HANDBOOK OF SCIENCE & ENGINEERING OF GREEN CORROSION INHIBITORS

HANDBOOK OF SCIENCE & ENGINEERING OF GREEN CORROSION INHIBITORS

Modern Theory, Fundamentals & Practical Applications

CHANDRABHAN VERMA
Post-Doctoral Fellow, Department of Chemistry, King Fahd University of Petroleum and Minerals, Saudi Arabia

ELSEVIER

Elsevier
Radarweg 29, PO Box 211, 1000 AE Amsterdam, Netherlands
The Boulevard, Langford Lane, Kidlington, Oxford OX5 1GB, United Kingdom
50 Hampshire Street, 5th Floor, Cambridge, MA 02139, United States

Copyright © 2022 Elsevier Inc. All rights reserved.

No part of this publication may be reproduced or transmitted in any form or by any means, electronic or mechanical, including photocopying, recording, or any information storage and retrieval system, without permission in writing from the publisher. Details on how to seek permission, further information about the Publisher's permissions policies and our arrangements with organizations such as the Copyright Clearance Center and the Copyright Licensing Agency, can be found at our website: www.elsevier.com/permissions.

This book and the individual contributions contained in it are protected under copyright by the Publisher (other than as may be noted herein).

Notices
Knowledge and best practice in this field are constantly changing. As new research and experience broaden our understanding, changes in research methods, professional practices, or medical treatment may become necessary.

Practitioners and researchers must always rely on their own experience and knowledge in evaluating and using any information, methods, compounds, or experiments described herein. In using such information or methods they should be mindful of their own safety and the safety of others, including parties for whom they have a professional responsibility.

To the fullest extent of the law, neither the Publisher nor the authors, contributors, or editors, assume any liability for any injury and/or damage to persons or property as a matter of products liability, negligence or otherwise, or from any use or operation of any methods, products, instructions, or ideas contained in the material herein.

British Library Cataloguing-in-Publication Data
A catalogue record for this book is available from the British Library

Library of Congress Cataloging-in-Publication Data
A catalog record for this book is available from the Library of Congress

ISBN: 978-0-323-90589-3

For Information on all Elsevier publications visit our website at
https://www.elsevier.com/books-and-journals

Publisher: Susan Dennis
Acquisitions Editor: Anita Koch
Editorial Project Manager: Lena Sparks
Production Project Manager: Bharatwaj Varatharajan
Cover Designer: Christian J. Bilbow

Typeset by Aptara, New Delhi, India

Contents

About the Author ix
Preface xi

1. Corrosion: basics and adverse effects

1.1 Summary 7
1.2 Important links 8
References and further reading 8

2. Forms of corrosion

2.1 Uniform corrosion 11
2.2 Intergranular corrosion 12
2.3 Pitting corrosion 13
2.4 Galvanic corrosion 14
2.5 Crevice corrosion 15
2.6 Stress corrosion cracking 16
2.7 Selective leaching 17
2.8 Erosion corrosion 18
2.9 Top-of-line corrosion 18
2.10 Summary 18
2.11 Useful links 19
References 19

3. Basics and theories of corrosion: thermodynamics and electrochemistry

3.1 Thermodynamics of corrosion 21
3.2 Types of corrosion 21
3.3 Factors affecting corrosion and corrosion rate 27
3.4 Summary 29
3.5 Useful links/websites 30
References 30

4. Concept of green chemistry in corrosion science

4.1 Concept of green corrosion inhibition 31
4.2 Assessment of green corrosion inhibitors: OSPAR and TEACH commission 32
4.3 Green chemistry principles and green corrosion inhibition 33
4.4 Summary 38
4.5 Useful websites 39
References 39

5. Classification of corrosion inhibitors

5.1 Classification of inorganic corrosion inhibitors 41
5.2 Classification of organic corrosion inhibitors 43
5.3 Summary 47
5.4 Useful websites 47
References 47

6. Corrosion and corrosion inhibition in acidic electrolytes

6.1 Common acidic electrolytes 50
6.2 Corrosion protection in acidic electrolytes 54
6.3 Case studies 54
6.4 Summary 57
6.5 Useful websites 58
References 58

7. Corrosion and corrosion inhibition in alkaline electrolytes

7.1 Corrosion protection in basic electrolytes: case studies 60
7.2 Summary 65
7.3 Useful websites 66
References 66

8. Corrosion and corrosion inhibition in neutral electrolytes

8.1 Corrosion protection in neutral electrolytes 70
8.2 Organic corrosion inhibitors: case studies 71
8.3 Summary 76
8.4 Useful website 77
References 77

9. Corrosion and corrosion inhibition in sweet and sour environments

9.1 Sweet corrosion 79
9.2 Sour corrosion 81
9.3 Protection for sweet and sour corrosion 82
9.4 Summary 83
9.5 Useful websites 83
References 83

10. Weight loss method of corrosion assessment

10.1 Advantages and disadvantages of weight loss study 86
10.2 WL technique of corrosion monitoring: case studies 87
10.3 Summary 91
10.4 Useful websites 91
References 91

11. Electrochemical methods of corrosion assessment

11.1 Electrochemical impedance spectroscopy 93
11.2 Potentiodynamic polarization 95
11.3 Electrochemical impedance spectroscopy and potentiodynamic polarization studies of corrosion inhibitors: case studies 97
11.4 Summary 101
11.5 Useful websites 101
References 101

12. Computational methods of corrosion assessment

12.1 Density functional theory 103
12.2 Molecular dynamics and Monte Carlo simulations 109
12.3 Summary 112
12.4 Useful links 113
References 113

13. Ionic liquids as green corrosion inhibitors

13.1 Ionic liquids: property and application 115
13.2 Mechanism of corrosion inhibition using ionic liquids 117
13.3 Ionic liquids as corrosion inhibitors: case studies 118
13.4 Ionic liquids as corrosion inhibitors for iron alloys in H_2SO_4 125
13.5 Miscellaneous 125
13.6 Summary and outlook 129
13.7 Useful websites 129
References 129

14. Green corrosion inhibitors from one step multicomponent reactions

14.1 Multicomponent reactions as a green synthetic strategy 135
14.2 Corrosion inhibitors derived from multicomponent reactions 136
14.3 Case studies: green corrosion inhibitors derived from multicomponent reactions 137
14.4 Summary 143
14.5 Useful websites 145
References 145

15. Green corrosion inhibitors from microwave and ultrasound irradiations

15.1 Microwave heating 148
15.2 Ultrasound heating 150
15.3 Case studies 151
15.4 Summary 157
15.5 Useful websites 158
References 158

16. Green corrosion inhibitors using environmental friendly solvents

16.1 Literature survey: corrosion inhibitors using green solvents 165
16.2 Summary 170
16.3 Useful websites 170
References 170

17. Plant extracts as green corrosion inhibitors

17.1 Mechanism of corrosion inhibition using plant extracts 175
17.2 Plant extracts as corrosion inhibitors: literature survey 176
17.3 Summary 184
17.4 Important websites 184
References 184

18. Chemical medicines (drugs) as green corrosion inhibitors

18.1 Chemical medicines as corrosion inhibitors: literature survey   194
18.2 Emerging trends in corrosion inhibition using chemical medicines   200
18.3 Summary   201
References   201

19. Natural polymers as green corrosion inhibitors

19.1 Literature survey: natural polymers as green corrosion inhibitors   208
19.2 Summary   217
19.3 Useful links   217
References   218

20. Carbohydrates as green corrosion inhibitors

20.1 Carbohydrates as green corrosion inhibitors: literature survey   226
20.2 Summary   231
References   231

21. Amino acids as green corrosion inhibitors

21.1 Literature survey: amino acids as green corrosion inhibitors   234
21.2 Summary   241
21.3 Useful websites   241
References   241

22. Oleochemicals as corrosion inhibitors

22.1 Oleochemicals as corrosion inhibitors: literature survey   243
22.2 Summary   252
22.3 Important websites   252
References   252

23. High temperature corrosion and corrosion inhibitors

23.1 High temperature corrosion inhibitors: literature survey   256
23.2 Conclusion   259
23.3 Important websites   259
References   260

24. Nanomaterials as corrosion inhibitors

24.1 Nanomaterials as corrosion inhibitors: literature survey   262
24.2 Summary   267
References   267

Index   271

About the Author

Chandrabhan Verma works at the Interdisciplinary Center for Research in Advanced Materials King Fahd University of Petroleum and Minerals in Saudi Arabia. He received the PhD degree in Corrosion Science from the Department of Chemistry, Indian Institute of Technology (Banaras Hindu University) in Varanasi, India. He is a Member of the American Chemical Society (ACS). His research is mainly focused on the synthesis and designing of environmentally friendly corrosion inhibitors useful for several industrial applications. Dr. Verma is the author of several research and review articles published in various peer-reviewed international journals such as ACS, Elsevier, RSC, Wiley, Springer, and other platforms. He is among the most highly cited researchers and academicians working in the field of Corrosion Science Engineering and serves as an Editor and Board Member in various reputed journals. Currently, Dr. Verma is editing and writing several books. Dr. Verma has received several national and international awards for his academic achievements.

Preface

Recently, research and development on green corrosion inhibition is gaining particular attention because of increasing ecological awareness and strict environmental regulations. In the last few years, numerous environmentally friendly corrosion monitoring practices and corrosion inhibitors have been developed and implemented. The use of computational modeling is one of the most recent and effective methods of corrosion monitoring as it is used to predict the metal-inhibitor bonding without polluting the environment. Nowadays, several synthetic and natural alternatives are used in place of traditional toxic heterocyclic corrosion inhibitors. Similar to traditional corrosion inhibitors, the environmentally friendly corrosion inhibitors become effective by absorbing on the metallic surface using their electron-rich centers. The electron-rich centers include polar functional groups including $-NH_2$ (amino), $-OH$ (hydroxyl), $-COOH$ (carboxyl), $-OMe$ (methoxy), $-CN$ (nitrile) $-NO_2$ (nitro), etc., and multiple (double and triple) bonds. This book describes the recent advancements in science and engineering of green corrosion inhibition and green corrosion inhibitors. Throughout this book, it can be seen that various environmentally friendly corrosion inhibitors and practices have been developed, and using these practices 15% (USD 375) to 35% (USD 875) of the cost of corrosion can be reduced.

A book to cover the recent developments in the science and engineering of green corrosion inhibitors is well overdue. Chandrabhan Verma addresses the topic in a book that attends to the fundamental characteristics of green corrosion inhibitors, their synthesis and characterization, chronological growths, and their industrial applications. The corrosion inhibition using environmentally friendly alternatives, especially synthetic green compounds, is broad-ranging. This book is divided into numerous sections each containing several chapters. Section 1, "Overview of corrosion and green corrosion inhibitors," describes the basics, theories, forms, and thermodynamics of corrosion along with the concept of green chemistry in corrosion science and engineering. In this section, classification of corrosion inhibitors has also been described in Chapter 5. Section 2, "Corrosion Environments," describes the nature and properties of some common electrolytes. In this section, the mechanism of corrosion and its protection are also discussed. Section 3, "Corrosion investigation: analysis and assessment," describes the advantages and disadvantages of some common corrosion monitoring practices along with their theories.

Section 4, "Synthetic green corrosion inhibitors," describes the corrosion inhibition properties of green corrosion inhibitors derived from chemical synthesis. This includes ionic liquids and compounds synthesized from one-step multicomponent reactions, ultrasound, and microwave irradiations and using environmentally friendly solvents, such as water, ionic liquids, and polyethylene glycol. Section 5, "Natural Green Corrosion Inhibitors," gives a description of natural corrosion inhibitors which include plant extracts, chemical medicines (drugs), natural polymers, carbohydrates, amino acids, oleochemicals, and their derivatives. Section 6, "Emerging trends in corrosion protection," includes two chapters namely, "High temperature corrosion and corrosion inhibitors" and "Nanomaterials as corrosion inhibitors."

Overall, this book is written for scholars in academia and industry, those working in corrosion engineering, and students of materials science and applied chemistry. On behalf of Elsevier, Dr. Chandrabhan Verma is very thankful to Prof. M.A. Quraishi for his valuable guidance and support. Special thanks to Anita Koch (acquisitions editor) and Lena Sparks (managing editor) for their dedication and support during this project. Finally, a thank you to Elsevier for publishing the book.

Chandrabhan Verma, PhD.

Corrosion: basics and adverse effects

Corrosion is the steady degradation of metallic materials through chemical and/or electrochemical reactions with the surrounding [1,2]. It is a natural process through which highly reactive metals convert into their stable forms, such as oxides, hydroxides, and sulfides. Metals and their alloys are widely used as construction materials in industries as well as for household applications. However, their life span is regrettably cut short by corrosion. Corrosion imposes enormous security, financial and environmental challenges on industries and nation at large [3,4]. It could bring about an extremely expensive and hazardous damage to home appliance, drinking water system, oil and gas transport pipeline, railway tracks, public buildings, bridges, and automobiles. Talking of adverse effects of corrosion, the United State over the last 22 years has suffered around 52 major corrosion-enhanced climate-related disasters, such as floods, fires, droughts, hurricanes, tropical storms, freezes, tornadoes, etc., that resulted into an estimate of about US $380 billion overall, which is around US $17 billion loss per year [5]. On the other hand, annual costs of corrosion in the United States was put at US $276 billion and US $2.2 trillion in 1998 and 2011, respectively [5]. According to the recent report of the National Association of Corrosion Engineers, recent cost of corrosion is about US $2.5 trillion which is equivalent to 3.4% of the world's gross domestic product. Economic costs of corrosion may be divided into direct and indirect. The direct costs of corrosion include:

(i) Cost of labor due to corrosion maintenance activities.
(ii) Cost of the utensils required because of corrosion-related actions.
(iii) Cost connected with loss or disruption in the supply of product.
(iv) Cost related to loss of consistency.
(v) Loss of productivity due to equipment failure.

Direct cost of corrosion includes replacement of corroded metallic structures and machineries or their components including metallic roofing, condenser, valves, pipelines, etc. It also includes costs of materials inspections, repairing, and maintenance, such as painting, coating, cathodic protection, and use of corrosion inhibitors. In summary direct cost of corrosion includes:

(i) Cost of replacement of corroded metallic structures.
(ii) Cost of corrosion control, e.g., coating, painting, cathodic protection, corrosion inhibitors, etc.

(iii) Cost of materials inspection, repair, and maintenance.
(iv) Loss of product manufacturing time.

According to a break-through released (in 2002) by the US Federal Highway Administration in a study entitled "Corrosion cost and preventing strategies in the United State" the direct cost of corrosion was about US $276 billion [5]. The cost of corrosion in the United States was increased to US $2.2 trillion in 2011. It is important to mention that cost of corrosion is further expected to increase because of the increased consumption of metallic materials in industries as well as household applications. The US economy can be divided into five foremost sectors and numerous subsectors. The five sectors of the US economy include transportation, utilities, manufacturing, production, and infrastructure. Fig. 1.1 represents cost of corrosion in the United States with respect to its total economy and the relative contributions of different sectors of the US economy.

Obviously, transportation sector deals with metallic equipment such as rail cars, aircrafts, HAZMAT transport, and motor vehicles. As depicted in Fig. 1.1, transportation sector contributes US $29.7 billion to the cost of corrosion in the United States. It is important to mention that more than 200 million motor vehicles are registered to the US consumers, organizations, and government. Therefore, car companies put extra efforts to the use of corrosion resistance materials through proper engineering designing. In motor vehicle sector, estimated direct cost of corrosion was US $23.4 billion in which around US $14.46 billion loss was due to the corrosion-reinforced reduction of the motor vehicles [5]. About US $6.45 billion was spent

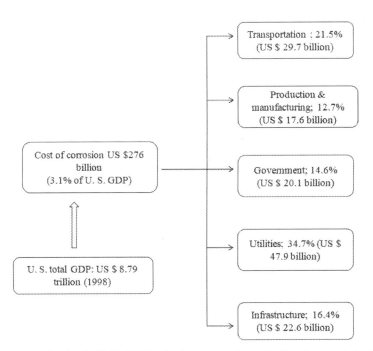

FIG. 1.1 Cost of corrosion in the United States in the transport, production & manufacturing, government, utilities, and infrastructure sectors.

for repairing and maintenance of the vehicles and their assets in response to the effects of corrosion. The use of corrosion resistance materials and engineering designing constitute another US $2.56 billion.

Along with vehicles, the US economy was badly affected by corrosion failure problems encounter in the railroad cars. Nearly, 1962 passenger and 1.3 million railroad cars are operating in the United States. Both internal and external corrosion failure has been encountered in the railroad with an estimated corrosion cost of US $0.5 billion. This corrosion cost is uniformly alienated into internal linings and coatings, and external coatings. Other units of the transportation sectors include ships and aircrafts. Corrosion cost of shipping industry in the United States was put at US $2.7 billion. This cost can be broken down into repair and maintenance costs (US $0.8 billion), corrosion-related down time (US $0.8 billion) and cost of construction of new ship (US $1.1 billion) [5]. In aeroplane industry, total annual costs of corrosion was estimated to be US $2.2 billion, which can be further divided into repair and maintenance (US $1.7 billion), engineering, designing, and manufacturing (US $0.2 billion) and downtime (US $0.3 billion) costs [5]. Similarly, hazardous materials transportation sector incurs corrosion cost of more than US $0.9 billion that includes costs of special packaging (US $0.5 billion) and transport vehicles [5]. A pictorial illustration of cost of corrosion in transportation sector is presented in Fig. 1.2.

Metallic materials are extensively used as constructional materials in production and manufacturing sectors such as oil and gas exploration (8%; US $1.4 billion), petroleum refineries (21%; US $3.7 billion), mining (1%; US $0.1 billion), paper and pulp (34%; US $6 billion), chemical, petrochemicals and pharmaceutical (10%; US $1.7 billion), food processing (12%; US $2.1 billion), and agricultural (6%; US $1.1 billion) [5]. The cost of corrosion is highly pronounced in these sectors. Fig. 1.3 represents the cost of corrosion in the numerous subdivisions of production and manufacturing sectors. It can be seen that paper & pulp (34%) and petroleum refineries (21%) contribute mainly into the corrosion costs of production and

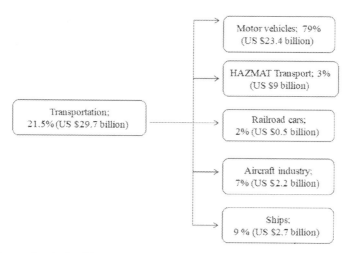

FIG. 1.2 Cost of corrosion in the US transportation sector.

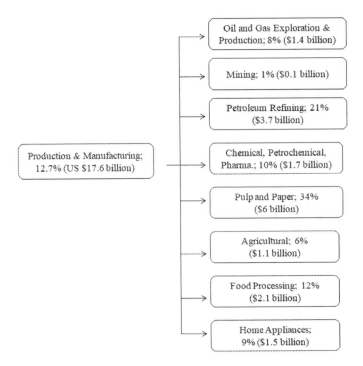

FIG. 1.3 Cost of corrosion in the US production and manufacturing sectors.

manufacturing. Generally, composition and chemistry of the crude oils are highly complicated and their corrosiveness depends upon the presence of total sulfur content, total acid number, moisture or water, salts, and dissolved gases. The economic adverse effects of the corrosion in utilities and infrastructure sectors are presented in Fig. 1.4. Cost of corrosion in infrastructure sector of the US economy can be divided into four subsectors namely, HAZMAT storage (31%; US $7 billion), waterways and ports (1%; US $0.3 billion), transmission pipelines (31%; US $7 billion), and highway bridges (37%; US $8.3 billion) [5]. Likewise, cost of corrosion in utilities sector can be categorized into three subsectors including electric utilities (14%; US $6.9 billion), gas distribution (10%; US $5 billion), and drinking water and sewer systems (75%; US $36 billion).

In view of this, it can be concluded that the US economics is greatly affected by corrosion. Apart from the US economics of the other developed as well as developing countries are also adversely affected by corrosion. Costs of corrosion of some of the major countries are presented in Table 1.1. Cost of corrosion is also highly pronounced in the case of metallurgical cleaning of the metallic ores using highly aggressive acidic solutions. Common cleaning processes are acid picking, acid descaling, and oil-well acidification. Generally, these processes are adopted to dissolve the surface impurities present in the metallic ores. However, these solutions also dissolve metallic components along with surface impurities. Nevertheless, several methods of corrosion mitigation have been developed depending upon the

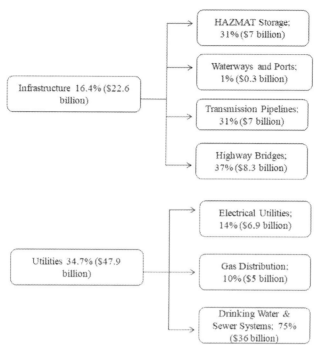

FIG. 1.4 Cost of corrosion in the US infrastructure (upper) and utilities (lower) sectors.

nature of metallic materials and environment. It is important to point out that corrosion cost can be cut down by 15% (US $375 billion) to 35% (US $875 billion) using previously developed methods of corrosion protection [6]. Some of the common methods of corrosion protection are coating, painting, galvanization, cathodic protection, and use of corrosion inhibitors [7]. Implementation of corrosion inhibitors is one of the most economic and commonly used methods [8]. Most of corrosion inhibitors become effective by adsorbing on metal surface and forming a surface protective film. Because of their ability to form surface protective film, organic corrosion inhibitors are also called filming or film-forming corrosion inhibitors [8].

It can be seen from Table 1.1 that costs of corrosion contribute greatly to the economics of developed and developing countries. Besides enormous economic losses, numerous accidents have been reported because of failure of metallic structures due to corrosion. Apart from financial and fatality losses, leakage of the pipelines of gas and liquid transportable materials can cause pollution and adversely affects the living organisms in the surrounding. For example, oil spill of Prudhoe Bay (in 2006) [6] caused leakage of around 1 million liter (267,000 gallon) of crude oil from corroded pipeline. Additionally, numerous accidents have been reported due to crevice and stress corrosion in chemical and nuclear reactors along with the collapse of several fighter planes. Numerous other such accidents have been reported previously and summary of some major accidents are listed in Table 1.2.

TABLE 1.1 Costs of corrosion in some major countries.

Country	Type of country	Direct cost	Gross domestic product (%)	Year	References
China	Developing	2127.8 billion RMB (~ 310 billion USD)	3.34	2015	[9]
USA	Developed	USD $5.5 billion 2.1		1949	[10]
		USD $70 billion	4.5	1978	[11]
		US $276 billion	3.1	1998	[12]
Australia	Developed	AUD $900 million	3.5	1972	[13]
		AUD $470 million	1.5	1974	[14]
		AUD $2 billion	1.5	1983	[15]
UK	Developed	£1365 billion pounds	3.5	1971	[16]
Japan	Developed	USD $9.2 billion	1.8	1977	[17]
		5.3 trillion Yen (Hoar)	1.02	1999	[18]
Kuwait	Developing	USD $1 billion	5.2	1995	[19]
Germany	Developed	USD $6 billion	3	1969	[20]
India	Developing	US $26.1 billion	2.4	2011–12	[21]

Apart from the above-mentioned accidents, numerous other accidents including EL AL Boeing 747 crash (October 4, 1992/Amsterdam, The Netherlands) [27], Silver Bridge Collapse of US Highway 35 bridge (December 15, 1967/United States) [28], Sinking of the Erika (December 12, 1999/Brittany, France) [29], etc., are reported because of corrosion failure. Therefore, it can be concluded that corrosion causes economic, environmental, and fatality losses. Adverse effects of corrosion can be summarized as follows:

(i) Damage and loss of metallic structures.
(ii) Reduction in the production, storage, and transport efficiency of plants.
(iii) Loss of valuable products through corroded structures.
(iv) Contamination of products.
(v) Environmental pollution due to leakage of liquids or gases being transported.
(vi) Reduction is the quality of materials being transported by contamination by corrosion products (rusts and scales).
(vii) Reduction in the transport efficiency due to blockage of the pipelines by corrosion products.
(viii) Increased possibility of localized corrosion around solid contaminants.
(ix) Cost of inspection, repairing and maintenance.
(x) Accidents due to mechanical damage of bridges, railways, motor vehicles, aircrafts, buildings and ships, etc.
(xi) Depletion of natural resources.
(xii) Difficulty in the efficiency of public transport due to corrosion failure of bridges and railway tracks.

TABLE 1.2 Some corrosion-related accidents in history.

Accident/country	Date	Description	References
Aloha incident/USA	April 18, 1988	A major fraction of upper fuselage of a 19-year-old Boeing 737, operated through Aloha airlines was lost at the height of 24,000 feet. Though the pilot managed to land the flight in a Maui, Hawaii island; however, one of the flight attendants died in the accident. The accident is resulted due to the fatigues cracks on the upper row of the rivets	[22]
Bhopal accident/India	December 2-3, 1984	In night of December 2-3, 1984, around 500 liter water inadequately inters into a MIC storage thank where 40 metric tons of MIC was stored in Union Carbide India Limited company, Bhopal (MP) India. By the chemical reaction of water and MIC (methylisocyanate) in the presence of iron as catalyst, sidewall of the storage tank get corroded that allowed the leakage of entire MIC. In this accident, 500,000 peoples were exposed to highly toxic MIC and around 3000 peoples were died. In another report, it was reported that the accident causes 558,125 injuries and death of 8000 peoples in next two weeks. The factory was closed after the accident	[23]
Carlsbad pipeline explosion/New Mexico	August 19, 2000	A natural gas transmission pipeline having diameter of 30-inch was ruptured due to corrosion in New Mexico. The released gas ignited and burned for 55 minute that caused death of 12 peoples and burning of three vehicles. In this accident, two nearby bridges made of steel were badly damaged. According to one report, the accident results into the total economic loss of US $998,296	[24]
Guadalajara sewer explosion/Mexico	April 22, 1992	At least nine corrosion explosions that killed 215 peoples were heard starting at around 10:30 AM (local time). Besides the fatality losses the corrosion explosions causes 1500 injuries and damage of more than 1600 buildings. This accident caused economic loss of more than US $75 million	[25]
Explosion by process chemicals/Louisiana	October 23, 1995	A yellow–brown vapor of nitrogen tetraoxide that is a liquefied mixture of water, poisonous gas, and oxidizer began leaking from the dome of the DOT class 105A railroad tank car UTLX 82329 at the Gaylord Chemical Corporation plant in Bogalusa, Louisiana. In this accident, about 3000 peoples were evacuated from the vapor cloud. The 4710 peoples were treated in local hospitals and 81 were admitted	[26]

1.1 Summary

Corrosion is a highly damaging and challenging phenomenon that has been identified with huge safety threat and economic losses. Both developed and developing countries are challenged by the loss of corrosion. According to the National Association of Corrosion Engineers, corrosion causes loss of about 3.5% of world's gross domestic product. Recent global annual cost of corrosion is about US $2.5 trillion. The cost of corrosion can be divided into direct and indirect. There are several well-known accidents that have been reported at different parts of the world because of corrosion failure. In view of the very high safety and economic losses, several methods of corrosion prevention and mitigation have been developed and being used. Among all the developed methods, use of organic compounds as corrosion inhibitors is one of the most effective, popular, and economic means of stemming corrosion.

1.2 Important links

http://impact.nace.org/documents/ccsupp.pdf
https://www.g2mtlabs.com/corrosion/cost-of-corrosion/
https://www.toppr.com/ask/question/what-are-the-adverse-effects-of-corrosion/

References and further reading

[1] A. Balamurugan, S. Rajeswari, G. Balossier, A. Rebelo, J. Ferreira, Corrosion aspects of metallic implants—an overview, Mater. Corros. 59 (2008) 855–869.
[2] M.G. Fontana, N.D. Greene, Corrosion Engineering, McGraw-Hill, New York, USA, 2018.
[3] B.N. Popov, Corrosion Engineering: Principles and Solved Problems, Elsevier, UK, 2015.
[4] H. Kaesche, Corrosion of Metals: Physicochemical Principles and Current Problems, Springer Science & Business Media, Springer-Verlag Berlin Heidelberg, 2012.
[5] G.H. Koch, M.P. Brongers, N.G. Thompson, Y.P. Virmani, J.H. Payer, Corrosion Cost and Preventive Strategies in the United States, Federal Highway Administration, USA, 2002.
[6] A.S. Williamson, P. Eng, Building a corrosion management system via material sustainability and stewardship.7/1/2020.
[7] M. Aliofkhazraei, Developments in Corrosion Protection, BoD–Books on Demand, InTechOpen, London, UK, 2014.
[8] M.A. Quraishi, D.S. Chauhan, V.S. Saji, Heterocyclic Organic Corrosion Inhibitors: Principles and Applications, Elsevier, UK, 2020.
[9] B. Hou, X. Li, X. Ma, C. Du, D. Zhang, M. Zheng, W. Xu, D. Lu, F. Ma, The cost of corrosion in China, npj Mater. Degr. 1 (2017) 1–10.
[10] H.H. Uhlig, The cost of corrosion to the United States, Corrosion 6 (1950) 29–33.
[11] L.H. Bennett, Economic Effects of Metallic Corrosion in the United States: a Report to the Congress, The Bureau, USA, 1978.
[12] G.H. Koch, M.P. Brongers, N.G. Thompson, Y.P. Virmani, J.H. Payer, Cost of corrosion in the United States. In: Handbook of environmental degradation of materials, Elsevier, UK, 2005, pp. 3–24.
[13] E.C. Potter, E.G. Potter The corrosion scene in Australia, Aust. Corros. Eng. (1972) 21–29.
[14] R. Revie, H. Uhlig, Cost of corrosion to Australia, J. Inst. Eng. Australia 46 (1974) 3–5.
[15] B.W. Cherry, B.S. Skerry, Corrosion in Australia: the report of the Australian National Centre for corrosion prevention and control feasibility study. In: Department of Materials Engineering, Monash University, Clayton, Vic: Dept. of Materials Engineering, Monash University, c1983, 1983.
[16] T. Hoar, Review lecture—corrosion of metals: its cost and control, Proceedings of the Royal Society of London. A. Mathematical and Physical Sciences, 348 (1976) 1–18.
[17] Committee on Corrosion Loss in Japan, Report on corrosion loss in Japan, Boshoku-Gijutsu (Corros. Eng.), 26 (1977) 401–512.
[18] Committee on Corrosion Loss in Japan, Survey of corrosion cost in Japan, Zairyoto-Kankyo (Corros. Eng.), 50 (2001) 490–512.

[19] F. Al-Kharafi, A. Al-Hashem, F. Martrouk, Economic Effects of Metallic Corrosion in the State of Kuwait, Final Report No. 4761, KISR Publications, December 1995.
[20] D. Behrens, Research and development programme on 'corrosion and corrosion protection' in the German federal republic, Br. Corros. J. 10 (1975) 122–127.
[21] R. Bhaskaran, L. Bhalla, A. Rahman, S. Juneja, U. Sonik, S. Kaur, J. Kaur, N. Rengaswamy, An analysis of the updated cost of corrosion in India, Mater. Performance 53 (2014) 56–65.
[22] W.R. Hendricks, The Aloha Airlines Accident—A New Era for Aging Aircraft, Structural Integrity of Aging Airplanes, Springer, Berlin, Heidelberg, 1991, pp. 153–165.
[23] P.K. Mishra, R.M. Samarth, N. Pathak, S.K. Jain, S. Banerjee, K.K. Maudar, Bhopal gas tragedy: review of clinical and experimental findings after 25 years, Int. J. Occup. Med. Environ. Health 22 (2009) 193.
[24] K.D. Koper, T.C. Wallace, R.C. Aster, Seismic recordings of the Carlsbad, New Mexico, pipeline explosion of 19 August 2000, Bull. Seismol. Soc. Am. 93 (2003) 1427–1432.
[25] S. Ranieri, Explosions In Guadalajara Caused By Gas Leaks In Sewer Lines: Summary Of Events, 1992.
[26] S.B. Presser, The Bogalusa explosion, single business enterprise, alter ego, and other errors: academics, economics, democracy, and shareholder limited liability: back towards a unitary abuse theory of piercing the corporate veil, Nw. UL Rev. 100 (2006) 405.
[27] P.U. De Haag, R. Smetsers, H. Witlox, H. Krüs, A. Eisenga, Evaluating the risk from depleted uranium after the Boeing 747-258F crash in Amsterdam, 1992, J. Hazard. Mater. 76 (2000) 39–58.
[28] S.G. Bullard, B.J. Gromek, M. Fout, R. Fout, The Silver Bridge Disaster of 1967, Arcadia Publishing, West Virginia, USA, 2012.
[29] I. No, M. Casualties, Report of the enquiry into the sinking of the ERIKA off the coasts of Brittany on 12 December 1999, Commentary, 1999.

CHAPTER 2

Forms of corrosion

Valuable information regarding the form of corrosion can be derived through careful observation of the corroded metallic structures. Corrosion can be classified into different classes and subclasses based on the nature in which the corrosion process manifests. Classification of corrosion as presented by ASM (the American Society of Metals), as shown in Fig. 2.1. Major forms of corrosion are briefly described in the following sections.

2.1 Uniform corrosion

Uniform corrosion, also known as general corrosion, is one of the most common forms of corrosion. Usually, uniform corrosion proceeds over the entire exposed metals surface. Uniform corrosion causes foremost destruction of metal on a tonnage basis. Although, this form of corrosion causes continuous thinning of the metallic objects; however, catastrophic breakdown of this form of corrosion is moderately rare. Uniform corrosion is relatively easy to observe, predict, and measure. Though it is not very dangerous, its continuous inspection and maintenance are highly essential as continuous corrosion of metal surface would result in roughness of the surface due to the accumulation of rusts and scales (if not soluble) that can even lead to other kinds of corrosion [1,2]. There are two important aspects of uniform corrosion, namely susceptibility and corrosion rate (C_R). Corrosion susceptibility or propensity is the ability of any particular material to undergo corrosive disintegration, and it is a function of thermodynamics. Generally, susceptibility determines whether the material will corrode or not under a given condition. Whereas, corrosion rate determines how much (quantitatively) a material will corrode in a given time interval.

Mostly, uniform corrosion is expressed as a weight or mass loss per unit area per unit time, e.g., mm (millimeter)/year. Corrosion rate (C_R) can be calculated as [3]:

$$C_R = \frac{KW}{At\rho} \tag{2.1}$$

where W is the weight or mass loss in gram, A is the surface area in cm^2, K is a constant 8.76×10^4 (mm/year), t is the temperature, and ρ is the density of the metal or alloy (in gram/cm^3).

Corrosion monitoring as expressed in Eq. (2.1) could be achieved via weight loss measurements. Weight loss experiment has several advantages. It is easy to perform, highly

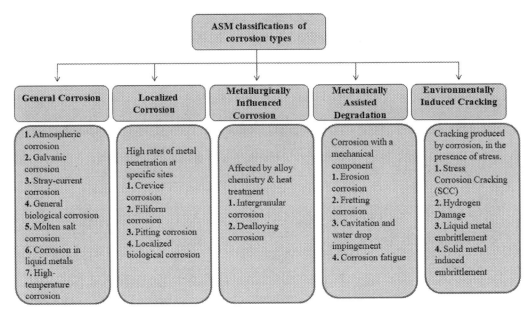

FIG. 2.1 ASM classification of corrosion types.

reproducible, and accurate. However, it generally takes a longer time to acquire results. This deficiency of weight-loss method can be overcome by determining the corrosion rate using electrochemical experiments. Some common examples of uniform corrosion are:

(a) Rusting of metals and alloys,
(b) Tarnishing of silver,
(c) Fogging of nickel,
(d) High-temperature oxidation, and
(e) Metal surface corrosion in aqueous electrolytes.

As uniform corrosion occurs evenly over the entire surface of the metal, it is relatively easy to control. It can be practically controlled using the following methods:

(a) Use of relatively thicker metallic materials for corrosion allowance.
(b) Cathodic protection through impressed current or sacrificial anode.
(c) Anodic protection.
(d) Use of corrosion inhibitors (in solution).
(e) Modifying the environment (by external additives).
(f) Painting and nonmetallic coatings.
(g) Metallic coatings, e.g., plating, anodizing or galvanizing.

2.2 Intergranular corrosion

Microstructure of metals and their alloys is constrained by grains that are separated by grain boundaries [4]. Intergranular corrosion (IGC) which is also known as intergranular attack, intercrystalline corrosion, or interdendritic corrosion, is a form of localized corrosion where corrosion occurs along the grain boundaries or just adjacent to the grain boundaries leaving most of the bulk of grains unaffected. IGC is principally associated with the effect of chemical segregation or precipitation of specific phase on or around the grain boundaries [4,5]. For example, chromium is added in austenitic stainless steel (18Cr–8Ni) and nickel alloys to improve their corrosion resistance property [6]. Generally, chromium becomes effective by forming the surface protective film of chromium oxide. However, at higher temperature (538–927°C) chromium precipitates as chromium carbide ($Cr_{23}C_6$) at the grain boundaries that results into the formation of chromium-depleted zones. The phenomenon of formation of such depleted zones at the grain boundaries is called sensitization.

It is important to mention that carbon remains immobile below 538°C, and above 927°C the $Cr_{23}C_6$ is soluble. This type of depletion quickly decreases the concentration of chromium from a region just adjacent to the grain boundary [6]. This decreases the 18Cr–8Ni alloy's tendency to develop corrosion protective surface films of chromium oxide and leads to an improved vulnerability for localized attack around the grain boundaries or nearby region [7]. IGC initiates and propagates along with the grain boundaries and seriously affects the mechanical strength of the metallic structures. IGC is generally experienced inside the heat-affected region. One of the common examples is welding-based sensitization of the material. In general, IGC needs a strong oxidizing stipulation. Testing of IGC is carried out by determining the susceptibility of a material to heat sensitization, followed by exposing the material to a string oxidizing environment such as HNO_3 (nitric acid), H_2SO_4 (sulfuric acid), $C_2H_2O_4$ (oxalic acid), or a mixed solution of H_2SO_4 (sulfuric acid) and $CuSO_4$ (copper sulphate) [8,9]. The possibility and rate of IGC can be minimized as follows:

(a) By injecting reducing gases such as hydrogen (H_2).
(b) By operating at lower acidity.
(c) By decreasing the level of carbon loading. This can be achieved by using low-carbon steel alloys such as 304 L and 316 L.
(d) Using postweld heat treatment.
(e) By solution annealing.
(f) Using titanium (e.g., type 321) or niobium (e.g., type 3470) based alloys. Niobium and titanium are strong carbides formers.

2.3 Pitting corrosion

Pitting or pitting corrosion is one of the most extensively studied forms of corrosion that leads to the formation of small holes in the metal [10]. Lack of predictability is the strongest motivation behind studying this form of corrosion. In general, driving force for pitting corrosion is breakdown of the passivity, i.e., depassivation [11,12]. The small depassivated area behaves as anode (where oxidation takes place) and the remaining potentially difference area behaves as cathode (where reduction takes place). Mostly, initiation of pitting requires chemical or structural inhomogeneity to the surface of metal. After breakdown of the passivity, the pit grows and penetrates to the bulk of mass of the metal with limited diffusion of ions.

There are several theories proposed for the breakdown of the passivity. One of such accepted theories is penetration mechanism theory according to which transfer of the aggressive ions occurs through passive layer to the interface of metal oxide [11]. This type of transfer generates potential difference that leads to excessive stress and rupturing of the passive film. According to the adsorption mechanism [13], pitting initiates with the generation of complexes that leads to an increase in the transfer of ions between electrolyte and metal-oxide interface. Pitting corrosion is measured using weight loss or gravimetric (coupon) testing method that involves the immersion of metallic coupons for a specific time. After the immersion time is elapsed, the coupons are taken out and then examined for surface corrosion, washed, cleaned, and weighted to determine weight loss. This is a highly effective, accurate and easy method of pitting corrosion testing.

Another method of pitting corrosion testing is electrochemical cyclic polarization. In this testing metallic specimen is allowed to polarize anodically from its equilibrium corrosion potential (E_{corr}) to a potential where the beginning of localized corrosion is identified through sudden rise in the measured current. This starting (onset) potential is called as breakdown potential (E_{BD}) [14]. After that, polarization process is then reversed (negative or cathodic direction) to measure the position at which the current returns to background. This position is called as repassivation potential (E_{RP}) [15]. The E_{RP} is the potential at which pits repassivate. Another analogous method of pitting testing is critical pitting temperature determination in which increase in temperature is measured instead of polarization of the specimen. Recently, electrochemical noise testing is gaining much attention in measuring pitting corrosion as it is a nondestructive and real-time monitoring technique.

2.4 Galvanic corrosion

Galvanic or two-metal corrosion occurs when two dissimilar metals are electrically connected while both are dipped in an electrolytic solution [16]. This is called as galvanic couple. In the galvanic couple, less noble materials will corrode in preference to the nobler one and behaves as the anode with respect to the other, which behaves as the cathode. The relative galvanic activity of a metal can be determined by its position in the galvanic series that provides rest potential of numerous metals in a specific environment (electrolyte). A general galvanic series of common metals and alloys from most reactive (least noble) to least reactive (most noble) is presented in Fig. 2.2. For example, if galvanic couple of aluminum and carbon steel is immersed in seawater, the aluminum will corrode more quickly than that of the carbon steel. In fact, carbon steel gets protected through anodic sacrificial mechanism of aluminum. It is important to mention that the greater the potential difference of the metals in a galvanic couple the greater the aggressiveness of (galvanic) corrosion. However, the aggressiveness of the galvanic couple can be more accurately determined by allowing for combination within mixed potential theory. Although, either (or both) metal(s) in the galvanic couple may or may not corrode itself (themselves), the self-corrosion rates of both metals will change. The concept of galvanic coupling serves as the basis for numerous corrosion protection techniques. Galvanic corrosion can be prevented as follows.

(a) By choosing metals of similar corrosion potential.
(b) By breaking the electrical connection between the galvanic couple.
(c) By separating the metals of the couple by inserting a properly sized spacer.

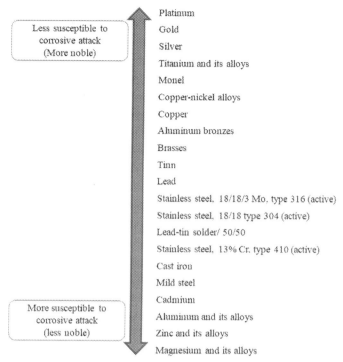

FIG. 2.2 Galvanic series of common metals and alloys from the most reactive (least noble) to least reactive (most noble).

(d) Installing an alternative anode that acquires greater corrosion potential than both metals of the couple.
(e) Applying coating on both metals.
(f) Adding corrosion inhibitors to the electrolyte.
(g) By keeping the anode to cathode area large.

2.5 Crevice corrosion

Crevice corrosion refers to the localized attack or corrosion on a surface of metal at or immediately adjacent to the crevice or gap between two joining surfaces [17,18]. The crevice may be produced between two metallic materials or one metallic and one nonmetallic materials. It is important to mention that without or outside the crevice, both materials are resistant to corrosive damage. Crevice corrosion identifies by the damages caused by corrosion of either of the two metals within or closed to the joining surfaces. Obviously, crevice corrosion is initiated and developed by differences in the concentration of some chemical species that result in setting up of electrochemical concentration cell. The electrochemical concentration cell developed through the concentration difference of oxygen is referred to as different aeration cell. Mechanism of crevice corrosion can be divided into four stages, namely,

(a) deoxygenation, (b) increasing localized concentration of salts and acids, (c) depassivation, and (d) propagation of crevice. After initiation, crevice corrosion propagates similarly to the pitting corrosion.

Major factors that influence the crevice corrosion are:

(a) Crevice type: metal-to-metal, metal-to-nonmetal.
(b) Crevice geometry or morphology: surface roughness, gap size, and depth.
(c) Nature of materials: structure and composition.
(d) Environment: temperature, oxygen, halide ions, and pH.
(e) Depletion of inhibitor (chemical compounds) in crevice.

The following practices and engineering designs can be adopted to minimize the possibility and rate of crevice corrosion.

- Avoid bolted joints and use welded butt joints.
- Avoid crevices through continuous soldering and welding.
- Insure inclusive drainage in vessel.
- Avoid use of water-absorbent gaskets, such as Teflon.
- Use of higher corrosion-resistant materials.
- Keep away from generating stagnant conditions.

2.6 Stress corrosion cracking

Stress corrosion cracking (SCC) is the cracking of metallic structures induced through the combination of corrosive environment and tensile stress [19,20]. The impact of SCC generally lies between fatigue threshold and dry cracking of that material. The stress in the metallic materials can be induced using welding, cold forming and cold reforming, grinding, and heat treatment. The significance and degree of stress is often underestimated. Development of the corrosion products in the restricted spaces can also generate significant stress that cannot be overlooked. Generally, SCC occurs in certain alloys in the suitable environmental-stress combinations. In SCC, most of the metal surface remains unaffected and unchanged; however, fine cracks penetrate bulk into the material. Recognition of cracks in SCC is extremely difficult. Numerous well-known accidents such as EL AL Boeing 747 crash in Amsterdam [21] are documented in literature because of the corrosion failure of metallic structures through SCC.

Various mechanisms for SCC are proposed in the literature. These include:

(a) *Adsorption model*: This model assumes that a specific chemical adsorbs on the crack surface and decrease the fracture stress.
(b) *Embrittlement model*: This model proposes that a specific species or chemical diffuses to the crack tip and embrittle the metal. Embrittlement of steel and titanium alloys by hydrogen is one of the common examples.
(c) *Film rapture model*: This model postulates that stress raptures the surface passive film and build-up a new active-passive cell. Under similar conditions, newly developed cell raptures again and this cycle continues until failure.

The SCC can be prevented by using the following methods.

(a) Avoid the use of chemical species that causes SCC or prevent their contact with metallic surface.
(b) Stress level and hardness of the material should be properly controlled as harder materials are relatively more prone to SCC.
(c) Encourage the use of materials that are not prone to crack formation in the specific environment.
(d) Use of self-healing surface coatings.
(e) Control by applying suitable operating temperature.
(f) Control by applying suitable electrochemical potential.
(g) Controlled heat treatment of the metallic structures.
(h) Controlled cold forming and cold reforming.

2.7 Selective leaching

Selective leaching is also known as parting, dealloying, selective attack, selective corrosion, and demetalification. Selective leaching is a form of corrosion in which a component of alloys is preferably leached out from the material under suitable condition [22]. Obviously, less noble (more reactive) metal is leached out from the alloys by a microscopic-scale corrosive dissolution mechanism [23]. The most susceptible alloys for selective leaching are those that contain metals with widely different corrosion potentials, i.e., metals that are situated widely apart from each other in the galvanic series. Iron, chromium, zinc, cobalt, aluminum, etc., are among the common metals susceptible to selective leaching. One of the common examples of selective leaching is dezincification of brass, which occurs when zinc is leached out from the alloy. Two mechanisms of selective leaching are expressed below.

(a) The two metals in an alloy dissolve and one of them deposited on the surface, e.g., dezincification of brass.
(b) One of the metals preferably dissolves and leached out leaving the other metal behind, e.g., selective leaching of molybdenum (Mo) from nickel alloys in molten NaOH (sodium hydroxide) solution.

In some selected alloys, leaching occurs through either mechanism varying upon the temperature as well as flow rate and concentration of the corrodent. The alloys exhibiting selective leaching become porous and lost their mechanical properties such as hardness, ductility and strength. Selective leaching can be prevented by the following methods.

(a) Selection of materials that are resistant to dealloying, e.g., (1) inhibited brass is comparatively more resistant to dezincification than alpha brass, and (2) ductile iron is less susceptible to corrosion than gray cast iron.
(b) Environmental modification that minimizes selective leaching.
(c) Cathodic protection using sacrificial anodic mechanism.
(d) Cathodic protection using impressed current.

2.8 Erosion corrosion

Erosion corrosion refers to the degradation of materials by the combined action of erosion and corrosion in the flowing corrosive liquids or the metal component moving through the fluid [24,25]. Erosion corrosion is mostly found around pump impellers, tube inlet ends, or tube blockages at high flow rates. Cavitation represents a special class of erosion-corrosion resulting from water bubbles formed by a high-speed impellers which when collapse cause pits on the metal surface [26,27]. Erosion corrosion can be controlled by:

(a) Using more resistant materials.
(b) Reduce turbulence.
(c) Using corrosion inhibitors.
(d) Cathodic protection using impressed current and sacrificial anodization.

2.9 Top-of-line corrosion

TLC is a form of corrosion that is motivated by the condensation of water in the inner wall of gas and liquids transport pipelines [28]. Literature study showed that numerous cases have been reported especially in the CO_2-dominated system. The pH of the water droplets drops down become of the dissolution of gases (CO_2 and H_2S) and organic acids that are naturally present in the gas stream and petroleum fluids. The condensed water droplets at low pH become corrosive. Presence of bicarbonate decreases the rate of TLC, whereas presence of acetic acid (CH_3COOH) increases the rate of TLC. There are numerous factors such as rate of water condensation, partial pressure of H_2S and CO_2, amount and aggressiveness of organic acids, fluids and gases velocity and temperature that affect the rate of water condensation and rate of TLC. The following approaches could be used to prevent TLC.

(a) Mostly water condensation occurs through heat exchange; therefore TLC can be prevented by applying local heat insulator on the inner wall of transport pipeline.
(b) Applying water-miscible corrosion inhibitors through batch injection mechanism.
(c) Using corrosion-resistant materials.
(d) Using materials coated with water repellents.
(e) By neutralizing the organic acids through chemicals such as N-methyl diethanolamine (MDEA).
(f) Through cathodic protection using a suitable galvanic couple and using impressed current.
(g) By sending a spray pig through the transport pipelines to distribute the corrosion inhibitor from flowing liquids to the top of inner wall of the line.

2.10 Summary

Corrosion can be divided into different forms on carefully observing the corroded metallic surface. Some common forms of corrosion include uniform corrosion, IGC, pitting corrosion, galvanic corrosion, and SSC, selective leaching, erosion corrosion, and TLC. Each form of corrosion occurs in a specific condition. Uniform corrosion occurs uniformly throughout the metallic structures whereas IGC which is also known as intergranular attack, intercrystalline corrosion occurs around the grain boundaries. Each form of corrosion requires a specific corrosion monitoring practices.

2.11 Useful links

https://www.nace.org/resources/general-resources/corrosion-basics/group-1/eight-forms-of-corrosion
https://cathwell.com/forms-of-corrosion/
https://www.gibsonstainless.com/types-of-corrosion.html
https://link.springer.com/chapter/10.1007%2F1-4020-7860-9_1

References

[1] Y. Zhao, A.R. Karimi, H.S. Wong, B. Hu, N.R. Buenfeld, W. Jin, Comparison of uniform and non-uniform corrosion induced damage in reinforced concrete based on a Gaussian description of the corrosion layer, Corros. Sci. 53 (2011) 2803–2814.
[2] C. Jirarungsatian, A. Prateepasen, Pitting and uniform corrosion source recognition using acoustic emission parameters, Corros. Sci. 52 (2010) 187–197.
[3] M.G. Fontana, Corrosion Engineering, Tata McGraw-Hill Education, New Delhi, India, 2005.
[4] E. Arzt, Size effects in materials due to microstructural and dimensional constraints: a comparative review, Acta Mater. 46 (1998) 5611–5626.
[5] M. Shimada, H. Kokawa, Z. Wang, Y. Sato, I. Karibe, Optimization of grain boundary character distribution for intergranular corrosion resistant 304 stainless steel by twin-induced grain boundary engineering, Acta Mater. 50 (2002) 2331–2341.
[6] C. Tedmon Jr, D. Vermilyea, J. Rosolowski, Intergranular corrosion of austenitic stainless steel, J. Electrochem. Soc. 118 (1971) 192.
[7] B. Wilde, Influence of silicon on the intergranular corrosion behavior of 18Cr-8Ni stainless steels, Corrosion 44 (1988) 699–704.
[8] J. Armijo, Intergranular corrosion of nonsensitized austenitic stainless steels, Corrosion 24 (1968) 24–30.
[9] T. Amadou, H. Sidhom, C. Braham, Double loop electrochemical potentiokinetic reactivation test optimization in checking of duplex stainless steel intergranular corrosion susceptibility, Metall. Mater. Trans. A 35 (2004) 3499–3513.
[10] G. Frankel, Pitting corrosion of metals: a review of the critical factors, J. Electrochem. Soc. 145 (1998) 2186.
[11] Y. Xu, M. Wang, H. Pickering, On electric field induced breakdown of passive films and the mechanism of pitting corrosion, J. Electrochem. Soc. 140 (1993) 3448.
[12] S. Lenhart, M. Urquidi-Macdonald, D. Macdonald, Photo-inhibition of passivity breakdown on nickel, Electrochim. Acta 32 (1987) 1739–1741.
[13] H.H. Uhlig, The adsorption theory of passivity and the flade potential, Zeitschrift für Elektrochemie, Berichte der Bunsengesellschaft für physikalische Chemie 62 (1958) 626–632.
[14] M. Wilms, V. Gadgil, J. Krougman, F. Ijsseling, The effect of σ-phase precipitation at 800 C on the corrosion resistance in sea-water of a high alloyed duplex stainless steel, Corros. Sci. 36 (1994) 871–881.
[15] N. Sridhar, G. Cragnolino, Applicability of repassivation potential for long-term prediction of localized corrosion of alloy 825 and type 316L stainless steel, Corrosion 49 (1993) 885–894.
[16] M. Tavakkolizadeh, H. Saadatmanesh, Galvanic corrosion of carbon and steel in aggressive environments, J. Compos. Constr. 5 (2001) 200–210.
[17] J. Oldfield, W. Sutton, Crevice corrosion of stainless steels: I. A mathematical model, Br. Corros. J. 13 (1978) 13–22.
[18] I. Rosenfeld, I. Marshakov, Mechanism of crevice corrosion, Corrosion 20 (1964) 115t–125t.
[19] M. Kimura, N. Totsuka, T. Kurisu, K. Amano, J. Matsuyama, Y. Nakai, Sulfide stress corrosion cracking of line pipe, Corrosion 45 (1989) 340–346.

[20] V. Raja, T. Shoji, Stress Corrosion Cracking: Theory and Practice, Woodhead Publishing, Elsevier, UK, 2011.
[21] H.F. Crombag, W.A. Wagenaar, P.J. Van Koppen, Crashing memories and the problem of 'source monitoring', Appl. Cogn. Psychol. 10 (1996) 95–104.
[22] C. Schroer, O. Wedemeyer, J. Novotny, A. Skrypnik, J. Konys, Selective leaching of nickel and chromium from Type 316 austenitic steel in oxygen-containing lead–bismuth eutectic (LBE), Corros. Sci. 84 (2014) 113–124.
[23] M.G. Fontana, N.D. Greene, Corrosion Engineering, McGraw-hill, New York, USA, 2018.
[24] R. Barik, J. Wharton, R. Wood, K. Stokes, R. Jones, Corrosion, erosion and erosion–corrosion performance of plasma electrolytic oxidation (PEO) deposited Al2O3 coatings, Surf. Coat. Technol. 199 (2005) 158–167.
[25] A.V. Levy, Solid Particle Erosion and Erosion-Corrosion of Materials, ASM International, Materials Park, Ohio, 1995.
[26] C. Kwok, F. Cheng, H. Man, Synergistic effect of cavitation erosion and corrosion of various engineering alloys in 3.5% NaCl solution, Mater. Sci. Eng.: A 290 (2000) 145–154.
[27] B. Vyas, I.L. Hansson, The cavitation erosion-corrosion of stainless steel, Corros. Sci. 30 (1990) 761–770.
[28] D. Larrey, Y.M. Gunaltun, Correlation of cases of top of line corrosion with calculated water condensation rates, Corrosion, NACE International, Orlando, Florida, March 2000.

CHAPTER 3

Basics and theories of corrosion: thermodynamics and electrochemistry

3.1 Thermodynamics of corrosion

Corrosion is a natural, spontaneous, and electrochemically favored phenomenon through which metals return to their natural oxidation states. Corrosion involves the oxidation-reduction reaction in which metals are oxidized by certain components of their surroundings such as water and oxygen. Through corrosion metals attempt to decrease their energy by spontaneously reacting with the components in their surroundings to form thermodynamically more stable compounds or solutions [1,2]. The driving force for corrosion is the change in the standard Gibb's free energy (ΔG) which is defined as the change in the free energy of the metal brought about by the corrosion (Fig. 3.1). For any process such as corrosion, if the ΔG value is negative then the process will be spontaneous and thermodynamically favored otherwise it will be nonspontaneous. Generally, ΔG is presented in calories per mole (cal/mol) or recently as joule per mole (J/mol).

The driving force for corrosion can be more commonly expressed in volts (V) which can be derived using the following equation [1,2]:

$$E = \frac{-\Delta G}{nF} \quad (3.1)$$

In the above equation, E is the driving force or driving energy (in volts, V) for the corrosion process, F is a constant, known as "faraday" and as the "electrical charge carried by one mole of electrons (or 96,490 C)," and n is the number of moles of electrons per mole of metal involved in the corrosion (oxidation-reduction) process. It can be concluded from Eq. (3.1) that in the case of ΔG of negative magnitude and minus sign, the spontaneous process always has a positive voltage E.

3.2 Types of corrosion

Corrosion can be categorized into two broad categories, chemical and electrochemical corrosion. A detailed description on chemical and electrochemical corrosion is given below:

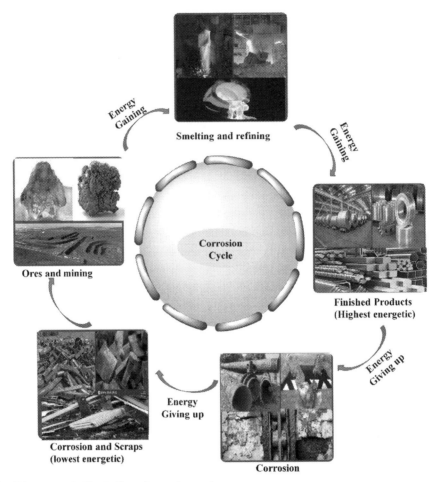

FIG. 3.1 Diagrammatic illustration of corrosion cycle.

3.2.1 Chemical corrosion or dry corrosion

Chemical or dry corrosion occurs through the chemical reactions between metal surfaces and chemicals and gases present in the surroundings including oxygen, hydrogen sulphide, sulfur dioxide, hydrogen, nitrogen, or anhydrous inorganic acids in a moisture-free state. Generally, dry corrosion is not as damaging as wet corrosion however dry corrosion is very sensitive to temperature. Increase in the temperature results in subsequent increase in the rate of dry corrosion. Dry corrosion is a relatively slow process that proceeds through the accumulation or collection of solid corrosion products that provide slight protection from further corrosive damage of the metal surface.

The inhibitive or protective nature of the corrosion products can be derived through *Pilling–Bedworth* ratio, Md/nmD, where d and m are the density and the atomic mass of the metal, respectively [3,4]; D and M are the density and the molecular weight of the corrosion product, respectively; while n represents the count of metal atom in the molecular formula of the corrosion product. For example, in Fe_2O_3, Fe_3O_4, and Al_2O_3, n values are 2, 3, and 2, respectively. If the volume of the corrosion product is smaller than that of the metal from which it is formed, then the corrosion product is expected to contain pores and cracks and to be nonprotective. In this situation, the magnitude of *Pilling–Bedworth* ratio will be less than unity, i.e., $Md/nmD < 1$. Conversely, the corrosion product would be protective when its volume is greater than the volume of the metal from which it is formed. In this condition, the corrosion product is expected to form a relatively more compact and compressed layer on the metal surface. In this condition, the *Pilling–Bedworth* ratio will be more than unity, i.e., $Md/nmD > 1$. However, in certain cases, the formation of volatile and soluble corrosion products takes place and are easily removed from the surface, exposing new metallic surface for further corrosion. Dry corrosion can also occur deep into the metallic materials through the penetration of corrosive species from porous corrosion products. This is called active corrosion. Dry corrosion occurs on both hetero- and homo-geneous metallic surfaces. Tarnishing of silver and copper and atmospheric oxidation (formation of oxides) are well-known examples of dry corrosion. Dry corrosion can be further divided into the following types:

(i) Oxidative dry-corrosion

Oxidative dry corrosion occurs due to the direct action of oxygen on metal (oxidation) in the absence of moisture. Dry corrosion is most common at room temperature [5]. Dry corrosion takes the following basic steps:

$$M \leftrightarrow M^{n+} + ne^- \quad \text{(Loss of electron)} \tag{3.2}$$

$$nO_2 + ne^- \leftrightarrow 2nO^- \tag{3.3}$$

Most metals, including Al, Sn, Pb, Cu, and Pt, form highly stable oxides at the metal-to-atmosphere interface, limiting further corrosion. If the surface oxide layer is porous, diffusion of cations occurs more rapidly as they are smaller in size than anions. This type of diffusion allows metals to corrode further by providing a new metallic surface. The nature of the oxide layer plays an extremely significant role in oxidative dry corrosion. The role of the oxide film can be summarized as follows:

1. The oxide layer acts as protective coating when it is stable and tightly adhered.
2. Oxidative corrosion does not occur when the oxide layer is unstable. In such situation, the unstable oxide layer decomposes back to metal and oxygen.
3. Oxidative corrosion will continue as long as the oxide film is volatile, as it removes from and exposes new metal surfaces to corrosion.
4. Oxidative corrosion will constantly occur if the oxide film is adequately porous.

(ii) Oxidative corrosion by gases other than oxygen

This kind of corrosion occurs through the reaction of a metal surface with gases other than oxygen, including CO_2, SO_2, H_2, H_2S, NO_2, Cl_2, and F_2. Similar to the oxidative dry corrosion, the corrosion products here may or may not be protective. The nature of the corrosion products depends on the nature of the metal and the atmosphere. For example, the corrosion of silver in the presence of chlorine (Cl) results in the formation of a highly stable, impervious, and adhering silver chloride (AgCl) film that provides excellent protection from further corrosion. Whereas, the corrosion of tin (Sn) in the presence of chloride results in the formation of a highly volatile and porous layer of $SnCl_4$.

(iii) Liquid metal corrosion

In numerous industries, liquefied (molten) metals such as Cd, Pb, Hg, Zn, etc., are transported through pipelines made-up of stainless steel and aluminum and result in their brittle failure due to internal penetration. Liquid metal corrosion is also known as liquid metal embrittlement and liquid metal cracking. Generally, liquid metal corrosion occurs at high temperature and mainly found in the devices employed in nuclear power plants. The liquid metal corrosion occurs in a specific combination of liquefied metals and stressed-out metals or their alloys that can direct to disastrous intergranular cracking. Some common examples of this combination include:

- Stainless steel and carbon steel are prone to liquid metal corrosion by zinc and lithium.
- Aluminum and its alloys are prone to liquid metal corrosion by zinc and mercury.
- Copper and its alloys are prone to liquid metal corrosion by lithium and mercury.

Because of this, use of mercury-based materials is highly restricted in all airplanes as they are mainly made-up of aluminum based of materials that are highly susceptible for liquid metal corrosion by mercury. Liquid metal corrosion may occur through the following types:

1. Through the mass transfer.
2. Through the formation of intermetallic compounds.
3. Solution of the structural metal.
4. Diffusion of liquid metal into solid metal.

Possibility of the liquid metal corrosion can be minimized by avoiding the contamination or contact of the cracks-causing liquids metals and avoiding the use of metals near or at their melting points. Liquid metal corrosion can also be prevented through barrier protection using a protective coating.

3.2.2 Electrochemical or wet corrosion

Electrochemical corrosion is one of the most common types of corrosion in which corrosion involves the transfer of electrons between a metal and electrolytic solution. Electrochemical corrosion or wet corrosion essentially requires four components listed below:

1. *Anodic sites*: where oxidation and corrosion occur through the discharge of electrons.
2. *Cathodic sites*: where reduction and consumption of electrons occur through various modes depending upon the nature of electrolyte.
3. *An electrolyte*: an electrolytic solution in which cathodic and anodic areas are resided.
4. *Potential difference*: both anode and cathode must have a different voltage (potential).
5. *Electrical connection*: anode and cathode have to be coupled electrically.
6. *An exposed metal surface*: the metal immersed in electrolyte should not be coated or protected.

According to Cushman (1907), the overall electrochemical corrosion proceeds through the formation of electrochemical cell that can be divided into two half-cell reactions, anodic and cathodic half-cell reactions.

3.2 Types of corrosion

Anodic half-cell reaction: the anodic half-cell reaction involves the oxidation and liberation of electrons. Anodic half-cell reaction of mono-, di-, and tri-valent metals can be presented as follows:

$$M \leftrightarrow M^+ + e^- \tag{3.4}$$

$$M' \leftrightarrow M'^{2+} + 2e^- \tag{3.5}$$

$$M'' \leftrightarrow M''^{3+} + 3e^- \tag{3.6}$$

where M, M', and M'' are the mono- (such as Na and Li), di-(such as Mg and Ca), and tri-valent (such as Al) metals.

Cathodic half-cell reaction

Cathodic reactions involve the consumption of electrons liberated at the anode. Depending upon the environmental conditions, cathodic half-cell reactions may proceed through the following mechanisms.

1. *Hydrogen evolution mechanism*: generally, hydrogen evolution mechanism operates in the acidic medium and in the absence of oxygen.

$$H^+ + e^- \rightarrow \frac{1}{2}H_2 \tag{3.7}$$

2. *Reduction of oxygen in acidic condition*: this mechanism operates with the formation of water by the consumption of electrons in oxygen reduction.

$$O_2 + 4H^+ + 4e^- \rightarrow 2H_2O \tag{3.8}$$

3. *Reduction of oxygen in neutral aqueous condition*: this mechanism operates with the formation of hydroxide by the consumption of electrons in oxygen reduction.

$$O_2 + 2H_2O + 4e^- \rightarrow 4OH^- \tag{3.9}$$

4. *Metal deposition*: in this type of cathodic reaction, metal cations reduce and deposit irrespective of the nature of the environment. Metallic deposition of the mono-, di-, and tri-valent metallic cations are presented below:

$$M^+ + e^- \leftrightarrow M \tag{3.10}$$

$$M'^{2+} + 2e^- \leftrightarrow M' \tag{3.11}$$

$$M''^{3+} + 3e^- \leftrightarrow M'' \tag{3.12}$$

where M, M', and M'' are the mono- (such as Na and Li), di-(such as Mg and Ca), and tri-valent (such as Al) metals.

5. *Reduction in the oxidation number of cations*: in this type of cathodic reaction, metal cations oxidation state reduces by the intake of electron(s) as presented below:

$$M'^{2+} + e^- \leftrightarrow M'^+ \tag{3.13}$$

$$M''^{3+} + e^- \leftrightarrow M''^{2+} \tag{3.14}$$

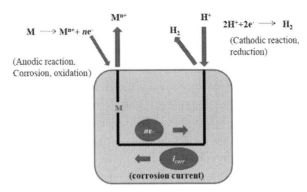

FIG. 3.2 A diagrammatic illustration of anodic and cathodic reaction, flow of charges (ne^-), and flow of current (i_{corr}) of a metal piece emerged into acidic electrolyte.

A diagrammatic illustration of anodic and cathodic reaction, flow of charges (e) and flow of current of a metal piece emerged into acidic electrolyte is shown in Fig. 3.2. At anode, oxidation of a metal (M) discharges electrons (ne^-) that pass through the electrical connection and reach to the cathodic site. As it is well-established, in acidic medium reduction of hydrogen is the main cathodic reaction, discharge of hydrogen gas occurs by the reduction of protons (H^+). In this process, flow of electrons (charge) occurs from anode to cathode and current flows (presented in the form of corrosion current) from cathodic to anodic direction.

One of the common examples of electrochemical corrosion which involves the formation of two half-cell reactions, i.e., anodic and cathodic half-cell reactions are rusting of iron. Anodic half-cell reaction results in the oxidation of iron to ferrous (Fe^{2+}) and ferric (Fe^{3+}) ions and discharge of the electrons. These electrons consume at cathode for the reduction of atmospheric oxygen into the water (Eq. 3.8). Under the same circumstances, water reduces to form hydroxide ions as presented in Eq. (3.9). The ferrous (Fe^{2+}) and ferric (Fe^{3+}) ions react with hydroxide ions to form corresponding ferrous [$Fe(OH)_2$] and ferric [$Fe(OH)_3$] hydroxide, i.e., rust.

$$Fe^{2+} + 2OH^- \rightarrow Fe(OH)_2 \tag{3.15}$$

$$Fe^{3+} + 3OH^- \rightarrow Fe(OH)_3 \tag{3.16}$$

Ferrous hydroxide, $Fe(OH)_2$ can be converted into $Fe(OH)_3$ in the presence of oxygen and moisture (water).

$$4Fe(OH)_2 + O_2 + 2H_2O \rightarrow 4Fe(OH)_3 \quad (Rust) \tag{3.17}$$

The $Fe(OH)_3$ so formed can be breaks in hydrated oxide forms (more common form of the rust according to the following equations:

$$4Fe(OH)_3 \rightarrow 2Fe_2O_3 \cdot 6H_2O \quad (Rust) \tag{3.18}$$

Anodic and cathodic half-cell reactions and rusting of iron in an open atmosphere is presented in Fig. 3.3.

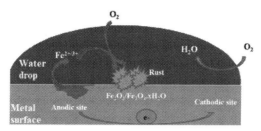

FIG. 3.3 Diagrammatic illustration of rusting of iron in the presence of oxygen and moisture.

Electrochemical or wet corrosion can be categorized into three classes depending upon the nature of anodic and cathodic sites developed over the metallic surface.

1. *Separable anode/cathode type (Sep. A/C)*: in this type of electrochemical reactions, a certain discrete area of the metal surface behaves as principally cathodic or anodic in nature. Generally, separation of these areas is very difficult and can be identified using only experimental studies as this ranges in fraction of a millimeter (mm). In this type of corrosion reaction, there will be a macroscopic flow of charge (electrons). Some common examples of Sep. A/C electrochemical reactions include: "long line" corrosion of buried transport pipelines, crevice corrosion, and iron corrosion in the presence of porous magnetite and oxygenated water.
2. Inseparable anode/cathode type (Insep. A/C): in this type of electrochemical reactions, though the survival of anodic and cathodic areas can be hypothetically predicted but they cannot be recognized by means of experimental techniques. Some common examples of Insep. A/C include: uniform corrosion of metals in alkaline (basic), neutral, acidic, and fused non-aqueous solutions, e.g., corrosion of zinc or iron in hydrochloric acid solution and corrosion if iron in oxygenated water or in reducing acids.
3. Interfacial anode/cathode type (interfacial A/C): in this type of electrochemical reactions, one complete interface acts as anode and the other interface behaves as a cathode. Some common examples of interfacial A/C include reaction of iron with oxygen at high temperature, reaction of iron with water at high temperature, and reaction of copper with SO_2 in carbon dioxide.

3.3 Factors affecting corrosion and corrosion rate

Since corrosion results through the interactions between the metal and environments therefore factors affecting corrosion and corrosion rate can be categorized into two categories. First category of factors is associated with metal, and the second category of factors is connected with the surrounding environment.

3.3.1 Factors related to metal

Factors related to the nature of metal that affect corrosion include the following:

(i) *Position in galvanic series*: difference metals can be arranged in their increasing or decreasing order of oxidation potential. A metal with higher value of oxidation potential is expected to corrode more quickly with the metal with lower value of oxidation potential. If two metals having different oxidation potentials are combined together then metal with higher oxidation potential behaves as anode while other metal acts as cathode in the couple.

(ii) *Over voltage (η)*: potential developed over the metallic surface with respect to the standard hydrogen electrode potential is known as electrode potential, E. Generally, difference in the electrode potential (E) and equilibrium electrode potential (E_0) is referred to as overvoltage (η). For the case where hydrogen evolution is a major cathodic reaction, lowering in the hydrogen overvoltage is consistent with the high corrosion rate as in such situations discharge of the hydrogen will be easy. An increase in the cathodic reaction causes subsequent increase in the rate of anodic reaction and ultimately corrosion.

(iii) *Cathode/anode ratio*: rate of corrosion greatly depends upon the ratio of cathode and anode. Generally, a higher value of cathodic/anodic ratio is consistent with high corrosion rate, i.e., high anodic area and/or lower cathodic area favors the corrosion.

$$\text{Corroion rate } (C_R) \propto \frac{\text{cathodic area}}{\text{anodic area}} \tag{3.19}$$

(iv) *Purity of metals*: generally, possibility and rate of corrosion increase in the presence of impurity. Presence of impurities makes the metallic surface inhomogeneous (or heterogeneous) help in setting down of the local electrochemical cells over the metallic surface. This type of setting of electrochemical cells results into the initiation and propagation of corrosion. For instance, corrosion of zinc in the presence of Pb or Fe impurity proceeds through the formation of electrochemical galvanic cells. Obviously, the rate of this type of corrosion is expected to increase on increasing the exposure time and extent of impurity.

(v) *Metallic physical states*: corrosion rate highly depends upon the physical state of the metal, such as stress, crystal structure, and grain size. In general, a smaller grain size is consistent with higher solubility and corrosion rate. More so, areas under stress become anode where rapid degradation of metal occurs through corrosion.

(vi) *Nature of corrosion products*: corrosion products developed over the metallic surface may be or may not be protective in nature. The behavior of corrosion products can be derived through "Pilling-Bedworth ratio" or "specific volume ratio" as described earlier. Generally, volatile and soluble corrosion products do not provide protection as they are continuously removed from the surface and allow newer areas to interact and corrode.

3.3.2 Factors related to environment

Factors related to the nature of environment that affect corrosion include the following:

(i) Temperature: generally, increase in temperature causes subsequent increase in the corrosion rate. Nevertheless, the rate of the wet corrosion decreases on increasing the temperature which is attributed due to the decrease in the solubility of the dissolved gases on elevating the temperature. It is important to mention that intergranular corrosion occurs only at high temperature. There is a thrum rule that the rate of corrosion becomes double for every rise of 10°C. The intergranular corrosion takes place only at higher temperature.

(ii) pH: pH exerts a significant effect on the corrosion rate. Relative to basic and neutral electrolytes, rate of corrosion is higher in acidic electrolytes, i.e., lowering in the pH causes an increase in the corrosion rate. Therefore, the corrosion rate of the metal can be decreased by increasing pH (pH > 7). For instance, the corrosion rate for iron corrosion increases on decreasing the pH. However, corrosion rate of few metals especially amphoteric metals such as Al, Zn, and Pb increases on increasing pH. Increased corrosion rate of these metals on increasing the pH proceeds through the formation of hydroxide complexes.

(iii) Oxygen concentration: generally, an increase in the oxygen concentration results in the increase in the corrosion rate. This is attributed due to the setting of differential aeration concentration cells. Increase in the concentration of oxygen also results in the increase in the rate of cathodic half-cell reaction according to Eqs. (3.8) and (3.9) that ultimately increases the overall rate of corrosion.

(iv) Fluid velocity: the rate of corrosion is expected to increase on increasing the fluid velocity. At high fluid velocity, rate of corrosion increases because of the combined effect of mechanical erosion-corrosion and dissolution of corrosion protective surface films and corrosion products.

(v) Suspended particles in atmosphere: corrosion rate is expected to increase in the presence both chemically inactive, such as charcoal and chemically active such as $(NH_4)_2SO_4$ and $NaCl$ suspended particles. Generally, chemically active suspended particles increase corrosion rate by making the atmosphere more corrosion because of their aggressiveness and through the adsorption of corrosive gases. On the other hand, chemically inactive suspended particles absorb the corrosive gases and make the electrolyte relatively more corrosive.

(vi) Conducting behavior of the electrolyte: since corrosion involves the charge transfer process therefore possibility and rate of corrosion in conducting electrolytes will be higher than that of the nonconducting electrolytes. For instance, the rate of soil corrosion is always higher in the presence of salts and moisture than in their absence.

(vii) Nature of counter ions of electrolytes: counter ions play a significant role in determining the corrosiveness of the medium. Some common counterions are Cl^-, SO_4^-, and $^-NH_2$, etc., and they are expected to increase the corrosion rate in their presence.

3.4 Summary

Corrosion can be classified into different forms on the basis of how corrosion manifests itself. One of the most common forms of corrosion is uniform corrosion where metallic surfaces corrode uniformly. Corrosion is the natural tendency of metallic materials to react with the constants of the surrounding environment. Corrosion may be classified as dry or wet corrosion. Generally, dry or chemical corrosion occurs in moisture-free conditions by the attacks of gases like O_2, CO_2, SO_2, H_2, H_2S, NO_2, Cl_2, F_2, etc. Dry corrosion in the presence of oxygen is known as oxidative dry corrosion. On the other hand, dry corrosion by remaining gases is known as nonoxidative dry-corrosion. Wet or electrochemical corrosion occurs in the presence of moisture and proceeds through electrochemical mechanisms. There are numerous factors that affect the rate of dry- and wet-corrosion. These factors can be divided into two classes; one that related to the metal and second that related to the environment. Temperature is the most important environmental factor that affects the rate of corrosion. Generally, every 10°C rise in temperature results in an increase of corrosion rate by two times.

3.5 Useful links/websites

https://corrosion-doctors.org/Corrosion-Thermodynamics/Introduction.htm
https://www.accessengineeringlibrary.com/content/book/9780071482431/chapter/chapter4
https://corrosion-doctors.org/Electrochemistry-of-Corrosion/Introduction.htm
https://www.sciencedirect.com/topics/materials-science/electrochemical-corrosion

References

[1] M. Pourbaix, Thermodynamics and corrosion, Corros. Sci. 30 (1990) 963–988.
[2] J. Van Muylder, Thermodynamics of corrosion. In: Electrochemical Materials Science, Springer, Boston, MA, 1981, pp. 1–96.
[3] R.W. Revie, Corrosion and Corrosion Control: An Introduction to Corrosion Science and Engineering, John Wiley & Sons, USA, 2008.
[4] C. Verma, E.E. Ebenso, M. Quraishi, Ionic liquids as green and sustainable corrosion inhibitors for metals and alloys: an overview, J. Mol. Liq. 233 (2017) 403–414.
[5] J. Colson, J. Larpin, High-temperature oxidation of stainless steels, MRS Bull. 19 (1994) 23–25.

CHAPTER 4

Concept of green chemistry in corrosion science

4.1 Concept of green corrosion inhibition

Corrosion is a highly damaging and challenging phenomenon; therefore various methods have been developed for its mitigation depending upon the nature of metal and surrounding environment. In view of high economic and safety risks the use of corrosion inhibitors is one of the most effective and economic methods of corrosion protection. Corrosion inhibitors are the chemicals or their mixture that minimize or prevent the corrosion. One of the oldest (before 1960) was use of inorganic compounds, including borates, molybdates, silicates, zinc salts, chromates, nitrates, etc. Inorganic inhibitors can be divided into two categories, depending upon their major influence on anodic or cathodic half-cell reaction. Most of the inorganic inhibitors become effective by forming a passive (oxide) film. These are called passivators or anodic-type corrosion inhibitors. However, some of them become effective by precipitating in the form of their oxides at the cathodic site. These are called precipitating or cathodic-corrosion inhibitors. However, inorganic compounds are relatively expensive and toxic; they were replaced by relatively less expensive polyphosphonates, surface activated chelates, carboxylates, phosphonates, gluconates, polyacrilates, and phosphonic acid.

Nevertheless, recent use of these chemicals is strictly restricted because of their toxicity. These chemicals are highly toxic therefore adversely affect the aquatic and soil life after their discharge. More so, some of the inorganic compounds possess the ability of bioaccumulation in living organisms including humans. Because of the increasing ecological awareness and strict environmental regulations, use of the chemicals has been totally stopped and they are replaced by relatively more eco-friendly organic corrosion inhibitors. Nowadays, the use of organic compounds has been associated with the concept of E4, i.e., economy, ecology, efficiency, and eco-friendly. Eco-friendly corrosion inhibitors may be of natural or designing origin. Natural eco-friendly corrosion inhibitors include plant extracts, bio- and natural-polymers, amino acids (AAs) and proteins, tannins and vitamins. Organic compounds derived through proper designing that avoids the use of toxic chemicals or solvents using energy-efficient heatings are also regarded as eco-friendly corrosion inhibitors. These include compounds derived through solid state reactions (SSRs), multicomponent reactions (MCRs), mechanochemical mixing (MCMs), and ultrasound (US) and microwave (MW) irradiations.

Obviously, these synthetic approaches are associated with several advantages including high synthetic efficiency, i.e., high atom economy, shorter reaction time, and ease to perform and lower number works-up and purification steps.

4.2 Assessment of green corrosion inhibitors: OSPAR and TEACH commission

Researches on the use of environmental friendly corrosion inhibitors and practices are gaining particular attention because of the high demands of "green chemistry" and ecological consciousness. The environmental friendly property of the corrosion inhibitors can be determined by assessing their toxicity for living organisms and to the environment, bioaccumulation, and biodegradability. Oslo Paris commission (OSPAR) and registration, evaluation, authorization and restriction (REACH) signify the international commissions that are used to appraise the above parameters. REACH represents a European Union regulation which was held on December 18, 2006, and was subjected to implement on June 7, 2007 [1,2]. An 849 pages report of European Union regulation is connected with the assessment of designing, production, and use of chemical substances and their impact of the human health and environment. OSPAR was subjected to implementation on September 22, 1992, which as set-up a number of guiding principles and parameters to assess the effect of chemical substances in the term of their toxicity, biodegradability, and bioaccumulation. These guiding principles also help in the assessment of the effect of the chemical substances on the environment and human health.

Toxicity a chemical can be determined using the effect of that chemical on a cell, or a tissue or a living organism. It is well established that toxicity of any chemical depends upon the gender, body weight, and age; therefore a population-level study is generally used to determine the toxicity of the chemical. Toxicity of a chemical can be defined in the term of its EC_{50} and LC_{50} concentrations. Generally, EC_{50} (effective concentration) represents the concentration or dose required to adversely affect the growth of 50% of a specific animal population. On the other hand, LC_{50} (lethal concentration) or LD_{50} (lethal dose) represents the concentration of the chemical which is required to kill 50% of a specific animal population. Obviously, a high value of EC_{50}/LD_{50} is consistent with lower toxicity and vice versa. The chemical substances having EC_{50}/LD_{50} values greater than 10 mg/Kg can be treated as nontoxic. EC_{50}/LD_{50} values vary depending upon the route of administration and animal species. For example, acute oral LD50 value for wormseed oil in mice is 380 mg/Kg and 255 mg/Kg in rate. Various chemical substances can be presented in the form of toxicity rating. The toxicity rating of dichlorvos for its single oral dose to rats is presented in Table 4.1.

Most of the chemical substances naturally undergo degradation or breakdown through microorganisms such as bacteria, fungi, and s protozoa. These microorganisms are called decomposers. Generally, the process of natural decomposition is very slow and takes several days, months, and sometimes years. Chemical substances can be divided into environmental friendly or toxic depending upon the time required for their decomposition. Generally, a chemical substance that decomposes up to 60% or more in 28 days can be treated as environmental friendly. Bioaccumulation of a chemical is defined as the ability of a chemical substance to accumulate in the living organism whose source of chemical is water. Bioaccumulation

TABLE 4.1 Toxicity rating of dichlorvos for its single oral dose to rate.

Toxicity rating	Commonly used term	LD50
1	Extremely toxic	1 or less
2	Highly toxic	1–50
3	Moderately toxic	50–500
4	Slightly toxic	500–5000
5	Practically nontoxic	5000–15,000
6	Relatively harmless	15,000 or more

is measured in the term of partition coefficient which is the ratio of concentrations of a chemical substance in a mixture of two immiscible solvents at equilibrium. Generally, partition coefficient is presented in the form of P_{OW} or K^{OW}. Most commonly used solvent mixture is water (hydrophilic) and octanol (hydrophobic). Therefore, partition coefficient defines how hydrophilic or hydrophobic a chemical is. Partition coefficient (P_{OW}) is different from the distribution coefficient (D_{OW}). Generally, partition coefficient is defined as the distribution of a unionized chemical in a mixture of two immiscible liquids whereas distribution coefficient defines the distribution of both ionized and unionized chemical substances in a mixture of two immiscible mixtures. The P_{OW} and D_{OW} can be presented as follows [3,4]:

$$\log P_{OW} = \left(\frac{[\text{solute}]_{\text{octane}}^{\text{un-ionized}}}{[\text{solute}]_{\text{water}}^{\text{un-ionized}}} \right) \quad (4.1)$$

$$\log D_{OW} = \left(\frac{[\text{solute}]_{\text{octane}}^{\text{un-ionized}} + [\text{solute}]_{\text{octane}}^{\text{ionized}}}{[\text{solute}]_{\text{water}}^{\text{un-ionized}} + [\text{solute}]_{\text{water}}^{\text{ionized}}} \right) \quad (4.2)$$

For an environmental friendly compound, P_{OW} or K_{OW} value to should less than 3.

4.3 Green chemistry principles and green corrosion inhibition

In simple terms, green chemistry is designing and modification of chemical processes that minimizes or retards the implementation or production of toxic materials. Green chemistry is also known as sustainable chemistry. Generally, green chemistry involves a set of principles and covers the synthesis, designing, use, and disposal of a chemical product, i.e., green chemistry:

- Minimizes environmental pollution at the molecular level.
- Includes all branches of chemistry not a single discipline of chemistry.
- Reduces the adverse effect of processes and chemical products on human health and on the environment.
- Sometimes minimize the discharge of toxic chemicals from previously existing products and processes.
- It also includes modifications and designing of existing products and processes to diminish their intrinsic hazards.

For the green chemist, the sequence looks like as follows:

- Designing of chemical products that exert less hazardous effect on the environment and human health.
- Making the feedstock of the solvents, chemicals, and reagents that are less hazardous effect on the environment and the human health.
- Designing of the processes and practices that avoid or minimize the formation of chemical waste.
- Designing of the processes and practices that avoid or minimize the use of energy or solvents.
- Use of chemicals products derived from natural and renewable resources or collected (extracted) from the waste.
- Recycling and reuse of chemical products and chemicals.

Recently, corrosion scientists and engineers developed numerous environmental friendly approaches for corrosion inhibition. The attempts of green corrosion inhibition can be categorized as follows:

(a) Synthetic green corrosion inhibitors.
(b) Green corrosion inhibitors through proper designing.
(c) Green compounds through natural resources.
(d) Green corrosion inhibition practices.

4.3.1 Synthetic green corrosion inhibitors

It is recalled that synthetic compounds represent as one of the most effective and economic classes of corrosion inhibitors. The cost-effectivity of the synthetic corrosion inhibitors is due to their ease of synthesis using inexpensive starting materials in high yield. However, not all synthetic approaches can be treated as equally important as some of them are associated with the lower yield and use of expensive starting chemicals, solvents and catalysts. Mainly, compounds synthesized using multistep reactions (MRs) are associated with several drawbacks including tedious, time-consuming, and low synthetic yield properties. Because of their multistep nature, these reactions are associated with the numerous works-up and purification processes that ultimately reduce their yields. The MRs are also associated with huge discharge of toxic chemicals, solvents, and catalysts into the surrounding environment due to multiple purification steps. This type consumption and discharge of expensive chemicals and solvents along with adversely affecting the environment make the reaction highly costly.

Unlike to this, MCRs that combine three or more reactant molecules in a single step have several advantages and emerged as one of the greenest synthetic protocols. The MCRs have the following advantages:

(i) Ease to perform due to one-step nature.
(ii) High synthetic yield, i.e., high atom economy due to reduction in the number of purification and works-up processes.
(iii) Avoid or reduce in the use of toxic and expensive solvents as the majority of the MCRs are carried out in solid phase.
(iv) Inexpensive and time-saving.

(v) Reduced environmental pollution risks as MCRs are associated with minimum production and discharge of side (waste) products.
(vi) High selectivity.
(vii) Facile automation.
(viii) Ease purification.

Because of the association of MCRs with the above advantages, compounds derived from MCRs are regarded as sustainable compounds. It is important to mention that several classes of organic compounds derived through MCRs are used as effective green corrosion inhibitors for different metals and alloys in various electrolytes [5].

Corrosion inhibitors derived from energy-efficient MW and US irradiations represent another class of green corrosion inhibitors. The advantages of these nonconventional irradiations include [6]:

- Rapid and uniform (homogeneous) heatings and activation of reactant molecules.
- High yield (atom economy) and minimum waste (side) products formation.
- High selectivity and possibility of modification of selectivities.
- Simplification of purification and work-up procedures.
- Reduced environmental pollution as MW and US irradiations as connected with production of formation of side (waste) products.
- Inexpensive and time-saving.
- Improved synthetic efficiency.

Obviously, because of their ease of handling and efficient heatings, MW and US irradiations are time-saving and cost-effective [6]. Clark reported an 85-fold increase in the rate of Suzuki coupling reaction catalyzed by MW irradiation with respect to the conventional heating [7]. It is important to mention that MW and US irradiations fulfill several requirements of the green chemistry and 12 principles of green chemistry. In fact, MW and US heatings in combination with MCRs represent one of the greenest synthetic approaches. Apart from the compounds derived from MCRs with and without MW and US heatings, solid supported synthesis, MCM, and solid phase reactions represent some other common alternative synthetic route for the production of green corrosion inhibitors. These syntheses are connected with lower waste generation and use and reduced use and discharge chemicals and solvents.

Water and supercritical CO_2 are established as environmental friendly solvents for the synthesis of numerous biologically and industrially useful products. Compounds synthesized in the water and CO_2 can be treated as environmentally sustainable. These solvents, particularly water, acquire several benefits including inexpensivity, free, and commercial availability, nontoxic and nonhazardous, unique redox stability and nonflammable. Supercritical CO_2 satisfies most of the characteristics of an ideal green solvent (Table 4.2). It can be seen from Table 4.2 that, similar to water, CO_2 fulfills all the basic requirements of an ideal green solvent.

4.3.2 Green corrosion inhibitors through proper designing

There are numerous classes of compounds that are environmentally friendly in nature because of their properties. Ionic liquids (ILs) represent the best example of this type of

TABLE 4.2 Comparison of common characteristics supercritical CO_2 with an ideal green solvent.

S. No.	Characteristics of an ideal green solvent	Supercritical CO_2
1	Easy to prepare/synthesis	Yes
2	Abundant/availability	Yes
3	Nontoxic	Yes
4	Nonflammable and easy to handle	Yes
6	Renewable and reuse	Yes
7	Eco-friendly	Yes
8	Easy to remove from a product	Yes
9	Does not contribute to global warming	Yes. As supercritical CO_2 can suitably be processed with minimum discharge
10	Noneutrophying	Yes
11	Does not participate to fog	Yes

compound. ILs are salts in liquid state and they exist in liquid state below 100°C. ILs are also called as ionic melts, liquid electrolytes, ionic glasses, fused salts, ionic salts, or liquid fluids. Generally, ILs consist of organic cations and inorganic anions. Imidazolium, pyridinium, tetra-ammonium, and phosphonium represent the major cations present in the ILs. Some of the common anions present in the ILs and their abbreviations are given in Table 4.3.

Because of their lower vapor pressure, ILs are considered as environmental friendly alternatives to be used for numerous industrial and biological applications including in the field of corrosion inhibition. Because of their several fascinating properties such as nonhazardous, low volatility, nonflammability, and high thermal, as well as chemical stabilities, ILs have been widely used for different applications. More so, properties of ILs can be suitably tailored by suitably varying the combination of cation and/ or anion. Literature study shows that several classes of ILs are extensively used as green and effective corrosion inhibitors for various metals and their alloys in almost all kinds of electrolytes [8]. Practically it has been observed that ILs show reasonably good corrosion inhibition efficiency. This is attributed due

TABLE 4.3 Name, chemical formula, and abbreviation of major anions present in ionic liquids.

S. No.	Full name	Chemical formula	Abbreviation
1	Nitrate	NO_3^-	$[NO_3]$
2	Acetate	CH_3COO^-	[Ac]
3	Hexaflurophosphate	PF_6^-	$[PF_6]$
4	Trifluoroacetate	CF_3COO^-	[TFA]
5	Tetrafluoroborate	BF_4^-	$[BF_4]$
6	Trifluromethylsulfonate	$CF_3SO_4^-$	[TfO]
7	Methylsulfate	$CH_3SO_4^-$	$[MeSO_4]$
8	Bis[(Trifluoromethyl) sulfonyl] amide	$(CF_3SO_2)_2N^-$	$[Tf_2N]$

to the presence of several electron-rich adsorption centers as well as due to their high solubility in the polar electrolytes. Protection efficiency of ILs can be suitably enhanced for different metals and electrolyte systems by varying the nature of cation and anion. Another example of this type of compound which is extensively used as green corrosion inhibitors for different metals and alloys is polyethylene glycol (PEG) [9,10]. PEG is a well-established inexpensive and green chemical to be used for various industrial and biological applications.

4.3.3 Green corrosion inhibitors derived through natural resources

Chemical species derived from natural resources always preferred to be used as environmental friendly alternatives for several industrial and biological purposes. Chemical species derived from natural resources are widely used as corrosion inhibitors for different metals and alloys. One of the most common examples of natural corrosion inhibitors is plant extract [11,12]. Because of their natural and biological origin, plant extracts are considered as environmental friendly and suitable materials to be used as corrosion. Extracts derived through different parts of the plants, such as leaves, barks, fruits, flowers, and peels, etc., are extensively used as corrosion inhibitors for different metal-electrolyte systems. Generally, each extract contains various phytochemicals that possess electron rich centers such as polar functional groups and aromatic ring(s) through which they adsorb effectively and exhibit metallic corrosion. It has also been observed that leaves extracts show better corrosion protection as compared to the other extracts.

Another class of green corrosion inhibitors derived from natural resources is chemical medicines (drugs). It is important to mention that there are numerous drugs, especially ayurvedic medicines derived from plants and other natural resources. Chemical medicines are widely used as corrosion inhibitors because of their high protection effectiveness [13,14]. The high inhibition efficiency of the drug molecules is attributed due to their complex structures that acquire several electron-rich centers including polar functional groups such as $-OH$, $-NO_2$, $-COOH$, $-CONH_2$, $-COCl$, and $-COOC_2H_5$, etc., and aromatic ring(s) through which they can easily get adsorbed. Though, use of the medicines derived from natural resources should be encouraged however implementation of the synthetic medicines cannot be treated as environmental friendly. Generally, synthesis of medicines requires an extremely high level of work-up and ultra-purification that make their synthesis expensive. More so, synthesis of the chemical medicines requires MRs that use and discharge huge amount of toxic solvents and chemicals into the surrounding environment and pollute it.

Carbohydrates, AAs, and their derivatives are also widely used as corrosion inhibitors for different meta-electrolyte combinations. Because of their natural and biological origin, carbohydrates and AAs along with their derivatives are treated as environmental friendly alternatives to be used as corrosion inhibitors. Despite of their complex molecular structures, carbohydrates, and their derivatives are highly soluble in polar electrolytes which is attributed due to their polar functional groups such as $-OH$, $-OMe$, $-CONH_2$, $-NH_2$, etc. Due to their complex structure, carbohydrates and their derivatives provide excellent metallic surface coverage that comes out in the form of their high inhibition efficiency [15]. Literature study showed chitosan and cellulose are the most frequently used carbohydrates based polymers that are used as corrosion inhibitors most extensively. Along with polysaccharides, mono- and oligo-saccharides have also been used extensively as corrosion inhibitors.

Similarly, AAs and its derivatives including heterocyclic compounds and proteins are extensively used as corrosion inhibitors in different metal-electrolyte systems.

Likewise, oleochemicals that are the chemicals derived from natural oils and fats can also be treated as natural resources to be used as corrosion inhibitors. Generally, these chemicals possess polar functional groups (hydrophilic) through which they easily get adsorbed and their nonpolar hydrophobic chain(s) act as water repellent, i.e., oleochemicals behave as surfactants. Literature study shows that several reports are published in which oleochemicals or their derivatives are used as corrosion inhibitors [16].

4.3.4 Green corrosion inhibition practices

Nowadays, computational modelings have emerged as environmental practices for the designing of effective corrosion inhibitors. The greenness of these modelings is based on the fact that various reactivity informations as well as reactivity parameters of a molecule can be determined before their synthesis. Generally, synthesis of organic compounds is always an expensive and nonenvironmental friendly approach. Therefore, computational simulations that provide useful reactivity parameters are gaining particular attention. These simulations provide several parameters in the term of which relative chemical reactivity including corrosion inhibition effect can be suitably described. It is well established that metal-inhibitors interaction involves charge (electrons) sharing. The molecular sites responsible for electron donation and acceptance can be derived through density functional theory [17,18]. Obviously, in the frontier molecular orbitals, highest occupied molecular orbital (HOMO) is responsible for electron donation whereas lowest unoccupied molecular orbital (LUMO) is responsible for electron acceptance. Therefore, higher values of E_{HOMO} and lower values of E_{LUMO} are consistent with high inhibition efficiency. Several other reactivity parameters such as electronegativity (χ), hardness (η), softness (σ), electrophilicity (ω), nucleophilicity (ε), fraction of electron transfer (ΔN), etc., can also be derived using the values of E_{HOMO} and E_{LUMO}.

Molecular dynamics (MD) and Monte Carlo (MC) simulations are common computational modelings through which orientation of the inhibitor molecules on metallic surface can be derived [19]. Generally, an inhibitor molecule with flat or horizontal orientation would cover a larger metallic surface and would behave as a better corrosion inhibitor as compared to the compound having vertical orientation. Along with orientation, MD and MC simulations provide information whether the interaction between inhibitor molecules is spontaneous or not. Generally, negative value of adsorption energy (E_{ads}) indicates that inhibitor spontaneous interacts with the metallic surface. In most of the cases, values of E_{ads} for metal–inhibitor interaction are found to be negative. Artificial neural networks and quantitative structure-activity relationships are two other commonly used computational modeling methods being used for the effectiveness of a compound toward its interaction with metallic surfaces.

4.4 Summary

Because of the increasing ecological awareness and strict environmental regulations, use of traditional corrosion inhibitors is highly restricted. Recently, corrosion scientists and engineers are trying to develop green corrosion inhibitors and green corrosion inhibition practices. Green corrosion inhibitors can be classified as natural, synthetic, or designed-type. Natural green corrosion inhibitors include plant extracts, natural medicines, carbohydrates, AAs, and their derivatives. Chemical species derived from raw natural materials can also be treated as green alternatives to be used as corrosion inhibitors. Compounds derived through MCRs with and without MW and US irradiations, MCMs, solid phase reactions, solid supported reactions (SSRs), etc., represent the major synthetic green corrosion inhibitors. More so, compounds synthesized in green solvents such as ILs, water, and supercritical CO_2 can also be considered as green. Designed green corrosion inhibitors are those whose properties including corrosion inhibition effectiveness and effect on the environment can be suitably modified. This class of corrosion inhibitors includes ionic liquids and PEG. Along with green corrosion inhibitors, corrosion monitoring with computational simulations including density functional theory and MD and MC simulations can be treated as environmental friendly approaches of the corrosion mitigation as they do not involve the use of any toxic chemicals and expensive instruments.

4.5 Useful websites

https://www.epa.gov/greenchemistry/basics-green-chemistry#:~:text=Green%20chemistry%20is%20the%20design,%2C%20use%2C%20and%20ultimate%20disposal

https://www.sigmaaldrich.com/chemistry/greener-alternatives/green-chemistry.html?gclid=CjwKCAiAkan9BRAqEiwAP9X6UWCBQmVvEzK1o-SqK3mv-wozo6Ow5fHLMOVseb3v3UMJQPAAPPpEyBoCzFYQAvD_BwE

References

[1] J. Zielonka, Europe as Empire: The Nature of the Enlarged European Union, Oxford University Press, UK, 2007.
[2] M.A. Schreurs, Y. Tiberghien, Multi-level reinforcement: explaining European Union leadership in climate change mitigation, Global Environ. Polit. 7 (2007) 19–46.
[3] C.T. Chiou, V.H. Freed, D.W. Schmedding, R.L. Kohnert, Partition coefficient and bioaccumulation of selected organic chemicals, Environ. Sci. Technol. 11 (1977) 475–478.
[4] W.B. Neely, D.R. Branson, G.E. Blau, Partition coefficient to measure bioconcentration potential of organic chemicals in fish, Environ. Sci. Technol. 8 (1974) 1113–1115.
[5] C. Verma, J. Haque, M. Quraishi, E.E. Ebenso, Aqueous phase environmental friendly organic corrosion inhibitors derived from one step multicomponent reactions: a review, J. Mol. Liq. 275 (2019) 18–40.
[6] C. Verma, M. Quraishi, E.E. Ebenso, Microwave and ultrasound irradiations for the synthesis of environmentally sustainable corrosion inhibitors: an overview, Sustain. Chem. Pharm. 10 (2018) 134–147.
[7] G. Stefanidis, A. Stankiewicz, Alternative Energy Sources for Green Chemistry, Royal Society of Chemistry, UK, 2016.
[8] C. Verma, E.E. Ebenso, M. Quraishi, Ionic liquids as green and sustainable corrosion inhibitors for metals and alloys: an overview, J. Mol. Liq. 233 (2017) 403–414.
[9] S. Umoren, O. Ogbobe, P. Okafor, E. Ebenso, Polyethylene glycol and polyvinyl alcohol as corrosion inhibitors for aluminium in acidic medium, J. Appl. Polym. Sci. 105 (2007) 3363–3370.
[10] M. Abdallah, H. Megahed, M. Radwan, E. Abdfattah, Polyethylene glycol compounds as corrosion inhibitors for aluminium in 0.5 M hydrochloric acid solution, J. Am. Sci. 8 (2012) 49–55.
[11] C. Verma, E.E. Ebenso, I. Bahadur, M. Quraishi, An overview on plant extracts as environmental sustainable and green corrosion inhibitors for metals and alloys in aggressive corrosive media, J. Mol. Liq. 266 (2018) 577–590.
[12] P.B. Raja, M.G. Sethuraman, Natural products as corrosion inhibitor for metals in corrosive media—a review, Mater. Lett. 62 (2008) 113–116.
[13] G. Gece, Drugs: a review of promising novel corrosion inhibitors, Corros. Sci. 53 (2011) 3873–3898.

[14] M. Abdallah, Rhodanine azosulpha drugs as corrosion inhibitors for corrosion of 304 stainless steel in hydrochloric acid solution, Corros. Sci. 44 (2002) 717–728.

[15] S.A. Umoren, U.M. Eduok, Application of carbohydrate polymers as corrosion inhibitors for metal substrates in different media: a review, Carbohydr. Polym. 140 (2016) 314–341.

[16] M. Quraishi, N. Saxena, D. Jamal, Inhibition of mild steel corrosion by oleochemical based hydrazides, Indian Journal of Chemical Technology (IJCT) 12 (2005) 220–224.

[17] I. Obot, D. Macdonald, Z. Gasem, Density functional theory (DFT) as a powerful tool for designing new organic corrosion inhibitors. Part 1: an overview, Corros. Sci. 99 (2015) 1–30.

[18] P. Udhayakala, T. Rajendiran, S. Gunasekaran, Theoretical approach to the corrosion inhibition efficiency of some pyrimidine derivatives using DFT method, J. Comput. Methods Mol. Des. 2 (2012) 1–15.

[19] C. Verma, H. Lgaz, D. Verma, E.E. Ebenso, I. Bahadur, M. Quraishi, Molecular dynamics and Monte Carlo simulations as powerful tools for study of interfacial adsorption behavior of corrosion inhibitors in aqueous phase: a review, J. Mol. Liq. 260 (2018) 99–120.

CHAPTER 5

Classification of corrosion inhibitors

Corrosion inhibitors are chemical species that inhibit or decrease the corrosion rate in their presence at low concentration. Corrosion inhibitors can be divided into various classes based on the nature, source of origin, mechanism of inhibition, effect on environment, and mode of adsorption on the metallic surface. Broadly corrosion inhibitors may be classified as organic and inorganic types. They can be further divided into different subclasses depending upon their various aspects. A broad classification of corrosion inhibitors is presented in Fig. 5.1. Some main organic corrosion inhibitors are heterocyclics, different families of organic compounds, polymers, and their composites, oligomers and their composites and nanomaterials. Major inorganic corrosion inhibitors are metal oxides and their organic and inorganic mixed composites.

5.1 Classification of inorganic corrosion inhibitors

Inorganic corrosion inhibitors can be divided into two classes, namely anodic- and cathodic-type depending upon their inhibitive influence on either of the corrosion reactions. Inorganic compounds that mainly affect the rate of anodic reactions are called as anodic corrosion inhibitors. Generally, these compounds become effective by forming the surface protective oxide film (passive film). The process of formation of oxide film generally referred as passive film or passive layer in known as passivation and such inhibitors are known as passivating inhibitors or simply passivators. Passivators may be further classified into oxidizing anions and un-oxidizing anions. Chromate, nitrite, and nitrite passivate the metallic surface even in the absence of oxygen and they are called oxidizing anions. On the other hand, tungstate, phosphate, and molybdate passivate the metallic surface only in the presence of oxygen. These anions are called nonoxidizing anions.

Cathodic corrosion inhibitors can be divided into three classes, namely, cathodic poisons, cathodic precipitates, and scavengers. Chemical species that slow down the rate of cathodic reactions are known as cathodic poisons. For example, antimony and arsenic create the connection of hydrogen more difficult and slow down the cathodic hydrogen evolution reaction. Some of the inorganic elements precipitate on cathodic sites in the form of their oxides and retard or slow down the rate of cathodic reactions. These elements are known as cathodic precipitates. For instance, magnesium, calcium, and zinc precipitate in the form of their

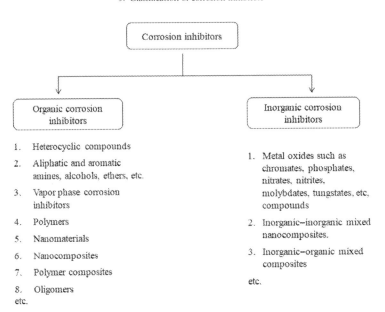

FIG. 5.1 Classification of corrosion based on their inorganic and organic nature.

oxides and act as cathodic precipitates. Generally, these precipitates avoid the diffusion of corrosive species, such as aggressive ions such as chloride and sulfate. It has been observed that the rate of cathodic reaction is faster in the presence of oxygen therefore cathodic reaction can be controlled by removing the oxygen. Some chemical species such as methyl sulfate and hydrazine are used to remove the oxygen from the corrosive environments. These chemicals are known as oxygen scavengers. A brief classification of inorganic corrosion inhibitors is outlined in Fig. 5.2.

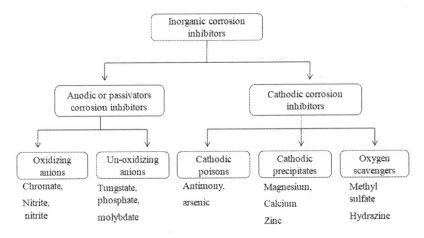

FIG. 5.2 Classification of inorganic corrosion inhibitors.

5.2 Classification of organic corrosion inhibitors

Various classifications of organic corrosion inhibitors have been proposed. Some of the major classifications are described below.

5.2.1 Classification of organic compounds based on the origin

Based on the origin organic compounds may be classified as synthetic corrosion inhibitors and natural corrosion inhibitors. Synthetic corrosion inhibitors are those that are synthesized in laboratories using different synthetic methods. Synthetic corrosion inhibitors include heterocyclics, nanomaterials, composites and nanocomposites, polymers and polymers composites, ionic liquids, different families of organic compounds, polyethylene glycol, etc. Natural corrosion inhibitors are derived through natural resources. Compounds derived from natural resources may be used with and without purification. Examples of natural corrosion inhibitors include plant extracts, natural medicines, carbohydrates, amino acids, oleochemicals, natural- and bio-polymers, natural gums and latex, etc. A brief classification of organic corrosion inhibitors based on their origin is summarized in Fig. 5.3.

5.2.2 Classification of organic compounds based on their nature

Based on their nature of chemical structure, organic corrosion inhibitors can be divided as aliphatic- and aromatic-types. Aliphatic corrosion inhibitors can be further classified as acyclic and alicyclic types. Acyclic compounds are those that do not contain cyclic ring(s) in their molecular structures. One of the most common examples of this category is acyclic

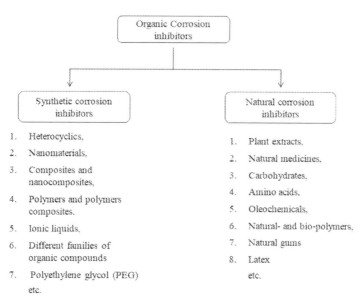

FIG. 5.3 A brief classification of organic corrosion inhibitors based on their origin.

amines with polar amino (–NH$_2$) functional group and a nonpolar hydrophobic hydrocarbon chain. Other examples include surfactants, polymers, alcohols, ethers, etc. Acyclic corrosion inhibitors are the compounds that possess one or more cyclic ring(s) in their molecular structures. Examples of this category of compounds include acyclic amines, cycloalkanes, cycloalkenes, cycloalkynes, and their derivatives.

Reports on the corrosion inhibition effect of acyclic and alicyclic compounds are relatively limited because of their low inhibition effect. However, corrosion inhibition effect of aromatic compounds especially for hetero-aromatic compounds is widely investigated. Benzene, aniline, phenol, naphthalene, anthracene, and their derivatives are reported as effective corrosion inhibitors for different metals and alloys in various electrolytes. Examples of the hetero-aromatic compounds include pyridine, quinoline, isoquinoline, and their derivatives. In numerous reports, corrosion inhibition effect of the five-membered heterocyclic compounds such as pyrrole, furan, thiophene, and their derivatives has also been reported. Obviously, hetero-aromatic compounds react with metallic surface using their nonbonding electrons of the heteroatoms (such as O, N, S, and S) and π-electrons of the multiple (double and triple) bonds. Generally, the corrosion inhibition effect of these compounds increases with increasing conjugation in the form polar functional groups and multiple bonds of the side chain(s). A brief classification of organic corrosion inhibitors based on their nature and chemical structures is summarized in Fig. 5.4.

5.2.3 Classification of organic compounds based inhibition mechanism

Organic corrosion inhibitors can also be classified on the basis of their mode of action. It has been observed that some of the organic compounds preferably inhibit either of the

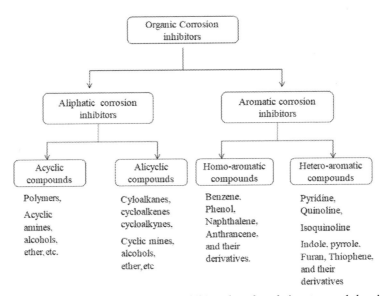

FIG. 5.4 **A brief classification of organic corrosion inhibitors based on their nature and chemical structures.**

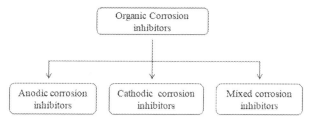

FIG. 5.5 Classification of the organic corrosion inhibitors based on their mechanism of action.

polarization (Tafel) reactions. Some of compounds mainly inhibit anodic Tafel reactions. These compounds are known as anodic corrosion inhibitors. Cathodic corrosion inhibitors are those that principally inhibit cathodic Tafel reaction. However, literature study shows that most of the organic compounds act as mixed-type corrosion inhibitors, i.e., these compounds adversely affect anodic as well as cathodic reactions. Classification of the organic corrosion inhibitors based on their mechanism of action is presented in Fig. 5.5.

5.2.4 Classification of organic compounds based molecular size

Organic corrosion inhibitors can also be divided into different classes depending upon their molecular size. Some main class of organic compounds based on their molecular size would be molecular, macromolecular, and nanosized organic corrosion inhibitors. Examples of the molecular corrosion inhibitors include different families of organic compounds. Macromolecular corrosion inhibitors can be categorized further into oligomeric and polymeric corrosion inhibitors. Literature study shows that natural and synthetic of oligomers and polymers are extensively investigated as corrosion inhibitors. Some of the well-known natural polymeric corrosion inhibitors include chitosan, chitin, cellulose, natural gums, proteins, and their derivatives. Synthetic polymers and oligomers are also reported as effective corrosion inhibitors. High inhibition efficiency of the macromolecular compounds is based on their extremely high molecular sizes that provide excellent surface coverage.

Among the nanosized organic corrosion inhibitors, graphene (G), graphene oxide (GO), carbon dots (CDs), carbon nanotubes (CNTs), and their derivatives and composites are extensively used. Generally, these nanomaterials show reasonably good anticorrosion effect at relatively low concentration which can be attributed due to their high surface areas as nanomaterials have extremely high surface area. Because of their high surface areas, nanomaterials provide excellent surface coverage and act as effective corrosion inhibitors. The classification of organic corrosion inhibitors based on their molecular size is presented in Fig. 5.6.

5.2.5 Classification of organic compounds based on their effect on the environment

Organic compounds may or may not have adverse effects on the surrounding environment. On this basis corrosion, inhibitors can be divided into two classes, namely toxic- and nontoxic or environmental friendly corrosion inhibitors. Most of the synthetic corrosion

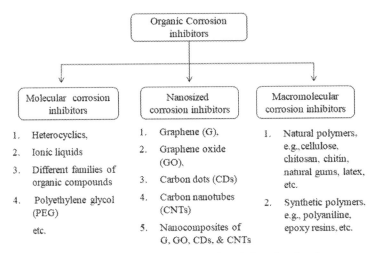

FIG. 5.6 Classification of organic corrosion inhibitors based on their molecular size.

inhibitors are toxic in nature as they use huge amount of toxic chemicals and solvents and expensive catalysts for their syntheses. However, synthetic corrosion inhibitors derived through multicomponent reactions with and without ultrasound and microwave irradiations, mechanochemical mixing and solid-phase reaction and solid supported synthesis can be treated as environmental friendly as these methods are associated with several advantages including high synthetic yield, ease to perform, production of less waste products, time-saving, less tedious, and many more [1,2]. More so, compounds synthesized in environmental friendly solvents, such as water and supercritical CO_2 can also be assumed as environmental friendly in nature.

5.2.6 Classification of organic compounds based on state of application

Organic corrosion inhibitors can also be divided into two classes, namely, aqueous phase corrosion inhibitors and coating phase corrosion inhibitors, based on their mode of application. Generally, coating phase corrosion inhibitors are used to provide long-term protection and they are used along with some polymer supports, such as epoxy resins mostly in coating and painting. However, aqueous phase corrosion inhibitors are used for the protection of metallic corrosion in aqueous electrolytes. It is important to mention that only water soluble compounds can be used as aqueous phase corrosion inhibitors.

5.2.7 Classification of organic compounds based on their mode of adsorption

Based on the mode of adsorption, organic compounds may be physical, chemical, and physiochemical (mixed) types. Generally, charged inhibitor molecules interact with the charged metallic surface through electrostatic force of attraction. In general, physical adsorption is consistent with the value of the standard Gibb's free energy (ΔG) of -20 kJ/mol or more positive [3,4]. It is important to mention that the extent of physical adsorption decreases on increasing the temperature. Chemical corrosion inhibitors are those that interact with metallic surfaces using a charge-sharing mechanism. Generally, chemical adsorption is consistent with the standard Gibb's free energy value of −40 kJ/mol or more negative [3,4]. Literature study shows that adsorption of most of the organic compounds follows physiochemisorption mechanism for which values of standard Gibb's free energy (ΔG) ranges in between −20 kJ/mol and −40 kJ/mol [5].

5.3 Summary

Corrosion inhibitors are chemical species that inhibit or decrease the corrosion rate in their presence at low concentration. They can be classified into different classes and categories depending upon their various aspects. Broadly corrosion inhibitors can be classified as organic- and inorganic-type corrosion inhibitors. Inorganic inhibitors can be further divided into anodic- and cathodic-type inorganic inhibitors. Some of the inorganic compounds passivate the metallic surface in the absence of oxygen. They are known as oxidizing anions. However, some anions essentially require the oxygen to passivate the metallic surface. These anions are known as un-oxidizing anions. Cathodic inorganic inhibitors can be classified as cathodic poisons that decrease the rate of cathodic reaction, cathodic precipitates that precipitate at cathodic site and avoid the diffusion of corrosion species and scavengers that react with the corrosive species, such as oxygen and decrease the rate of cathodic reaction. Various classifications have been proposed for organic corrosion inhibitors. One of the common classifications is based on their chemical nature, i.e., aromatic or aliphatic. Aromatic corrosion inhibitors can be divided as homo- and hetero-aromatic corrosion inhibitors. Organic compounds can also be classified as natural or synthetic types depending upon the origin. Organic corrosion inhibitors may be classified as anodic-, cathodic-, or mixed-type depending upon their major corrosion inhibition effect. Based on their mode of adsorption, they can be classified as physical, chemical, or physiochemical types. On the basis of their molecular size, organic corrosion inhibitors can be classified as molecular, nanosized, and macromolecular type. Based on their effect on the environment, organic corrosion inhibitors can be divided as environmental friendly or environmental toxic types.

5.4 Useful websites

https://www.dynagard.info/different-types-corrosion-inhibitors/
https://corrosion-doctors.org/Inhibitors/Introduction.htm
https://www.intechopen.com/books/corrosion-inhibitors/corrosion-inhibitors

References

[1] C. Verma, J. Haque, M. Quraishi, E.E. Ebenso, Aqueous phase environmental friendly organic corrosion inhibitors derived from one step multicomponent reactions: a review, J. Mol. Liq. 275 (2019) 18–40.

[2] C. Verma, M. Quraishi, E.E. Ebenso, Microwave and ultrasound irradiations for the synthesis of environmentally sustainable corrosion inhibitors: An overview, Sustain. Chem. Pharm. 10 (2018) 134–147.

[3] L. Li, Q. Qu, W. Bai, F. Yang, Y. Chen, S. Zhang, Z. Ding, Sodium diethyldithiocarbamate as a corrosion inhibitor of cold rolled steel in 0.5 M hydrochloric acid solution, Corrosion Sci. 59 (2012) 249–257.

[4] T. Yan, S. Zhang, L. Feng, Y. Qiang, L. Lu, D. Fu, Y. Wen, J. Chen, W. Li, B. Tan, Investigation of imidazole derivatives as corrosion inhibitors of copper in sulfuric acid: combination of experimental and theoretical researches, J. Taiwan Inst. Chem. Eng. 106 (2020) 118–129.

[5] C. Verma, M. Quraishi, A. Singh, 2-Amino-5-nitro-4, 6-diarylcyclohex-1-ene-1, 3, 3-tricarbonitriles as new and effective corrosion inhibitors for mild steel in 1 M HCl: Experimental and theoretical studies, J. Mol. Liq. 212 (2015) 804–812.

CHAPTER 6

Corrosion and corrosion inhibition in acidic electrolytes

Metallic materials are highly susceptible for corrosion in acidic electrolytes. Acidic solution of nitric acid (HNO_3), acetic acid (CH_3COOH), sulfuric acid (H_2SO_4), hydrochloric acid (HCl), hydrofluoric acid (HF), phosphoric acid (H_3PO_4), formic acid (HCOOH), perchloric acid ($HClO_4$), sulfamic acid (H_3NSO_3), etc., are extensively used as electrolytes for metallic corrosion. Mechanism and rate of corrosion vary depending upon the nature of electrolyte and metal. Generally, highly concentrated acidic electrolytes are used for industrial descaling, acidic cleaning, oil-well acidification, and acid pickling processes whereas less concentrated acidic electrolytes are used for academic and research purposes. In acidic electrolytes, the mechanism of metal corrosion involves two half-cell reactions, anodic oxidation and cathodic reduction reaction.

(a) *Anodic oxidation reaction*: at anode metal oxidation takes place through the release of electrons. Anodic reaction of a mono- (M: Na, Li, and K), di- (M′: Fe, Mg, and Ca), and tri-valent (Fe and Al) metals are presented below:

$$M \leftrightarrow M^+ + e^- \tag{6.1}$$

$$M' \leftrightarrow M'^{2+} + 2e^- \tag{6.2}$$

$$M'' \leftrightarrow M''^{3+} + 3e^- \tag{6.3}$$

(b) *Cathodic reduction reaction*: cathodic reaction involves the consumption of electrons released at the anode. In the acidic electrolytes, the nature of cathodic reaction depends upon the availability of oxygen. In the absence of oxygen hydrogen evolution or reduction of protons is the major cathodic reaction. However, formation of hydroxides from the reduction of oxygen has been reported as major cathodic reaction in the presence of oxygen.

$$H^+ + e^- \leftrightarrow \frac{1}{2}H_2 \quad \text{(Hydrogen evolution; absence of oxygen)} \tag{6.4}$$

$$2O_2 + 4H^+ + 4e^- \leftrightarrow 4OH^- \quad \text{(Hydroxide formation; presence of oxygen)} \tag{6.5}$$

Metal cations (M+, M′2+ and M″3+) can react with hydroxide ions to form the corresponding hydroxide corrosion products [e.g. $Fe(OH)_2$, $Mg(OH)_2$, $Fe(OH)_3$, and $Al(OH)_3$]. However, the formation of these products would be the last but not first product. In the acidic electrolytes, metal and metal oxides are expected to convert into their chlorides (in HCl), sulfates (in H_2SO_4), nitrates (in HNO_3), and phosphate (in H_2PO_4). Formation of the metal chlorides and sulfates in the presence of HCl and H_2SO_4 are presented below:

$$M + 2HCl \rightarrow MCl_2 + H_2 \tag{6.6}$$

$$2M' + 3HCl \rightarrow 2M'Cl_3 + 3H_2 \tag{6.7}$$

$$MO + 2HCl \rightarrow MCl_2 + H_2O \tag{6.8}$$

$$M_2O_3 + 6HCl \rightarrow 2MCl_3 + 3H_2O \tag{6.9}$$

$$M + H_2SO_4 \rightarrow MSO_4 + H_2 \tag{6.10}$$

$$MCl_2 + H_2SO_4 \rightarrow MSO_4 + 2HCl \tag{6.11}$$

Replacement of chlorides/sulfates with hydroxide ions, i.e., formation of final corrosion products can occurs through following mechanism:

$$MCl + OH^- \leftrightarrow M(OH) + Cl^- \tag{6.12}$$

$$MCl_2 + 2OH^- \leftrightarrow M(OH)_2 + 2Cl^- \tag{6.13}$$

$$MCl_3 + 3OH^- \leftrightarrow M(OH)_3 + 3Cl^- \tag{6.14}$$

$$M'SO_4 + 2OH^- \leftrightarrow M'(OH)_2 + SO_4^- \tag{6.15}$$

In acidic electrolytes, corrosion products are highly soluble and do not provide any protection from corrosion. Therefore, in such electrolytes, corrosion is expected to continue until the metallic materials completely corrode. An account of different kinds of acidic electrolytes is given below.

6.1 Common acidic electrolytes

6.1.1 Hydrochloric acid (HCl)

Hydrochloric acid which is also known as muriatic acid, hydronium chloride, sprite of salt and acidum salt represents a colorless inorganic chemical that acquires a characteristic smell. HCl is soluble in aqueous medium up to all proportions or compositions. Therefore, it is classified as a strongly acidic chemical. Technically grades HCl are available as 33%–37% that possess a slightly yellow color. An alchemist, J. ibn Hayyan discovered the HCl around 800 AD [1]. Different concentrations of HCl-based electrolytes are used for industrial and academic purposes. Generally, lower HCl concentrations are used for academic purposes and highly concentrated HCl solutions are used for industrial purposes. One of the most common industrial uses of HCl solution is acid pickling in which highly concentrated HCl solution is used to remove the surface impurities such as rusts and scales present over the metallic surface. Generally, acid pickling is achieved at low temperature, usually below 15°C.

Generally, acid pickling using HCl-based electrolytes at higher temperature requires specific equipment because at elevated temperatures HCl easily undergoes vaporization and adsorption of its vapor on metallic surface in open atmosphere is impossible. It is important to mention that pickling efficiency of the HCl-based electrolytes increases on increasing the

TABLE 6.1 EU 1272/2008 classification of HCl.

S. No.	HCl concentration	Specification
1	≤10%	The HCl solutions with range of concentration acquire the specification of H335 that causes target specific organ toxicity
2	10%–25%	The HCl solutions acquire specifications of H319 and H315 that define serious eyes and skin irritation, respectively
3	≥25%	This defines a highly toxic range of HCl and grouped as 1B code. This grade of HCl causes skin corrosion and possesses the specification H314 characteristics for skin burns and eye damage

temperature. Every 10°C rise in temperature causes a two-times increase in the pickling efficiency. Though picking process requires lower temperature in the open system, in the closed system, HCl-based pickling can be achieved at as high as 80°C. In total, 5%–15% HCl solutions are widely employed for the pickling process. It is important to mention that chloride ion induces pitting corrosion, therefore, acid pickling cannot be used for stainless steel. Zinc is also highly susceptible to dissolve in HCl-based electrolytes therefore pickling cannot also be applied for zinced parts. It is important to mention that HCl solution of 15%–22% for picking time of 7–10 minute immersion in the temperature range of 30–40°C is the most appropriate selection. As per the EU 1272/2008 that defines packaging, classification, leveling of chemicals, and mixtures of chemicals, HCl can be classified into three classes (Table 6.1).

Pickling conditions for different metals and alloys in HCl-based electrolytes are presented in Table 6.2. It can be clearly seen that lower HCl concentrations (5%–10%) are mainly used for the acid descaling process whereas highly concentrated HCl solutions (10%–25%) are used for acid pickling. HCl solutions of very low concentration (<5%) are widely used as corrosive electrolytes for academic purposes.

6.1.2 Sulfuric/sulphuric acid (H_2SO_4)

Sulfuric acid (H_2SO_4) which is also called as oil of vitriol is a highly useful acid and its different concentrations are widely used for acid picking. Technical grade of sulfuric acid is 90% which is slightly brown in color. Similar to the pickling in HCl, pickling is greatly influenced by the change in temperature. Generally, sulfuric acid solution causes serious skin and eye irritations. EU 1272/2008 classification of sulfuric acid is given in Table 6.3.

TABLE 6.2 Pickling conditions for various metals and alloys in HCl-based electrolytes.

S. No.	Temperature of pickling bath	HCl concentration	Example
1	RT	1%–5%	Steel deoxidation
2	RT	3.8%	Neutralization of alkali residues
3	RT to 60°C	5%–20%	Cu and Cu alloys
4	RT to 80°C	10%–25%	Carbon steel
5	RT to 80°C	Various	Stainless steel

TABLE 6.3 EU 1272/2008 classification of H_2SO_4.

S. No.	Specification code	H_2SO_4 concentration	Description
1	1A	≥15 %	Skin corrosion
2	H314	≥15 %	Skin burns and eye damage
3	H315	5%–15 %	Skin irritation
4	H319	5%–15 %	Eye irritation

TABLE 6.4 Pickling conditions for different metals and alloys in H_2SO_4-based electrolytes.

S. No.	Temperature of pickling bath	H_2SO_4 concentration	Example
1	RT	2%–10%	Steel deoxidation
2	RT	3%–6%	Neutralization of alkali residues
3	60–100°C	5%–25%	Steel alloys
4	50–100°C	10%–25%	Carbon steel
5	50–90°C	10%–60%	Cu and Cu alloys
6	RT to 90°C	96%	Glass and dull burning in a mixture of sulphuric and nitric acid

Generally, lower concentrations of H_2SO_4 are used for neutralization of the alkali residues and higher H_2SO_4 concentrations are mainly used for pickling of steel, copper, and their alloys. Pickling conditions for different metals and alloys in H_2SO_4-based electrolytes are presented in Table 6.4.

6.1.3 Phosphoric acid (H_3PO_4)

Phosphoric acid (H_3PO_4) is a colorless syrupy liquid. It is also known as orthophosphoric acid and its available analytical grade is 85%. In pure form phosphoric acid exists in the solid-state. As compared to HCl and H_2SO_4, phosphoric acid is a weak acid. It is one, two, and three hydrogen that can be successively replaced to form $H_2PO_4^-$ (dihydrogen phosphate), HPO_4^{2-} (hydrogen phosphate) and PO_4^{3-} (phosphate), respectively. As phosphoric acid is less aggressive than sulfuric and hydrochloric acids, its high concentrations (15%–70%) act as effective oxide dissolver. At low concentration, phosphoric acid forms iron phosphate that provides slight corrosion. Similar to H_2SO_4, phosphoric acid causes serious skin and eye irritations. EU 1272/2008 classification of the phosphoric acid is presented in Table 6.5.

TABLE 6.5 EU 1272/2008 regulation classification of H_3PO_4.

S. No.	H_3PO_4 concentration	Specification code	Description
1	≥15%	1A	Skin corrosion
2	≥15%	H314	Eye damage and skin burns
3	5%–15%	H315	Skin irritation
4	5%–15%	H319	Eye irritation

TABLE 6.6 Pickling conditions for different metals and alloys in H_3PO_4-based electrolytes.

S. No.	Temperature of pickling bath	H_3PO_4 concentration	Example
1	Variable	2%–10%	Phosphating solution
2	Variable	2%–20%	Cleaning solutions
3	40–80°C	2%–20%	Steel
4	40–80°C	5%–30%	Al and Al alloys
5	40–80°C	10%–40%	Pickling of Cu and Cu alloys
6	40–80°C	10%–60%	H_3PO_4 mixture with other acids
7	40–80°C	15%–60%	Burning of Cu and Cu alloys

Phosphoric acid solutions are extensively used for the acid pickling process. Pickling conditions for different metals and alloys in H_3PO_4-based electrolytes are presented in Table 6.6. Inspection of the table reveals that phosphoric acid solutions are widely used for different metals and alloys including iron, copper and aluminum. Generally, lower acid concentrations are useful for acid cleaning processes.

6.1.4 Other acidic electrolytes

Apart from HCl, H_2SO_4, and H_3PO_4, several other electrolytes such as nitric acid (HNO_3) formic acid (HCOOH), acetic acid (CH_3COOH), sulfamic acid (H_3NSO_3), oxalic acid [$(COOH)_2$] are extensively used as electrolytes for different industrial and academic applications. Pickling conditions of different metals and alloys in HNO_3-based electrolytes are presented in Table 6.7. It is important to mention that formic, oxalic, and acetic acids are used as electrolytes in combination with other strong acids such as sulfuric and hydrochloric acid. Among the tested organic acids, oxalic acid is relatively more effective, cheap than other organic acids. Sulfamic acid is generally used to carbonate scales and metal oxides.

TABLE 6.7 Pickling conditions of different metals and alloys in HNO_3-based electrolytes.

S. No.	Temperature of pickling bath	HNO_3 concentration	Example
1	RT to 60°C	1%–30%	Al, Mg, Zn, Cu and other metals
2	RT to 60°C	3%–25%	Still alloys and other metals in mixture of HF and HNO_3
3	RT to 60°C	5%–35%	Other substances
4	RT to 60°C	10%–65%	Etching of different metals
5	RT to 50°C	25%–35%	Passivation
6	60–65%	RT to 60°C	Heat treatment of Cu and Cu-based alloys
7	60–65%	RT to 60°C	Matte and gloss burning in H_2SO_4 and HNO_3 mixture

6.2 Corrosion protection in acidic electrolytes

As described above, acidic solutions are widely used in acid cleaning and acid pickling processes. Generally, these processes are associated with the loss of metallic components because of the corrosion. The use of organic compounds is the most effective, economic, and popular method of acid corrosion mitigation. These compounds adsorb on metal surface through their polar electron-rich centers and form inhibitive film which isolates metal from the aggressive acidic solution. Adsorption of corrosion inhibitors may involve physisorption, chemisorption, or physiochemisorption (mixed) adsorption. Generally, physisorption occurs by the electrostatic interactions between charged metallic surface and charged inhibitor molecules, whereas chemisorption occurs by the charge sharing between them.

Generally, in acidic solutions heteroatoms of organic corrosion inhibitors get protonated and converted in their cationic (protonated) form. On the other hand, the metallic surface becomes negatively charged because of the adsorption of counter ions of electrolytes. These opposite charged species attracted each other through electrostatic (physisorption) force of attraction. However, the protonated heteroatoms can return in their neutral form by accepting the electrons released at cathode. Heteroatoms with their free unshared electron pairs interact with metallic surfaces using electron transfer mechanisms. This is known as chemisorption of the inhibitor molecules. It is important to mention that metals are already electron-rich species therefore transfer of electrons from inhibitor to metal (donation) causes interelectronic repulsion which enables metals to transfer their electrons into the antibonding molecular orbitals of the inhibitor molecules. This process is known as retro-donation or back-donation. Both donation and retro-donation strengthen each other through a phenomenon known as synergism.

Corrosion inhibition effect of organic compounds greatly depends upon the nature of substituents. In general, electron-donating substituents such as –OH, –Me, –NH$_2$, –NHMe, and –NMe2 enhance the electron density at active sites and increase the inhibition power of organic compounds. Unlikely, electron-withdrawing substituents such –CN, –COOH, and –NO$_2$ decrease the electron density at the active sites and therefore decrease the inhibition effect of organic compounds. Nevertheless, several exceptions have been reported in the literature. The effect of substituents can be best explained using Hammett substituent constant values. Substituents with negative sign of Hammett substituent constant (σ) are expected to increase the protection effectiveness and converse is true for electron-withdrawing substituents with the positive sign of Hammett substituent constant (σ). Different families of organic compounds have been extensively used as corrosion inhibitors in acidic electrolytes. It has been observed that inhibition power of these compounds increases on increasing their concentration. Most of the organic compounds acted as mixed-type corrosion inhibitors and their adsorption obeyed the Langmuir adsorption isotherm.

6.3 Case studies

Corrosion inhibition effects of organic compounds are widely investigated and reported for different metals and alloys in acidic solutions. However, HCl and sulfuric acid based electrolytes are most frequently utilized as media for corrosion inhibition. Although other acidic electrolytes are also used extensively, the present chapter describes the case studies related only with the hydrochloric acid and sulfuric acid solutions.

6.3.1 Corrosion inhibition in HCl-based electrolytes

HCl-based aggressive solutions ranging from 0.1M to 15% are widely used as electrolytes for corrosion inhibition study. In these electrolytes, organic compounds are used as the first line of defense. Yıldız et al. [2] reported the corrosion inhibition property of 4-Amino-3-hydroxynaphthalene-1-sulphonic acid designated as 4A3H1S for mild steel in 0.1 M HCl using various electrochemical and surface investigation methods. Results showed that 4A3H1S acted as a good corrosion inhibitor and its protection efficiency increased with its concentration. The 4A3H1S exhibited maximum inhibition efficiency of 94.7% at 10 mM concentration. Polarization studies showed that 4A3H1S acted as mixed-type corrosion inhibitors as its presence adversely affected the rate of both anodic as well as cathodic reaction. It was observed that 4A3H1S inhibited corrosion by adsorbing on the metallic surface using the Langmuir adsorption isotherm mechanism. Surface study carried out through scanning electron microscopy (SEM) method showed that 4A3H1S inhibited corrosion by adsorbing on the metallic surface. A significant improvement in the surface morphology of the metal surface was observed in the presence of 4A3H1S. The studies of metallic corrosion and corrosion protection in 0.1M HCl electrolyte are also reported in other studies.

2-butyl-hexahydropyrrolo[1,2-b][1,2]oxazole designated as BPOX was tested as corrosion inhibitors for mild steel in hydrochloric acid solution of 0.5M HCl using various experimental methods [3]. Inhibition effect of the BPOX was treated at its different concentrations at various temperatures. It was observed that inhibition power of the BPOX increases with its concentration and decreases with the raise in temperature. The BPOX showed the highest protection power of more than 95% at 10^{-3} M concentration (at 40°C). Potentiodynamic polarization study showed that BPOX acted as a mixed-type corrosion inhibitor. Presence of BPOX in the corrosive electrolyte causes increase in the value of charge transfer resistance. Thermodynamic analyses showed that inhibition of metallic corrosion is attributed due to the adsorption of BPOX on the metallic surface. The values of standard Gibbs free energy (ΔG) ranged in between -20 kJmol^{-1} and -40 kJmol^{-1} indicating that adsorption of BPOX followed the physiochemisorption mechanism. Baeza and coworkers demonstrated the corrosion inhibition effect of 1,3,4-thiadiazole-2,5-dithiol (bismuthiol) on copper in 0.5M HCl solution [4]. Results showed that bismuthiol inhibits corrosion by forming surface protective complexes. Adsorption of bismuthiol on copper surface in 0.5M HCl followed the Langmuir adsorption isotherm model.

Shalabi et al. [5] reported the corrosion inhibition power of some phenyl sulfonylacetophenoneazo derivatives (PSAAD) for aluminum in 0.5M HCl using experiential and computational methods. Analyses showed that PSAADs acted as effective corrosion inhibitors and their inhibition effectiveness followed the order: PSAAD-I (75.4%) < PSAAD-II (84.2%) < PSAAD-III (88.2%). Potentiodynamic polarization measurements revealed that PSAADs acted as mixed type corrosion inhibitors with slight cathodic predominance. Adsorption of PSAAD-I, PSAAD-II and PSAAD-III on aluminum surface obeyed the Langmuir adsorption isotherm model. Density functional theory (DFT)-based theoretical study showed that PSAAD-I, PSAAD-II and PSAAD-III interact with metallic surfaces using donor acceptor interactions in which electron rich centers behave as adsorption centers. A good agreement in the results of experimental and density functional; theory measurements were observed.

Similarly, corrosion inhibition effect of organic compounds has also been investigated in other HCl-based electrolytes such as in 1–5M and 15% HCl solution. In fact, 1 and 2 M HCl solutions are the most widely used HCl-based electrolytes. Several reports are published on describing the anticorrosion effect of organic compounds in these electrolytes. In total, 15% HCl represents the highly corrosive HCl-based electrolyte and it is mainly used in industry for acid pickling process. The literature study showed that several reports published in which organic compounds are effectively used as corrosion inhibitors in 15% HCl solution. Ansari et al. demonstrated the corrosion inhibition effect of two pyrazolone derivatives designated as PZ-1 and PZ-2, respectively, for N80 steel corrosion in 15% using experimental and computational methods [6]. Analyses showed that inhibition effect of tested compounds was substituent dependent and PZ-1 showed better protection effectiveness than that of the PZ-2. Polarization analyses showed that PZ-1 and PZ-2 acted as mixed-type corrosion inhibitors with slight cathodic predominance. Adsorption of both corrosion inhibitors on the metallic surface followed the Langmuir adsorption isotherm model. Quantum chemical calculations analysis showed that protonated forms of the inhibitor molecules behaved as more effective inhibitors than that of their nonprotonated-form.

6.3.2 Corrosion inhibition in H_2SO_4-based electrolytes

Similar to HCl, H_2SO_4-based electrolytes are widely used as electrolytes for corrosion inhibition studies. For this purpose, various concentrations of the sulfuric acid have been used. It is recalled that less concentrated sulfuric acid solutions are useful for academic purposes and highly concentrated sulfuric acid solutions are used in industrial cleaning processes including acid pickling and oil-well acidification. Zhou and coworkers reported the corrosion inhibition power of imidazole-based organic compounds for mild steel in acidic solution of 0.1M H_2SO_4 [7]. Among the evaluated compounds, omeprazole showed the highest protection effectiveness of 98.72% at 10^{-3}M concentration. Polarization measurements showed that tested imidazole-based compounds adversely affect the rate of cathodic as well as anodic reaction and behaved as mixed-type corrosion inhibitors. The change in the shape of polarization curves was consistent with their concentration. All tested imidazole-based compounds acted as interface-type corrosion inhibitors as their presence increases the values of the charge transfer process. The DFT-based study showed that imidazole-based compounds interact with metallic surface through donor acceptor process. Lastly, orientation of the tested compounds was measured using molecular dynamics simulations. It was observed that imidazole-based compounds adsorbed using their flat orientations and therefore behaved as effective corrosion inhibitors. Corrosion inhibition effect of other organic compounds has also been investigated in 0.1M H_2SO_4 medium.

Inhibition of metallic corrosion in 0.5 M H_2SO_4 is widely reported using organic compounds. Omotosho and coworkers described the corrosion inhibition effect of aniline (Ph-NH_2) for stainless steel corrosion in 0.5M H_2SO_4 [8]. The study showed that aniline showed the highest protection effectiveness of 96.68% at 0.043M concentration at 60°C. In another study, phthalocyanine blue, an organic dye, was evaluated as a corrosion inhibitor for carbon steel in 0.5M H_2SO_4 medium [9]. Study was conducted using weight loss, electrochemical, and SEM methods. Weight loss results showed that inhibition efficiency of the phthalocyanine blue increases with increasing its concentration and maximum inhibition efficiency of 35.052% was observed at 800 ppm concentration (25°C). Potentiodynamic polarization studies showed that phthalocyanine blue inhibits both cathodic hydrogen evolution and anodic oxidation reactions and acted as mixed type corrosion inhibitor. Electrochemical impedance spectroscopy study revealed that presence of the phthalocyanine blue increases the resistance for corrosion process thereby it behaved as interface-type corrosion inhibitors. Inhibition effect of several other organic compounds has been reported in 0.5M H_2SO_4 medium.

Metallic corrosion is also widely studied in other concentrations of the H_2SO_4 electrolyte. Between 1 and 2M sulfuric acid solutions are the most commonly used H_2SO_4-based electrolytes. Abdulridha et al. [10] investigated the inhibition power of a new Schiff's base designated as AS for carbon steel corrosion in 1M HCl using weight loss, electrochemical, SEM, and AFM methods. Tween-80 was added along with AS to increase its solubility in the tested electrolyte. Protection effectiveness of the AS increases with its concentration and it showed highest protection effectiveness of 91.32% at 0.08 mM concentration. Polarization studies revealed that AS became effective by inhibiting anodic as well as cathodic reaction, i.e., AS behaved as a mixed-type corrosion inhibitor. SEM and AFM studies showed that morphology of the metallic surfaces was greatly improved in the presence of AS indicating that it inhibits corrosion of metal by adsorbing and forming a surface protective film. DFT analyses showed that AS interacts strongly with metallic surfaces using donor–acceptor interactions. Adsorption of the AS on metallic surface followed the Langmuir adsorption isotherm model. Inhibition effect of organic corrosion inhibitors in 1M H_2SO_4 medium is widely investigated.

Highly concentrated H_2SO_4 solutions are mostly used for industrial cleaning processes. Solomon and coworkers reported the corrosion inhibition power of the chitosan (Chi), a polysaccharide and its composite with silver (AgNPs-Chi) a steel alloy (St37) in 15% H_2SO_4 medium [11]. The study showed that chitosan showed inhibition efficiency of nearly 45%; however, its composite with silver showed as high as 94% of inhibition efficiency. Both Chi and AgNPs-Chi inhibit corrosion by adsorbing on the metallic surface and their adsorption follows the Langmuir adsorption isotherm model. Adsorption mechanism of corrosion inhibition of AgNPs-Chi was further studied by surface analyses. Other reports are extensively reported in literature in which organic compounds are used as effective corrosion inhibitors in highly concentrated H_2SO_4.

6.4 Summary

Acidic solution of phosphoric acid (H_3PO_4), sulfuric acid (H_2SO_4), hydrochloric acid (HC), nitric acid (HNO_3), acetic acid (CH_3COOH), sulfuric acid (H_2SO_4), hydrochloric acid (HCl), hydrofluoric acid (HF), phosphoric acid (H_3PO_4), formic acid (HCOOH), perchloric acid ($HClO_4$), sulfamic acid (H_3NSO_3), etc., are extensively used as electrolytes for metallic corrosion. Generally, highly concentrated acidic solutions are used for industrial purposes and less concentrated acidic solutions are used for academic purposes. Most commonly used acidic solutions are phosphoric acid (H_3PO_4), sulfuric acid (H_2SO_4), hydrochloric acid (HC), nitric acid (HNO_3). Generally, organic acids such as acetic acid (CH_3COOH) and formic acid (HCOOH) are used in a mixture of strong acids. Sulfamic acid (H_3NSO_3) solutions are mostly used to remove the carbonate-based scales and cleaning of the metal oxides present over the metallic surface. Metallic corrosion in acidic electrolytes involves two types of cathodic reactions depending upon the availability of oxygen. In the absence of oxygen hydrogen evolution is the main cathodic reaction whereas in its presence major cathodic reaction is reduction of oxygen. The use of organic corrosion inhibitors is one of the most effective, popular, and economic methods of corrosion mitigation in acidic electrolytes. Organic compounds become effective by forming the surface protective film in which polar functional groups and multiple bonds of the aromatic rings act as adsorption centers.

6.5 Useful websites

https://www.hindawi.com/journals/ijc/2017/9425864/
https://www.efunda.com/materials/corrosion/corrosion_basics.cfm
https://www.corrosionpedia.com/definition/440/electrolyte-corrosion

References

[1] A.E. Williams-Jones, R.J. Bowell, A.A. Migdisov, Gold in solution, Elements 5 (2009) 281–287.
[2] R. Yıldız, T. Doğan, İ. Dehri, Evaluation of corrosion inhibition of mild steel in 0.1 M HCl by 4-amino-3-hydroxynaphthalene-1-sulphonic acid, Corrosion Sci. 85 (2014) 215–221.
[3] G. Moretti, F. Guidi, F. Fabris, Corrosion inhibition of the mild steel in 0.5 M HCl by 2-butyl-hexahydropyrrolo[1, 2-b][1, 2] oxazole, Corrosion Sci. 76 (2013) 206–218.
[4] H. Baeza, M. Guzman, P. Ortega, L. Vera, Corrosion inhibition of copper in 0.5 M hydrochloric acid by 1, 3, 4-thiadiazole-2, 5-dithiol, J. Chil. Chem. Soc. 48 (2003) 23–26.
[5] K. Shalabi, Y. Abdallah, A. Fouda, Corrosion inhibition of aluminum in 0.5 M HCl solutions containing phenyl sulfonylacetophenoneazo derivatives, Res. Chem. Intermed. 41 (2015) 4687–4711.
[6] K. Ansari, M. Quraishi, A. Singh, S. Ramkumar, I.B. Obote, Corrosion inhibition of N80 steel in 15% HCl by pyrazolone derivatives: electrochemical, surface and quantum chemical studies, RSC Adv. 6 (2016) 24130–24141.
[7] Y. Zhou, L. Guo, S. Zhang, S. Kaya, X. Luo, B. Xiang, Corrosion control of mild steel in 0.1 M H_2O_4 solution by benzimidazole and its derivatives: an experimental and theoretical study, RSC Adv. 7 (2017) 23961–23969.
[8] O.A. Omotosho, J.O. Okeniyi, E.I. Obi, O.O. Sonoiki, S.I. Oladipupo, T.M. Oshin, Inhibition of stainless steel corrosion in 0.5 M H_2SO_4 in the presence of $C_6H_5NH_2$, TMS 2016 145th Annual Meeting & Exhibition, Springer, 2016, pp. 465–472.
[9] J. Valle-Quitana, G. Dominguez-Patiño, J. Gonzalez-Rodriguez, Corrosion inhibition of carbon steel in 0.5 M H2SO4 by phtalocyanine blue, International Scholarly Research Notices, 2014, Hindawi Publishing Corporation, 2014, pp. 1–8.
[10] A.A. Abdulridha, M.A.A.H. Allah, S.Q. Makki, Y. Sert, H.E. Salman, A.A. Balakit, Corrosion inhibition of carbon steel in 1 M H2SO4 using new Azo Schiff compound: electrochemical, gravimetric, adsorption, surface and DFT studies, J. Mol. Liq. 315 (2020) 113690.
[11] M.M. Solomon, H. Gerengi, T. Kaya, S.A. Umoren, Enhanced corrosion inhibition effect of chitosan for St37 in 15% H2SO4 environment by silver nanoparticles, Int. J. Biol. Macromol. 104 (2017) 638–649.

CHAPTER

7

Corrosion and corrosion inhibition in alkaline electrolytes

Similar to acidic electrolytes, basic electrolytes mainly sodium hydroxide (NaOH) and potassium hydroxide (KOH) are extensively used in corrosion studies. Metals are highly susceptible to corrosion in the basic solution. In alkaline solution, anodic metallic solution reaction involves the similar mechanism as was in the case of acidic electrolyte. However, the mechanism of anodic reaction is entirely different. In alkaline electrolytes formation of hydroxide complexes is possible that ultimately converts into passive film of metal oxides (passivation). This explains why the rate of metallic corrosion is relatively lesser in the alkaline electrolytes as compared to the acidic electrolytes. The anodic dissolution of mono- (M: Na, Li, and K), di- (M': Fe, Mg, and Ca), and tri-valent (Fe and Al) metals are presented below:

$$M \leftrightarrow M^{+} + e^{-} \tag{7.1}$$

$$M' \leftrightarrow M'^{2+} + 2e^{-} \tag{7.2}$$

$$M'' \leftrightarrow M''^{3+} + 3e^{-} \tag{7.3}$$

These cations react with hydroxide ions (counter ions of alkaline solutions) to form corresponding hydroxide complexes.

$$M^{+} + \mathrm{OH}^{-} \rightarrow M(\mathrm{OH}) \tag{7.4}$$

$$M'^{2+} + 2\mathrm{OH}^{-} \rightarrow M'(\mathrm{OH})_2 \tag{7.5}$$

$$M''^{3+} + 3\mathrm{OH}^{-} \rightarrow M''(\mathrm{OH})_3 \tag{7.6}$$

Wang et al. [1,2] proposed the stepwise formation of the hydroxide complexes of aluminum as shown below:

$$\mathrm{Al} + \mathrm{OH}^{-} \rightarrow \mathrm{Al(OH)_{ads}} + e^{-}$$

$$\mathrm{Al(OH)_{ads}} + \mathrm{OH}^{-} \rightarrow \mathrm{Al(OH)_{2,ads}} + e^{-}$$

$$\mathrm{Al(OH)_{2,ads}} + \mathrm{OH}^{-} \rightarrow \mathrm{Al(OH)_{3,ads}} + e^{-}$$

Handbook of Science & Engineering of Green Corrosion inhibitors.
DOI: https://doi.org/10.1016/B978-0-323-90589-3.00022-7

Though in highly concentrated alkaline solution, formation of hydroxide complexes may form in a single step. This was proposed following steps for the passivation of zinc in alkaline solutions [3]:

$$Zn + OH^- \rightarrow Zn(OH)^-_{ads} \quad (7.7)$$

$$Zn(OH)^-_{ads} \rightarrow Zn(OH)_{ads} + e^- \quad (7.8)$$

$$Zn(OH)_{ads} + OH^- \rightarrow Zn(OH)_2 + e^-$$

$$Zn(OH)_2 + OH^- \rightarrow ZnO + H_2O + e^- \quad (7.9)$$

Similarly, passivation of the other common metals in the alkaline solution can be presented as follows:

$$Ni(OH)_2 + OH^- \rightarrow NiO + H_2O + e^- \quad (7.10)$$

$$Fe(OH)_2 + 2OH^- \rightarrow Fe_2O_3 + 2H_2O + 2e^- \quad (7.11)$$

$$2Al(OH)_2 + 2OH^- \rightarrow Al_2O_3 + 3H_2O + 2e^- \quad (7.12)$$

$$2Al(OH)_3 \rightarrow Al_2O_3 + 3H_2O \quad (7.13)$$

In alkaline electrolytes, passivation of the metal surface provides slight protection from corrosion. However, in aqueous alkaline solution surface oxide or hydroxide layers become unstable (charged) through the following amphoteric reactions:

$$M-OH + H^+ \leftrightarrow M-OH \quad (7.14)$$

$$M-OH + OH^- \leftrightarrow M-O^- + H_2O \quad (7.15)$$

$$M-OH \leftrightarrow M-O^- + H^+ \quad (7.16)$$

This explains why metals undergo corrosion in alkaline electrolytes in spite of their ability of passivation, i.e., formation of protection metal oxide film.

7.1   Corrosion protection in basic electrolytes: case studies

Theoretically, various types of alkaline electrolytes are possible; however, the two most commonly studied alkaline electrolytes are NaOH and KOH. It is important to mention that the use of organic compounds is one of the most effective and popular methods of corrosion mitigation in alkaline electrolytes. In the literature, several reports are published in which plant extracts are used as effective corrosion inhibitors against metallic corrosion in alkaline media. Generally, each plant extracts contain numerous phytochemicals (organic compounds) that contain various electron-rich polar functional groups such as –OH, –COOC$_2$H$_5$, –CONH$_2$, –COCl, –NH$_2$, etc. and multiple bonds through which they easily get adsorb and exhibit high protection effectiveness. Nevertheless, use of synthetic compounds as corrosion inhibitors gained highest priority among the available methods of corrosion mitigation in alkaline corrosive solutions. Some important case studies on corrosion inhibition in NaOH and KOH solutions are described below.

7.1.1 Corrosion protection in NaOH-based electrolytes

In several studies, corrosion inhibition of different metals and alloys including steel, aluminum, copper, zinc, and nickel using organic corrosion inhibitors is widely studied. Fawzy et al. [4] reported the corrosion inhibition of (sodium 2-(dodecylamino)-5-guanidinopentanoate) for SABIC iron in different concentrations of HCl, NaOH, and NaCl media. The (sodium 2-(dodecylamino)-5-guanidinopentanoate) showed a maximum inhibition effect of 70% in NaOH solution at its 900 ppm concentration. It was observed that among the tested electrolytes, inhibition power of the tested compounds followed the order: NaOH < NaCl << HCl. It was estimated the tested compounds inhibited metallic corrosion by adsorbing on the metal surface. Adsorption of the compound on metallic surface obeyed the Langmuir adsorption isotherm model. Polarization study showed that investigated compounds acted as mixed-type corrosion inhibitors and their presence adversely affected the anodic and cathodic half-cell reactions. This group of authors [5] also reported the corrosion inhibition power of amino acid based surfactants for carbon steel corrosion in 0.5M NaCl and 0.5M NaOH solution. Authors observe that in NaOH medium, among the tested three amino acid based surfactants N-dodecyltryptophan (TS) showed highest protection effectiveness of 84% at its 900 ppm concentration. The amino acid based surfactants inhibit corrosion by adsorbing on the metallic surface that obeyed the Langmuir adsorption isotherm model. Polarization studies showed that all tested amino acid based surfactants acted as mixed-type corrosion inhibitors.

Organic compounds are widely studied as corrosion inhibitors for aluminum in NaOH media. Wang et al. [1] reported the corrosion inhibition of an aluminum alloy (AA5052) in a highly alkaline solution of 4M NaOH using L-cysteine. Study showed that L-cysteine exhibited highest protection effectiveness of 93.3% at 30 mM concentration. Adsorption of the cysteine on metallic surface obeyed the Langmuir adsorption isotherm model. Potentiodynamic polarization studies showed that cysteine affects both anodic as well as cathodic reactions and act as mixed-type corrosion inhibitor. However, cysteine mainly acted as cathodic-type inhibitors. Quantum chemical calculation studies showed that cysteine interacted with the metallic surface using donor-acceptor interactions. In another study, effect of substituents was demonstrated for the corrosion of aluminum in 0.5M NaOH using 2-aminobenzene-1,3-dicarbonitriles (ABDNs) various experimental and computational methods were used to demonstrate the effect of substituents [6].

Both experimental and computational analyses showed that inhibition efficiencies of the ABDNs followed the order: ABDN-III (–OH+–OMe) > ABDN-II (–OH) > ABDN-I (–Me). The highest protection effectiveness of the ABDN-III is attributed due to its highest molecular size and presence of both methoxy (–OMe) and hydroxyl (–OH) substituents. Because of their electron donating nature, these substituents enhance electron density at the donor sites computational analyses showed that substituents greatly increased the binding affinity of the compounds tested as corrosion inhibitors. Adsorption of the ABDNs followed the Langmuir adsorption isotherm. Anodic and cathodic polarization curves for aluminum corrosion in 0.5M NaOH with and without ABDNs are shown in Fig. 7.1. From the polarization curves, it can be clearly observed that presence of the ABDNs adversely affect anodic as well as cathodic reactions. In the presence of ABDNs values of corrosion current densities (i_{corr}) are greatly reduced. This observation indicated that these compounds become effective by

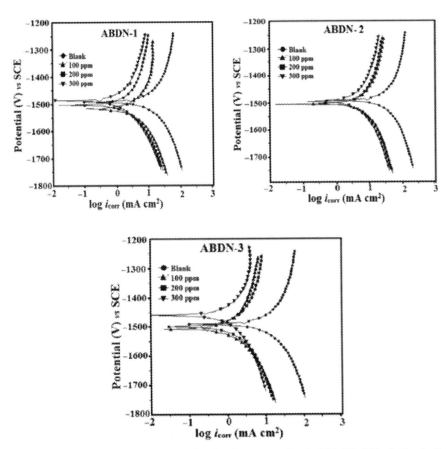

FIG. 7.1 Potentiodynamic polarization curves for aluminum corrosion in 0.5 M NaOH solution in the absence and presence of different concentrations of ABDNs [6]. (Reproduced with Permission@ Copyright: Elsevier). *ABDNs*, 2-aminobenzene-1,3-dicarbonitriles.

blocking the active sites responsible for corrosion through their adsorption. It was also observed that decrease in the i_{corr} values for different studied ABDNs was consistent with their order of inhibition efficiency.

EIS studies showed that the presence of ABDNs increases the values of charge transfer resistance. This observation showed that ABDNs behaved as interface-type corrosion inhibitors. Scanning electron microscopy (SEM) study carried out in conjunction with energy dispersive X-ray (EDX) showed that metallic surfaces protected by ABDNs were smoother than that of the unprotected metallic surface. This is attributed due to the adsorption of the inhibitor molecules on the metallic surface and formation of the protective film. SEM images of the inhibited and uninhibited metallic surfaces are presented in Fig. 7.2. It can be seen that inhibited metallic surfaces are much smoother than that of uninhibited metallic surface. The SEM observation of supported by detecting the elements present on the metallic surface through EDX studies. The corrosion inhibition or organic compounds for aluminum has also been reported in other studies. Organic compounds are also reported as corrosion inhibitors

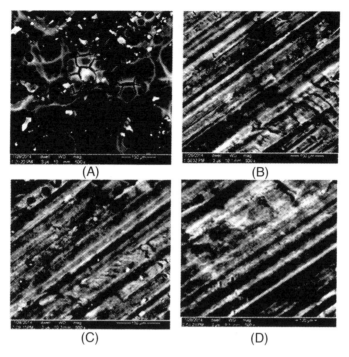

FIG. 7.2 SEM images of aluminum surface corroded in 0.5M HCl in the absence (A) and presence of ABDN-I (B), ABDN-II (C), and ABDN-III (d) [6]. (Reproduced with Permission@ Copyright: Elsevier). *ABDN*, 2-aminobenzene-1,3-dicarbonitriles.

for brass, zinc, and nickel. A summary of corrosion inhibition of some organic compounds for different metals and alloys in NaOH-based electrolytes is presented in Table 7.1.

7.1.2 Corrosion protection in KOH-based electrolytes

Corrosion inhibition of metals and alloys using organic corrosion inhibitors in KOH-based electrolytes has also been reported widely. Oguzie et al. [22] reported the inhibition power of congo red (CR) dye for aluminum corrosion in 2M KOH solution. Analyses showed that CR showed inhibition efficiency of 48.63% in the presence of KI. The CR adsorbed physically on the metallic surface. Adsorption of the CR on aluminum surface obeyed the Langmuir adsorption isotherm model. Effect of synergism was studied using different halides and it was observed that their inhibition effect followed the order: KI > KBr > KCl. In another study corrosion inhibition power of another dye, crystal violet (CV) was evaluated for aluminum in 1M HCl and M KOH solutions [23]. Results showed that CV showed highest protection effectiveness of 85.3%. Synergistic effect of potassium iodide (KI) was tested for corrosion inhibition of CV for aluminum in alkaline and it was observed in the presence of KI; inhibition efficiency of CV was greatly increased. Adsorption of the CV on the aluminum surface followed the Freundlich adsorption isotherm model. Similar observation was further observed by this group of authors for the inhibition effect of malachite green for aluminum corrosion

TABLE 7.1 Name, inhibition efficiency, concentration, metal, electrolyte, and mode of adsorption of some common compounds evaluated as corrosion inhibitors in sodium hydroxide based electrolytes.

Chemical name	Inhibition efficiency	Conc.	Metal	Electrolyte	Mode of adsorption	References
(sodium 2-(dodecylamino)-5-guanidinopentanoate)	70% (in 0.5M NaOH)	900 ppm	SABIC iron	0.1–1.0M HCl, NaCl, and NaOH	Langmuir isotherm, mixed-type	[4]
Amino acids-based surfactants	84%	900 ppm (TS; best)	Carbon steel	0.5 NaCl and 0.5 M NaOH	Langmuir isotherm, anodic-type	[5]
L-cysteine	93.3%	30 mM	AA5052 aluminum	4M NaOH	Langmuir isotherm, cathodic-type	[1]
2-aminobenzene-1,3-dicarbonitriles (ABDNs)	94.50%	300 ppm (ABDN-III)	Aluminum	0.5M NaOH	Langmuir isotherm, mixed-type	[6]
1-(4-aminophenyl)propan-1-one based azo-dyes	78.12% (compound-III; best)	11×10^{-6}M	Aluminum	0.1M NaOH	Temkin isotherm, mixed-type	[7]
Theophylline (expired drug)	90%	2.5%	Aluminum	7075 Aluminium Alloy	Mixed-type but cathodic predominant.	[8]
Quinol, phenol, p-cresol, o-cresol, pyrocatechol & o-nitrophenol.	97%	1:1 Glycerol +water	Aluminum	0.01 and 0.01M NaOH	Mixed-type	[9]
alginates (Alg) and pectates (Pect)	88.68%	1.6% (Pect.; best)	Aluminum	4M NaOH	–	[10]
2,6-dimethylpyridine	75.37%	3×10^{-3}M	Aluminum	1M HCl, NaCl and NaOH	Langmuir isotherm, mixed-type	[11]
Gum Arabic	74%	0.5g/L	Aluminum	1–2.5M NaOH	Langmuir adsorption isotherm	[12]
Bismarck brown (BB)	71%	0.025M	Aluminium (AA 1060)	0.1M NaOH	Chemisorption	[13]
Methyl cellulose	61.48%	2000 ppm	Al-Si alloy	0.1M NaOH	Freundlich isotherm	[14]
Polyvinyl Alcohol (PVA)	60.3%	3×10^{-4}M (PVA) +0.05KCl	Aluminum	0.1 M NaOH	Physisorption	[15]
3-Hydroxy flavones (3HF)	85.98%	4×10^{-5}M + QAI	Aluminum	1M NaOH	Langmuir isotherm	[16]
8-hydroxyquinoline	95%	46 mM	Al and Al-HO411 alloys	0.2 M NaOH	–	[17]
Sodium phytate	≈90%	0.01M	Copper	0.1M NaOH	–	[18]
Methionine	72.04%	10 mM	Brass	1M NaOH	Mixed-type	[19]
Coumarin derivatives	69.88%	11×10^{-6}M	Zinc	2M NaOH	Temkin isotherm, mixed-type	[20]
Natural Oils	72%	150 ppm	Nickel Electrode	0.01M NaOH	–	[21]

TABLE 7.2 Name, inhibition efficiency, concentration, metal, electrolyte, and mode of adsorption of some common compounds evaluated as corrosion inhibitors in sodium hydroxide based electrolytes.

Chemical name	Inhibition efficiency	Conc.	Metal	Electrolyte	Mode of adsorption	References
Congo Red" dye	48.63%	CR (0.005M) + KI	Aluminum	2M KOH	Physical, Langmuir isotherm	[22]
Crystal Violet (CV) Dye	85.3%	CV+10^{-6} KI	Aluminum	1M HCl & 0.5M KOH	Freundlich isotherm	[23]
Malachite green (MG) Dye	77.9%	MG+10^{-3} mM KI	Aluminum	1M HCl & 0.5M KOH	Flory–Huggins isotherm.	[24]
Urea and thiourea	60.8%	25 mM	Aluminum	5M KOH	Mixed-type inhibitors	[25]
Divinyl Sulfone -b-Cyclodextrin Polymer	93.61%	250 ppm	Zinc	3.5M KOH	Langmuir isotherm, mixed-type	[26]
Poly(ethylene glycol) and cetyltrimethylammonium bromide	96.8%	250 mg/L CTMAB	Zn and Zn–Ni	8M KOH	Langmuir isotherm, mixed-type	[27]
Polysorbate 20	86.8%	Indium + Tween 20	Zinc	3M KOH	Anodic-type	[28]
Lauroamide Propylbetaine	84.1%	800 ppm	Zinc	7.0M KOH	Langmuir adsorption	[29]

in the 1M HCl and 0.5M KOH [24]. However, its adsorption followed the Flory–Huggins adsorption isotherm model. Table 7.2 describes the results of some common studies conducted on corrosion inhibition of metals and alloys in KOH-based electrolytes.

7.2 Summary

Alkaline solutions are widely used as electrolytes for corrosion inhibition studies. Metallic corrosion in alkaline electrolytes proceeds through the formation of surface oxides and hydroxides complexes. The metal oxide film provides little bit corrosion protection however in the aqueous alkaline electrolytes; surface film becomes unstable due to the charge development. A huge amount of metallic components is lost during metallic corrosion in alkaline electrolytes. The use of organic compounds is the first-line defense against metallic corrosion in alkaline solutions. Sodium hydroxide and potassium hydroxide solutions are the most extensively used alkaline electrolytes. Several classes of organic compounds are used as corrosion for different metals and alloys in alkaline electrolytes. Organic compounds inhibit corrosion mainly adsorbing on the metallic surface and forming the surface protective film. Organic compounds containing polar functional groups and multiple bonds act as excellent corrosion inhibitors. These electron-rich centers help organic compounds to adsorb on the metallic surface.

7.3 Useful websites

https://www.sciencedirect.com/topics/medicine-anddentistry/alkali#:~:text=Alkalis%20are%-20very%20corrosive%20in,%2C%20lithium%2C%20and%20potassium%20hydroxides

https://blog.storemasta.com.au/examples-corrosive-substances-ph-levels

References

[1] D. Wang, L. Gao, D. Zhang, D. Yang, H. Wang, T. Lin, Experimental and theoretical investigation on corrosion inhibition of AA5052 aluminium alloy by l-cysteine in alkaline solution, Mater. Chem. Phys. 169 (2016) 142–151.

[2] D. Wang, H. Li, J. Liu, D. Zhang, L. Gao, L. Tong, Evaluation of AA5052 alloy anode in alkaline electrolyte with organic rare-earth complex additives for aluminium-air batteries, J. Power Sources 293 (2015) 484–491.

[3] A.A. Azim, L. Shalaby, H. Abbas, Mechanism of the corrosion inhibition of Zn Anode in NaOH by gelatine and some inorganic anions, Corrosion Sci. 14 (1974) 21–24.

[4] A. Fawzy, M. Abdallah, M. Alfakeer, H. Ali, Corrosion inhibition of SABIC iron in different media using synthesized Sodium N-dodecyl arginine surfactant, Int. J. Electrochem. Sci. 14 (2019) 2063–2084.

[5] A. Fawzy, M. Abdallah, I. Zaafarany, S. Ahmed, I. Althagafi, Thermodynamic, kinetic and mechanistic approach to the corrosion inhibition of carbon steel by new synthesized amino acids-based surfactants as green inhibitors in neutral and alkaline aqueous media, J. Mol. Liq. 265 (2018) 276–291.

[6] C. Verma, P. Singh, I. Bahadur, E. Ebenso, M. Quraishi, Electrochemical, thermodynamic, surface and theoretical investigation of 2-aminobenzene-1, 3-dicarbonitriles as green corrosion inhibitor for aluminum in 0.5 M NaOH, J. Mol. Liq. 209 (2015) 767–778.

[7] M. Abdallah, S. Atwa, I. Zaafarany, Corrosion inhibition of aluminum in NaOH solutions using some bidentate azo dyes compounds and synergistic action with some metal ions, Int. J. Electrochem. Sci. 9 (2014) 4747–4760.

[8] P. Su, L. Li, W. Li, C. Huang, X. Wang, Y. Liu, A. Singh, Expired drug theophylline as potential corrosion inhibitor for 7075 Aluminum Alloy in 1 M NaOH solution, Int. J. Electrochem. Sci. 15 (2020) 1412–1425.

[9] N. Subramanyan, V. Kapali, S.V. Iyer, Influence of hydroxy compounds on the corrosion and anodic behaviour of Al in NaOH solutions, Corrosion Sci. 11 (1971) 115–123.

[10] I. Zaafarany, Corrosion inhibition of aluminum in aqueous alkaline solutions by alginate and pectate water-soluble natural polymer anionic polyelectrolytes, Portugaliae Electrochim. Acta 30 (2012) 419–426.

[11] R. Padash, G.S. Sajadi, A.H. Jafari, E. Jamalizadeh, A.S. Rad, Corrosion control of aluminum in the solutions of NaCl, HCl and NaOH using 2, 6-dimethylpyridine inhibitor: Experimental and DFT insights, Mater. Chem. Phys. 244 (2020) 122681.

[12] S. Umoren, I. Obot, E. Ebenso, P. Okafor, O. Ogbobe, E. Oguzie, Gum Arabic as a potential corrosion inhibitor for aluminium in alkaline medium and its adsorption characteristics. In: Anti-Corrosion Methods and Materials, 53, Emerald Insight, 2006, pp. 277–282.

[13] E. Oguzie, B. Okolue, C. Ogukwe, C. Unaegbu, Corrosion inhibition and adsorption behaviour of bismark brown dye on aluminium in sodium hydroxide solution, Mater. Lett. 60 (2006) 3376–3378.

[14] S. Eid, M. Abdallah, E. Kamar, A. El-Etre, Corrosion inhibition of aluminum and aluminum silicon alloys in sodium hydroxide solutions by methyl cellulose, J. Mater. Environ. Sci. 6 (2015) 892–901.

[15] S. Umoren, E. Ebenso, P. Okafor, U. Ekpe, O. Ogbobe, Effect of halide ions on the corrosion inhibition of aluminium in alkaline medium using polyvinyl alcohol, J. Appl. Pol. Sci. 103 (2007) 2810–2816.

[16] J.M. Princey, P. Nagarajan, Corrosion inhibition of aluminium using 3-Hydroxy flavone in the presence of quarternary ammonium salts in NaOH medium, J. Korean Chem. Soc. 56 (2012) 201–206.

[17] H. Soliman, Influence of 8-hydroxyquinoline addition on the corrosion behavior of commercial Al and Al-HO411 alloys in NaOH aqueous media, Corrosion Sci. 53 (2011) 2994–3006.

[18] Y.-H. Wang, J.-B. He, Corrosion inhibition of copper by sodium phytate in NaOH solution: cyclic voltabsorptometry for in situ monitoring of soluble corrosion products, Electrochim. Acta 66 (2012) 45–51.

[19] J.F. Wu, Q. Wang, S.T. Zhang, L.L. Yin, Methionine as corrosion inhibitor of brass in O2-free 1M NaOH solution, Adv. Mater. Res. (2011) 241–245.

[20] M. Abdallah, O. Hazazi, B. ALJahdaly, A. Fouda, W. El-Nagar, Corrosion inhibition of zinc in sodium hydroxide solutions using coumarin derivatives, Int. J. Innov. Res. Sci. Eng. Technol. 3 (2014) 13802–13819.

[21] M. Abdallah, I. Zaafarany, S. Abd El Wanees, R. Assi, Corrosion behavior of nickel electrode in NaOH solution and its inhibition by some natural oils, Int. J. Electrochem. Sci. 9 (2014) 1071–1086.

[22] E. Oguzie, G. Onuoha, A. Onuchukwu, The inhibition of aluminium corrosion in potassium hydroxide by "Congo Red" dye, and synergistic action with halide ions, Anti-Corrosion Methods Mater. 52 (2005) 293–298.

[23] E.E. Oguzie, Inhibiting effect of crystal violet dye on aluminum corrosion in acidic and alkaline media, Chem. Eng. Commun. 196 (2008) 591–601.

[24] E. Oguzie, C. Akalezi, C. Enenebeaku, J. Aneke, Corrosion inhibition and adsorption behavior of malachite green dye on aluminum corrosion, Chem. Eng. Commun. 198 (2010) 46–60.

[25] Z. Moghadam, M. Shabani-Nooshabadi, M. Behpour, Electrochemical performance of aluminium alloy in strong alkaline media by urea and thiourea as inhibitor for aluminium-air batteries, J. Mol. Liq. 242 (2017) 971–978.

[26] H.M. Abd El-Lateef, M.A.E.A.A. Ali, Divinyl sulfone cross-linked β-cyclodextrin polymer as new and effective corrosion inhibitor for Zn anode in 3.5 M KOH, Trans. Indian Inst. Metals 69 (2016) 1783–1792.

[27] H.M. Abd El-Lateef, M. Elrouby, Synergistic inhibition effect of poly (ethylene glycol) and cetyltrimethylammonium bromide on corrosion of Zn and Zn—Ni alloys for alkaline batteries, Trans. Nonferrous Metals Soc. China 30 (2020) 259–274.

[28] X. Li, M. Liang, H. Zhou, Q. Huang, D. Lv, W. Li, Composite of indium and polysorbate 20 as inhibitor for zinc corrosion in alkaline solution, Bull. Korean Chem. Soc. 33 (2012) 1566–1570.

[29] L. Zhou, H. Liu, K. Liu, P. He, S. Wang, L. Jia, F. Dong, D. Liu, L. Du, Corrosion inhibition and passivation delay action of lauroamide propylbetaine on zinc in alkaline medium, Russian J. Electrochem. 56 (2020) 638–645.

CHAPTER 8

Corrosion and corrosion inhibition in neutral electrolytes

Neutral electrolytes include the synthetic and marine solutions of sodium chloride (NaCl). Different concentrations of NaCl solutions, especially 3% and 3.5% NaCl solutions are widely used as electrolytes for corrosion inhibition studies. Generally, sodium chloride solutions are highly aggressive and cause various corrosion-related failures and problems including loss or reduction in mechanical, corrosion resistance, weldability, and fatigue resistance of the metallic structures. Metallic corrosion in sodium chloride solutions involves anodic oxidation and cathodic oxygen reduction reactions.

$$Fe \rightarrow Fe^{2+/3+} + 2e^-/3e^- \quad \text{(Anodic oxidation)} \tag{8.1}$$

$$O_2 + 2H_2O + 4e^- \rightarrow 4OH^- \quad \text{(Cathodic reduction)} \tag{8.2}$$

Various reactive species can be formed by the reaction of ferrous and ferric ions with hydroxide and chloride ions [1,2]:

$$Fe^{2+} + 2Cl^- \rightarrow FeCl_2 \tag{8.3}$$

$$Fe^{3+} + 3Cl^- \rightarrow FeCl_3 \tag{8.4}$$

$$Fe^{2+} + 2OH^- \leftrightarrow Fe(OH)_2 \tag{8.5}$$

$$Fe^{3+} + 3OH^- \leftrightarrow Fe(OH)_3 \tag{8.6}$$

Formation of the ferrous and ferric hydroxides is also possible through hydroxide-chloride ions replacement reactions:

$$FeCl_2 + 2OH^- \leftrightarrow Fe(OH)_2 + 2Cl^- \tag{8.7}$$

$$FeCl_3 + 3OH^- \leftrightarrow Fe(OH)_3 + 3Cl^- \tag{8.8}$$

The ferrous and ferric chlorides can also undergo hydrolysis reaction to form corresponding hydroxides.

$$FeCl_2 + 2H_2O \rightarrow Fe(OH)_2 + 2HCl \tag{8.9}$$

$$FeCl_3 + 3H_2O \rightarrow Fe(OH)_3 + 3HCl \tag{8.10}$$

All these products, i.e., $Fe(OH)_2$, $Fe(OH)_3$, $FeCl_2$, and $FeCl_3$ are highly soluble in an aqueous medium therefore they would not provide protection from corrosion. In the aqueous solutions, presence of sodium chloride can play two significant roles.

(i) It can increase corrosion rate through forming the unstable chloride ($FeCl_2$ and $FeCl_3$) intermediates, and
(ii) As per Eqs. (8.9) and (8.10), sodium chloride can reduce the pH of aqueous solution through the formation of hydrochloric acid.

Aluminum and other metals undergo corrosion using similar mechanisms. The anodic and cathodic reaction for aluminum corrosion is shown below:

$$Al \rightarrow Al^{3+} + 3e^- \quad \text{(Anodic oxidation)} \tag{8.11}$$

$$O_2 + 2H_2O + 4e^- \rightarrow 4OH^- \quad \text{(Cathodic reduction)} \tag{8.12}$$

The aluminum cations so formed readily undergo hydration reaction to form various mononuclear (only one Al) and polynuclear (more than one Al) hydroxyl complexes [3].

$$Al^{3+} + H_2O \leftrightarrow Al(OH)^{2+} + H^+ \tag{8.13}$$

$$Al^{3+} + 2H_2O \leftrightarrow Al(OH)_2^+ + 2H^+ \tag{8.14}$$

$$Al^{3+} + 3H_2O \leftrightarrow Al(OH)_3 + 3H^+ \tag{8.15}$$

$$Al^{3+} + 4H_2O \leftrightarrow Al(OH)_4^- + 4H^+ \tag{8.16}$$

$$2Al^{3+} + 2H_2O \leftrightarrow Al_2(OH)_4 + 2H^+ \tag{8.17}$$

$$3Al^{3+} + 4H_2O \leftrightarrow Al_3(OH)_4^{5+} + 4H^+ \tag{8.18}$$

$$13Al^{3+} + 28H_2O \leftrightarrow Al_{33}O_4(OH)_{24}^{7+} + 32H^+ \tag{8.19}$$

The above hydroxide complexes react with chloride ions to form different chloride compounds. All these products are expected to be soluble and nonprotective.

8.1 Corrosion protection in neutral electrolytes

It is well-established that corrosion causes huge safety and economic losses. However, using proper corrosion monitoring techniques the loss of corrosion can be saved from 15% (US $875 billion) to 35% (US $375 billion). It is important to mention that corrosion results in the loss of more than 30% of marine metallic assets, such as ships and boats. In marine environments, several methods of corrosion protection such as painting and coating, alloying and de-alloying, and passivating the metallic surface using organic compounds are widely employed against metallic corrosion. Corrosion protection in sodium chloride solutions using organic compounds is one of the most effective and economic methods. Sometimes, the addition of some external chemical species of organic, inorganic, or their mixture needed for effective corrosion protection. These additives synergistically enhance the corrosion inhibition effectiveness of organic compounds.

Organic compounds inhibit corrosion by adsorbing on the metallic surface. Adsorption of the organic compounds on the metal surface may involve the physisorption, chemisorption, or mixed-(physiochemisorption) mechanism. Generally, physisorption occurs through electrostatic interactions among charged metallic surfaces and charged inhibitor molecules. On the other hand, chemisorption takes place through sharing of charges (electrons) between them. In most of the cases, adsorption of the organic compounds on metallic surface follows physiochemisorption mechanism. Value of the standard Gibbs free energy (ΔG) can be used as a gauge to determine the mode of adsorption of organic compounds on the metal surface. Generally, the Gibbs free energy (ΔG) values of -20 kJ mol^{-1} or less (in negative) are consistent with physisorption and its values of -40 kJ mol^{-1} are associated with chemisorption mechanism. Literature study shows that in most of the corrosion inhibition studies values of the Gibbs free energy (ΔG) stand in between -20 kJ mol^{-1} and -40 kJ mol^{-1}, which indicate that organic compounds interact with metallic surface through physiochemisorption mechanism.

Organic compounds contain various electron-rich polar functional groups such as $-OH$ (hydroxyl), $-NH_2$ (amino), $-CONH_2$ (amide), $-NHR/-NR_2$ (substituted amines), $-NO_2$ (nitro), $-CN$ (nitrile), $-COOC_2H_5$ (ester), $>C=O$ (carbonyl), $-N=N-$ (azo), $>C=N-$ (imine), $>C=S$ (thio-carbonyl), and $-COCl$ (acid chloride), etc., through which they get adsorbed on the metal surface. Along with the polar functional groups, organic compounds also transfer their π-electrons into the d-orbitals of the surface metallic atoms. The process of transfer of electrons from inhibitors to metal is known as donation. However, metals are electron-rich species, this type of electron donation causes interelectronic repulsion. This renders metals to transfer their electrons to the empty molecular orbitals of the organic molecules. This is known as retro- or back-donation. Greater is the donation greater will be retro-donation, i.e., donation and retro-donation strengthen each other through synergism. Adsorption of the organic corrosion inhibitors (CINHs) on the metallic surface can be illustrated as follows:

$$M \leftrightarrow M^{n+} + ne^-$$

$$M^{n+} + nCl \leftrightarrow [M(Cl)_n]_{ads}$$

$$[M(Cl)_2]_{ads} + CINH \leftrightarrow [M^{n+}(CINH)]_{ads} + nCl^-$$

However, if organic compounds contain acidic hydrogen then formation of neutral complex is also possible [4]:

$$MCl + CINH - H \leftrightarrow [M(CINH)]_{ads} + HCl$$

After getting adsorbed organic compounds build a hydrophobic film and that protects metals from corrosion.

8.2 Organic corrosion inhibitors: case studies

The use of organic compounds is one of the most effective and popular methods of corrosion mitigation in sodium chloride solutions. Organic compounds of natural (carbohydrates, amino acids, natural medicines, and plant extracts) and synthetic (different families of organic compounds) origin are effectively evaluated as corrosion in sodium chloride based electrolytes. Corrosion inhibition using organic compounds in sodium chloride solutions for different metals and alloys is described below.

8.2.1 Organic corrosion inhibitors for iron alloys in NaCl electrolytes

Because of their numerous useful properties, such as high tensile and mechanical strength, relatively higher ductility, malleability, conductivity and high melting point, steel alloys are widely used as constructional and building materials. The metallic equipment made-up of steel alloys includes medical tools and devices, magnetic cores, wires, and military assets. However, these materials are highly prone to corrosion when exposed to the environment especially in aqueous solution of the sodium chloride. Organic compounds are extensively used as corrosion inhibitors for steel alloys in sodium chloride. A description on the corrosion inhibition effect of some major organic compounds for steel alloys is given in Table 8.1. In another study, Amar and coworkers [5] described the inhibition power of two phosphonic acids, namely, piperidin-1-yl-phosphonic acid and (4-phosphono-piperazin-1-yl) phosphonic acid (PPPA) for iron corrosion in 3% sodium chloride solution. The study was conducted using various experimental methods including chemical, electrochemical, and surface studies methods. Among the tested compounds, PPPA acted as the best corrosion inhibitor and it showed the highest protection power of 93.2% at 5×10^{-3} mol/L M concentration. Both piperidin-1-yl-phosphonic acid and PPPA exhibited high corrosion inhibition effectiveness through their adsorption that obeyed the Langmuir adsorption isotherm model.

In another study, the effect of five imidazoles and triazoles compounds on steel corrosion in 2.5% and 3.5% NaCl solutions was studied using electrochemical and surface analyses [6]. Results demonstrated that adsorption of 3AMTA, 2ATA, and 2-ATDA on steel surface obeyed the Langmuir adsorption isotherm model, whereas adsorption of remaining compounds followed the Temkin adsorption isotherm model. Among the tested compounds, 2-ABA exhibited 94% and 88% inhibition efficiencies in 3.5% and 2.5% NaCl solutions, respectively. Electrochemical results were also supported by surface morphology study using scanning electron microscope (SEM). In the SEM study, metallic specimens were allowed to corrode free in the absence and presence of the best corrosion inhibitor (2-ABA at 2×10^{-2} M) for 30 minutes. SEM images of corroded and protected metallic specimens are illustrated in Fig. 8.1. It can be clearly seen that the surface of the nonprotected metallic specimens is highly corroded and damaged. This is attributed to the free aggressive attacks of electrolyte molecules. But in the presence of 2-ABA, the metal surface morphology improves significantly owing to the formation of protective surface film through the adsorption of inhibitor molecules (2-ABA).

Inhibition effect of four porphyrin derivatives was reported for N80 corrosion in 3.5% NaCl solution saturated with CO_2 [8]. Inhibition power four evaluated compounds followed the order: HPTB (85%) < THP (86%) TPP < (88%) < T4PP (91%). All tested compounds showed their maximum protection efficiency at 200 ppm concentration. Computational analyses provide good agreement to the results derived from experimental methods. Analyses of Table 8.1 showed that most of the evaluated organic compounds exhibit good corrosion inhibition effectiveness. Most of the organic compounds become effective by adsorbing on the steel alloys surface usually through Langmuir adsorption isotherm mechanism.

TABLE 8.1 Name, inhibition efficiency, concentration, metal, electrolyte, and mode of adsorption of some common compounds evaluated as corrosion inhibitors for steel alloys in NaCl-based electrolytes.

Chemical name	Inhibition efficiency	Conc.	Metal	Electrolyte	Mode of adsorption	References
Phosphonic acids	93.2% (PPPA)	5×10^{-3} M	Iron	3% NaCl	Langmuir adsorption, mixed-type	[5]
Cyclic nitrogen compounds	94% (2-ABA)	2×10^{-2} M	Steel	2.5 & 3.5% NaCl	Temkin/Langmuir, mixed-type	[6,7]
Porphyrins	91% (T4PP)	200 ppm	N80 steel	3.5% NaCl+CO_2	Mixed-type corrosion inhibitors	[8]
Benzotriazole	93%	5 mM	Carbon steel	3.5% NaCl+ H_2S	Interface-type inhibitor	[9]
Imidazoline amide (IM)	97%	20 ppm	Iron	3M NaCl +CO_2	Langmuir adsorption isotherm	[10]
Phosphonic acids	98.7% (TMPA)	5×10^{-3} M	Carbon steel	Sea water	Mixed-type inhibitors	[11]
Amines	78.2% (IMC-80-Q)	150 mg/L	N80 steel	3% NaCl+CO_2	Langmuir & Freundlich below and above 150 ppm	[12]
Piperidin-1-yl-phosphonic	89.9%	5×10^{-3} M +20% Zn^{2+}	Armco iron	3% NaCl	Predominantly cathodic-type inhibitor	[13]
Ethylamino imidazoline derivative	97.3%	80 ppm+ 2000 ppm KI	N80 steel	3% NaCl+CO_2	Temkin adsorption, mixed-type	[14]
3-amino-5-mercapto-1,2,4-triazole (AMTA)	97%	2 mM	Pure iron	3.5% NaCl	Mixed-type corrosion inhibitor	[1]
Benzimidazole	92%	100 ppm	Carbon steel	5% NaCl	Interface-type inhibitor	[15]
Imidazolinium Carboxylate Salt	86%	4 mM	Mild steel	0.01M NaCl	Mixed-type inhibitor	[16]
1H-benzotriazole, Na-molybdate & Na-phosphate	76% (S10)	–	Fe & Cu alloys/	Sea water	Mixed-type inhibitors	[17]
2-mercaptobenzoxazole	96.45%	0.5 mM +1.5 mM $ZnCl_2$	Mild steel	0.1M NaCl	Mixed- and interface-type inhibitor	[18]
Thiamine (vitamin B1) & biotin (vitamin B7)	91.42% (VB1)	200 ppm	Mild steel	400 ppm NaCl	Mixed-type inhibitors	[19]

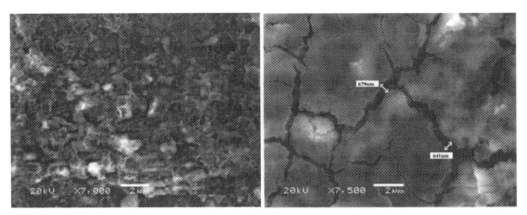

FIG. 8.1 SEM micrograph of steel surface corroded in 3.5% NaCl solution for 30 minutes in the absence (left-handside) and presence of 2×10^{-2} M concentration of 2-ABA (right-hand side). (Reproduced with permission@ Copyright Elsevier). *SEM*, scanning electron microscope.

Polarization studies suggest that most of the organic compounds adversely affect the anodic and cathodic reactions and act as mixed-type corrosion inhibitors. Presence of the organic compounds in the sodium chloride solutions increases the values of charge transfer resistance for metallic corrosion and behaves as interface-type corrosion inhibitors.

8.2.2 Organic corrosion inhibitors for aluminum alloys in NaCl electrolytes

Because of their very high thermal and electrical conductivity, aluminum alloys are extensively employed in automotive, electrical, and aviation industries. The aluminum-based alloys possess excellent corrosion resistance property which is attributed to the formation of passive film of aluminum oxide. Organic compounds are used as the first line of defense against corrosion of aluminum in sodium chloride solution. A description on the corrosion inhibition effect of some major organic compounds for aluminum alloys is given in Table 8.2. Organic compounds adsorb over the metallic surface and form corrosion inhibitive film. Polar functional groups and π-electrons of multiple bonds help in getting them adsorb on the aluminum surface. In a study, Zhu et al. [20] reported the inhibition effect of bis-[3-(triethoxysilyl)propyl-]tetrasulfide for AA 2024-T3 corrosion in 0.6M NaCl solution using surface and electrochemical analyses methods. Study showed that investigated compound inhibits corrosion by adsorbing on the metallic surface thereby it increased the values of charge transfer resistance. Adsorption mechanism of corrosion inhibition was supported by SEM study. Potentiodynamic polarization studies showed that presence of inhibitor shifted corrosion potential (E_{corr}) toward negative direction, therefore, acted as mainly cathodic-type inhibitor.

Corrosion inhibition effect of three surfactants for Al-brass alloy corrosion in 3.5% NaCl tested in seawater [21]. Inhibition effectiveness of the tested compounds followed the order: DPh(EO) < LAPACl < SDBS. Potentiodynamic studies showed that all tested compounds become effective in both electrolytes by forming different types of oxide complexes. In another report, corrosion inhibition power of 1,4-naphthoquinone (NQ) was evaluated for

TABLE 8.2 Name, inhibition efficiency, concentration, metal, electrolyte, and mode of adsorption of some common compounds evaluated as corrosion inhibitors for aluminum alloys in NaCl-based electrolytes.

Chemical name	Inhibition efficiency	Conc.	Metal	Electrolyte	Mode of adsorption	References
Bis-[3-(triethoxysilyl) propyl]tetrasulfide (bis-sulfur silane)	–	–	AA 2024-T3	0.6M NaCl	Cathodic-type inhibitor	[20]
Anionic, nonionic & cationic surfactants	91% (SDBS)	1000 ppm	Aluminium–brass	3.5%M NaCl and sea water	Mixed-type inhibitors	[21]
1,4-naphthoquinone (NQ)	98.2%	5×10^{-3} M	Aluminum	0.5M NaCl	Interface-type inhibitor	[22]
Triazole and triazole derivatives	–	–	2024 Al alloys/	0.005M NaCl	Mixed-type inhibitors	[23]
Sinapinic acid	65.4%	5×10^{-5}M	Al–2.5 Mg	0.5M NaCl	Cathodic-type, Freundlich isotherm	[24]
bis-[triethoxysilylpropyl] tetrasulfide and bis-[trimethoxysilylpropyl] amine	–	–	AA 2024-T3 and & galvanized Steel	0.6 M NaCl	Mixed-type corrosion inhibitors	[25]
8-hydroxy-quinoline and its derivative	–	–	Aluminum alloy 2024-T3	3.5 NaCl	Mixed type of corrosion inhibitors	[26]

aluminum in 0.5M sodium chloride solution using various electrochemical and surface analyses methods [22]. Analyses showed that the tested compound showed reasonably good corrosion inhibition effectiveness at relatively low concentration. EIS study showed that presence of the NQ in corrosive medium increases the values of charge transfer resistance thereby it acts as interface-type corrosion inhibitors. Inspection of Table 8.2 showed that most of the studied compounds acted as efficient corrosion inhibitors for aluminum alloys in sodium chloride based electrolytes. Mostly, organic compounds become effective by inhibiting the anodic and cathodic half-cell reactions and behaving as mixed-type corrosion inhibitors.

8.2.3 Organic corrosion inhibitors for copper alloys in NaCl electrolytes

Because of their excellent mechanical properties and resistance to corrosion, copper and its alloys are widely used as construction materials for industrial as well as household applications. Copper-based alloys are widely used to make ships and boats. These materials are widely used as power lines, electrical and thermal conductors, and heat exchangers. However, these materials are also very prone to corrosion especially in the sodium chloride solutions. Again application of organic compounds is one of the first lines of defense against metallic corrosion. Generally, copper forms different coordination compounds in its different oxidation states which act as a barrier for aggressive attacks for corrosive species. It is important to notice that organic compounds containing heteroatoms in the form of polar functional

TABLE 8.3 Name, inhibition efficiency, concentration, metal, electrolyte, and mode of adsorption of some common compounds evaluated as corrosion inhibitors for aluminum alloys in NaCl-based electrolytes.

Chemical name	Inhibition efficiency	Conc.	Metal	Electrolyte	Mode of adsorption	References
Sodium oleate (SO)	95.37%	200×10^5 M	Copper	3% M NaCl	Frumkin isotherm, mixed-type	[27]
Cysteine	84.13%	18 mM	Copper	0.6M NaCl &1M HCl	Langmuir isotherm, cathodic-type	[29]
	(1M HCl)					
Benzotriazole (BTAH), & 1-hydroxybenzotriazole (BTAOH).	89.32% (BTAH)	5 mM	Copper	3% NaCl	Mixed-type inhibitors	[30]
Sodium succinate (SS)	79.65%	0.01 M	Copper	5M NaCl	Mixed-type inhibitor	[31]
5-(Phenyl)-4H-1,2,4-triazole-3-thiol (PTAT)	90%	1500 ppm	Copper/	3.5% NaCl	Mixed-type inhibitor	[32]
Bis-(1-benzotriazolymethylene)-(2,5-thiadiazoly)-disulfide	86.7% (3% NaCl)	1×10^{-3} M	Copper	3% NaCl and 0.5 M HCl	Mixed type corrosion inhibitor	[33]

groups and multiple bonds act as ideal corrosion inhibitors. Literature study showed that several reports are published dealing with the corrosive effect of organic compounds in sodium chloride solution. A description on the corrosion inhibition effect of some major organic compounds for copper and copper alloys is given in Table 8.3.

Amin [27] investigated the corrosion inhibition of sodium oleate (SO) for copper corrosion in 3% NaCl solution. Polarization studies showed that SO acted as mixed-type corrosion inhibitors and its presence adversely affected the anodic as well as cathodic reactions. Adsorption of SO on copper surface obeyed the Frumkin adsorption isotherm model. Zucchi et al. [28] investigated the corrosion inhibition effect of different organic compounds namely, 5-n-hexylbenzotriazole, S-chlorobenzotriazole, 2,4-dimercaptopyrimidine, 5-methylbenzotriazole, and 2-amino-5-mercaptothiadiazole at their 0.1 and 0.5M concentrations for copper corrosion in 0.1M NaCl solution. Inhibition efficiencies of the tested compounds followed the order: 2ASMcTDZ < DTU < TTA < Cl - BTA < C6BTA. Inhibition efficiency of the different compounds was measured by evaluating the anodic and cathodic polarization curves. From Table 8.3, it can be derived that most of the evaluated organic compounds exhibit good corrosion inhibition effectiveness for copper in sodium chloride solutions. Most of the organic compounds become effective by adsorbing on the metallic surface. Mostly, organic compounds act as mixed-type corrosion inhibitors for copper corrosion in sodium chloride solution and they adversely affect the anodic as well as cathodic half-cell reactions.

8.3 Summary

Sodium chloride based solutions are widely used as electrolytes in corrosion studies. Corrosion inhibition of different metals and their alloys are extensively reported in sodium chloride solution. Sodium chloride solutions are extensively used as electrolytes for steel, aluminum, copper, and their alloys. Most of the metals undergo passivation through the formation of metal oxides. However, the passive film of the metal oxides is unstable due to the development of charge. The use of organic compounds is one of the most economic, popular, and effective methods in the sodium chloride solution. The compounds become effective by forming the surface protective film using their electron-rich polar functional groups and multiple bonds. It can also be seen that 3% and 3.5% NaCl solutions are most widely used in corrosion studies. Outcomes of the polarization studies showed that most of organic compounds become effective by inhibiting the both anodic and cathodic half-cell reaction. Mostly, adsorption of the organic compounds on metallic surface on metallic surface obeyed the Langmuir adsorption isotherm.

8.4 Useful website

https://onlinelibrary.wiley.com/doi/10.1002/9783527610426.bard040201

References

[1] E.-S.M. Sherif, R. Erasmus, J. Comins, In situ Raman spectroscopy and electrochemical techniques for studying corrosion and corrosion inhibition of iron in sodium chloride solutions, Electrochim. Acta 55 (2010) 3657–3663.
[2] W. Li, K. Nobe, A.J. Pearlstein, Potential/current oscillations and anodic film characteristics of iron in concentrated chloride solutions, Corrosion Sci. 31 (1990) 615–620.
[3] D. Zhu, W.J. van Ooij, Corrosion protection of AA 2024-T3 by bis-[3-(triethoxysilyl) propyl] tetrasulfide in neutral sodium chloride solution. Part 1: corrosion of AA 2024-T3, Corrosion Sci. 45 (2003) 2163–2175.
[4] M. Scendo, The effect of purine on the corrosion of copper in chloride solutions, Corrosion Sci. 49 (2007) 373–390.
[5] H. Amar, J. Benzakour, A. Derja, D. Villemin, B. Moreau, A corrosion inhibition study of iron by phosphonic acids in sodium chloride solution, J. Electroanalyt. Chem. 558 (2003) 131–139.
[6] M. Şahin, S. Bilgic, H. Yılmaz, The inhibition effects of some cyclic nitrogen compounds on the corrosion of the steel in NaCl mediums, Appl. Surface Sci. 195 (2002) 1–7.
[7] G. Gece, S. Bilgiç, Quantum chemical study of some cyclic nitrogen compounds as corrosion inhibitors of steel in NaCl media, Corrosion Sci. 51 (2009) 1876–1878.
[8] A. Singh, Y. Lin, M.A. Quraishi, L.O. Olasunkanmi, O.E. Fayemi, Y. Sasikumar, B. Ramaganthan, I. Bahadur, I.B. Obot, A.S. Adekunle, Porphyrins as corrosion inhibitors for N80 Steel in 3.5% NaCl solution: electrochemical, quantum chemical, QSAR and Monte Carlo simulations studies, Molecules 20 (2015) 15122–15146.
[9] A. Solehudin, Performance of benzotriazole as corrosion inhibitors of carbon steel in chloride solution containing hydrogen sulfide, Int. Refereed J. Eng. Sci 1 (2012) 21–26.
[10] X. Zhang, F. Wang, Y. He, Y. Du, Study of the inhibition mechanism of imidazoline amide on CO2 corrosion of Armco iron, Corrosion Sci. 43 (2001) 1417–1431.
[11] H. Amar, T. Braisaz, D. Villemin, B. Moreau, Thiomorpholin-4-ylmethyl-phosphonic acid and morpholin-4-methyl-phosphonic acid as corrosion inhibitors for carbon steel in natural seawater, Mater. Chem. Phys. 110 (2008) 1–6.
[12] X. Jiang, Y. Zheng, W. Ke, Effect of flow velocity and entrained sand on inhibition performances of two inhibitors for CO2 corrosion of N80 steel in 3% NaCl solution, Corrosion Sci. 47 (2005) 2636–2658.
[13] H. Amar, J. Benzakour, A. Derja, D. Villemin, B. Moreau, T. Braisaz, A. Tounsi, Synergistic corrosion inhibition study of Armco iron in sodium chloride by piperidin-1-yl-phosphonic acid–Zn2+ system, Corrosion Sci. 50 (2008) 124–130.
[14] P. Okafor, X. Liu, Y. Zheng, Corrosion inhibition of mild steel by ethylamino imidazoline derivative in CO2-saturated solution, Corrosion Sci. 51 (2009) 761–768.
[15] D.A. Lopez, S. Simison, S. De Sanchez, The influence of steel microstructure on CO2 corrosion. EIS studies on the inhibition efficiency of benzimidazole, Electrochim. Acta 48 (2003) 845–854.

[16] A.L. Chong, J.I. Mardel, D.R. MacFarlane, M. Forsyth, A.E. Somers, Synergistic corrosion inhibition of mild steel in aqueous chloride solutions by an imidazolinium carboxylate salt, ACS Sustain. Chem. Eng. 4 (2016) 1746–1755.
[17] K.S. Bokati, C. Dehghanian, S. Yari, Corrosion inhibition of copper, mild steel and galvanically coupled copper-mild steel in artificial sea water in presence of 1H-benzotriazole, sodium molybdate and sodium phosphate, Corrosion Sci. 126 (2017) 272–285.
[18] S. Alinejad, R. Naderi, M. Mahdavian, Effect of inhibition synergism of zinc chloride and 2-mercaptobenzoxzole on protective performance of an ecofriendly silane coating on mild steel, J. Ind. Eng. Chem. 48 (2017) 88–98.
[19] A. Aloysius, R. Ramanathan, A. Christy, S. Baskaran, N. Antony, Experimental and theoretical studies on the corrosion inhibition of vitamins–Thiamine hydrochloride or biotin in corrosion of mild steel in aqueous chloride environment, Egyp. J. Petr. 27 (2018) 371–381.
[20] D. Zhu, W.J. van Ooij, Corrosion protection of AA 2024-T3 by bis-[3-(triethoxysilyl) propyl] tetrasulfide in sodium chloride solution. Part 2: mechanism for corrosion protection, Corrosion Sci. 45 (2003) 2177–2197.
[21] M. Osman, Corrosion inhibition of aluminium–brass in 3.5% NaCl solution and sea water, Mater. Chem. Phys. 71 (2001) 12–16.
[22] E. Sherif, S.-M. Park, Effects of 1, 4-naphthoquinone on aluminum corrosion in 0.50 M sodium chloride solutions, Electrochim. Acta 51 (2006) 1313–1321.
[23] M. Zheludkevich, K. Yasakau, S. Poznyak, M. Ferreira, Triazole and thiazole derivatives as corrosion inhibitors for AA2024 aluminium alloy, Corrosion Sci. 47 (2005) 3368–3383.
[24] S. Lamaka, M. Zheludkevich, K. Yasakau, M. Montemor, M. Ferreira, High effective organic inhibitor corrosion for 2024 aluminium alloy, Electrochim. Acta 52 (2007) 7231–7247.
[25] D. Zhu, W.J. van Ooij, Enhanced corrosion resistance of AA 2024-T3 and hot-dip galvanized steel using a mixture of bis-[triethoxysilylpropyl] tetrasulfide and bis-[trimethoxysilylpropyl] amine, Electrochim. Acta 49 (2004) 1113–1125.
[26] S.-M. Li, H.-R. Zhang, J.-H. Liu, Corrosion behavior of aluminum alloy 2024-T3 by 8-hydroxy-quinoline and its derivative in 3.5% chloride solution, Trans. Nonferrous Metals Soc. China 17 (2007) 318–325.
[27] M.A. Amin, Weight loss, polarization, electrochemical impedance spectroscopy, SEM and EDX studies of the corrosion inhibition of copper in aerated NaCl solutions, J. Appl. Electrochem. 36 (2006) 215–226.
[28] F. Zucchi, G. Trabanelli, C. Monticelli, The inhibition of copper corrosion in 0.1 M NaCl under heat exchange conditions, Corrosion Sci. 38 (1996) 147–154.
[29] K.M. Ismail, Evaluation of cysteine as environmentally friendly corrosion inhibitor for copper in neutral and acidic chloride solutions, Electrochim. Acta 52 (2007) 7811–7819.
[30] M. Finšgar, A. Lesar, A. Kokalj, I. Milošev, A comparative electrochemical and quantum chemical calculation study of BTAH and BTAOH as copper corrosion inhibitors in near neutral chloride solution, Electrochim. Acta 53 (2008) 8287–8297.
[31] O.A. Hazzazi, Corrosion inhibition studies of copper in highly concentrated NaCl solutions, J. Appl. Electrochem. 37 (2007) 933–940.
[32] H.O. Curkovic, E. Stupnisek-Lisac, H. Takenouti, Electrochemical quartz crystal microbalance and electrochemical impedance spectroscopy study of copper corrosion inhibition by imidazoles, Corrosion Sci. 51 (2009) 2342–2348.
[33] D.-Q. Zhang, L.-X. Gao, G.-D. Zhou, Inhibition of copper corrosion by bis-(1-benzotriazolymethylene)-(2, 5-thiadiazoly)-disulfide in chloride media, Appl. Surface Sci. 225 (2004) 287–293.

CHAPTER 9

Corrosion and corrosion inhibition in sweet and sour environments

Sweet and sour corrosion mostly occurs in petroleum industries where metallic structures undergo corrosion failure in the presence of carbon dioxide (CO_2) and hydrogen sulfide (H_2S), respectively. Sweet and sour corrosion is separately described below.

9.1 Sweet corrosion

Sweet corrosion is also known as CO_2 corrosion which occurs in the presence of high CO_2 partial pressure and is devoid of high levels of hydrogen sulfide and other sulfides [1]. Generally, CO_2 converts into carbonic acid and causes corrosion. Sweet corrosion is one of the most common forms of corrosion encountered in the refinery, oil, gas, and petroleum industries. CO_2 causes major corrosion failure of transport pipelines and downhole tubing. Sweet corrosion mainly occurs in the presence of moisture. CO_2 dissolves in water and forms carbonic acid (H_2CO_3) aqueous [2,3]. Therefore, the rate and aggressiveness of the sweet corrosion depends upon the presence of moisture. Formation of carbonic acid (H_2CO_3) through the reaction of water and carbon dioxide is presented below:

$$CO_{2(gas)} \rightarrow CO_{2(aq)} \tag{9.1}$$

$$CO_{2(aq)} + H_2O \rightarrow H_2CO_3 \tag{9.2}$$

Although H_2CO_3 is a relatively weak acid, it undergoes dissociation in the aqueous solution to form bicarbonate and liberate hydrogen ions. The bicarbonate ions further undergo dissolution to form carbonate ions.

$$H_2CO_3 \rightarrow HCO_3^- + H^+ \tag{9.3}$$

$$HCO_3^- \rightarrow CO_3^{2-} + H^+ \tag{9.4}$$

Mechanism of sweet corrosion is very much similar to the mechanism of metallic corrosion in aqueous acidic solutions. A schematic illustration of sweet corrosion is presented in Fig. 9.1. At anode, rapid oxidation of iron results in the formation of ferrous ions and discharge of the electrons. These electrons are consumed at the cathode for the reduction of hydrogen ions.

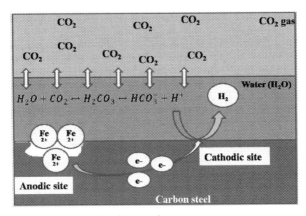

FIG. 9.1 Mechanism of sweet corrosion of carbon steel.

Ferrous sulphate forms as corrosion product (scale) may or may not be protective. In sweet environment, reactions taking place at anodic and cathodic sites can be presented as follows:

Reactions at anodic sites [4]:

$$Fe \leftrightarrow Fe^{2+} + 2e^- \tag{9.5}$$

$$Fe^{2+} + CO_3^{2-} \leftrightarrow FeCO_3 \tag{9.6}$$

$$Fe^{2+} + 2HCO_3^- \leftrightarrow Fe(HCO_3)_2 \tag{9.7}$$

$$Fe(HCO_3)_2 \leftrightarrow FeCO_3 + CO_2 + H_2O \tag{9.8}$$

Reactions at anodic sites [4]:

$$CO_2 + H_2O \leftrightarrow H_2CO_3 \tag{9.9}$$

$$2HCO_3^- + 2e^- \leftrightarrow H_2 + 2CO_3^{2-} \tag{9.10}$$

$$HCO_3^- \leftrightarrow H^+ + CO_3^{2-} \tag{9.11}$$

However, the overall sweet corrosion reaction can be presented as follows:

$$Fe + CO_2 + H_2O \rightarrow FeCO_3 + H_2 \tag{9.11}$$

The protectiveness of the $FeCO_3$ depends upon numerous factors, such as pH, temperature, and flow rate of the liquid. Following factors affect the rate of sweet corrosion:

(i) *Effect of CO_2 partial pressure*: obviously, an increase in CO_2 concentration causes increase in the corrosion rate. This is attributed to the increased dissolution of CO_2 at its high partial pressure. However, at very high CO_2 partial pressure, especially at high pH, increase in the CO_2 partial pressure would be beneficial for the corrosion protection. In such conditions, increase in the CO_2 partial pressure results in the increase in precipitation of $FeCO_3$. Therefore, under favorable conditions, increase in CO_2 partial pressure would be beneficial for sweet corrosion protection.

(ii) *Effect of temperature*: generally, an increase in temperature (up to 70°C) results in the corresponding increase in sweet corrosion rate. However, at high temperature (above 70°C), corrosion rate diminished due to the formation of highly protective $FeCO_3$.

(iii) *Effect of pH*: it is extensively reported that rate of sweet corrosion increases on decreasing pH and vice versa. At low pH, cathodic reduction rate would be high. However, at high pH, solubility of the $FeCO_3$ would be low as it will precipitate as solid scale at cathodic sites. This type of precipitation results in the formation of protective film of $FeCO_3$.

(iv) *Effect of flow*: The corrosion rate would be high in the flowing liquids. This is due to the fact that the solid corrosion scale of $FeCO_3$ would be continuously removed and new surfaces would be exposed for further corrosion.

9.2 Sour corrosion

Corrosion resulted through hydrogen sulfide (H_2S) is known as sour corrosion. Generally, crude oils and gases contain dissolved sulfur-containing compounds, especially H_2S that causes corrosion of the metallic structures. Sour corrosion is mainly encountered in petroleum industries. H_2S can undergo dissociation to give several reactive and corrosive species. In an aqueous medium, dissociation of H_2S occurs as follows [5,6]:

$$H_2S_{(g)} \leftrightarrow H_2S_{(aq)} \tag{9.12}$$

$$H_2S_{(aq)} \leftrightarrow H^+_{(aq)} + HS^-_{(aq)} \tag{9.13}$$

$$HS^-_{(aq)} \leftrightarrow H^+_{(aq)} + S^{2-}_{(aq)} \tag{9.14}$$

These species produce a more aggressive environment by changing the electrochemical nature of the environment. Generally, these species deposit as corrosion products, and that influence the localized corrosion. Various other sulphides including pyrite, pyrrhotite, cubic, and amorphous ferrous sulfide, mackinawite, troilite, etc., are formed from the dissolution of H_2S in the aqueous electrolytes. These species also greatly affect the corrosiveness of the environment. According to this mechanism, mackinawite forms as one of the major corrosion products in the sour corrosion of iron. Sun et al. showed that dissolution in H_2S medium occurs through the following series of reactions [7,8].

$$Fe + H_2S_{(free)} \leftrightarrow Fe:H_2S_{(ads)} \tag{9.15}$$

$$Fe:H_2S_{(ads)} \leftrightarrow Fe:HS^-_{(ads)} + H^+_{(ads)} \tag{9.16}$$

$$Fe:HS^-_{(ads)} + H^+_{(ads)} \leftrightarrow Fe:HS^-_{(ads)} + H_{(ads)} + e^- \tag{9.17}$$

$$Fe:HS^-_{(ads)} + H_{(ads)} + e^- \leftrightarrow FeS_{(ads)} + 2H_{(ads)} \tag{9.18}$$

$$2nFeS_{(ads)} \leftrightarrow nFe_2S_2 \leftrightarrow FeS(\text{mackinawide}) \tag{9.19}$$

Mechanism of anodic and cathodic reactions for iron corrosion in the presence of H_2S is presented in Fig. 9.2.

Following factors affect the rate of sour corrosion:

(i) *Effect of H_2S partial pressure*: effect of H_2S partial pressure in sour corrosion is similar to the effect of effect of CO_2 partial pressure in sweet corrosion. Generally, at low pH, the rate of corrosion increases with increasing H_2S partial pressure. However, if all

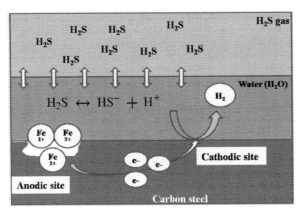

FIG. 9.2   Mechanism of sour corrosion of carbon steel.

conditions are favorable, the rate of sour corrosion decreases on increasing the high H_2S partial pressure. This is attributed due to formation of sulphide based scales at elevated temperature.

(ii) *Effect of temperature*: generally, rate of sour corrosion increases on increasing the temperature. This is because the sulphide films produced at low temperature are relatively more compact and less porous that provide good anticorrosive protection. However, sulphide scales produced at high temperature are highly porous and less compact and provide either no or very less anticorrosive protection.

(iii) *Fluid velocity*: low fluid velocity rate favors the sour corrosion as at stagnant or low fluid velocity allows the deposition of solid deposits where water accumulation can take place.

(iv) *Effect of dissolved salability/salts*: high chloride/salinity level favors the sour corrosion.

(v) *Effect of pH*: in general, increase in the pH or alkalinity diminishes the sour corrosion as high pH favors the precipitation of solid protective sulphide scales. However, this type precipitation of solid scales can induce localized (pitting) corrosion.

9.3   Protection for sweet and sour corrosion

The use of organic compounds, especially containing heteroatoms is established as one of the most effective methods of sweet and sour corrosion mitigation. Some of the commonly used organic compounds are amines, amides, carboxylic acids, nitrogen containing salts, and polyoxyalkylized imidazoline. It is proposed that most of these compounds become effective by adsorbing and forming the surface protective film. Presence of the polar functional groups and multiple bonded electron rich centers act as adsorption centers adsorption of organic compounds over the metallic surface in sweet and sour environments. It is important to mention that various natural and synthetic polymers are also employed as effective corrosion inhibitors, especially in the sweet environment. As the polymers contain several electron rich adsorption centers, their use as corrosion inhibitors should be preferred. Although several organic compounds are tested as sweet and sour corrosion inhibitors at laboratory scale, only few of them are used as anticorrosive formulations for industrial purposes.

9.4 Summary

Carbon dioxide (CO_2) and hydrogen sulfide (H_2S) are present as dissolved gases in petroleum-based oil and gases. These gases become highly corrosive especially in the presence of moisture. Corrosion caused by carbon dioxide (CO_2) is known as sweet corrosion, whereas corrosion resulting from hydrogen sulfide (H_2S) is known as sour corrosion. Generally, CO_2 becomes effective by forming the highly corrosive carbonic acid (H_2CO_3) through combination of CO_2 and H_2O. In both sweet and sour corrosion, anodic half-cell reaction involves the oxidative decomposition of metal and cathodic half-cell reaction involves the reductive hydrogen evolution. The rate sweet and sour corrosion is greatly depends upon the pH, temperature, CO_2 partial pressure (in case sweet corrosion), H_2S partial pressure (in case sour corrosion), nature of solid deposits, salts/salinity etc. Organic compounds containing polar functional groups and multiple bonds are used as the first line of defense against sweet and sour corrosion.

9.5 Useful websites

https://www.corrosionpedia.com/definition/6143/sour-corrosion
https://www.coursera.org/lecture/corrosion/sour-corrosion-RrEmS
https://www.hindawi.com/journals/isrn/2012/892385/
https://www.coursera.org/lecture/corrosion/sweet-and-sour-corrosion-AUorp

References

[1] Z. Yin, W. Zhao, Z. Bai, Y. Feng, W. Zhou, Corrosion behavior of SM 80SS tube steel in stimulant solution containing H2S and CO2, Electrochim. Acta 53 (2008) 3690–3700.
[2] G. Das, Precipitation and kinetics of ferrous carbonate in simulated brine solution and its impact on CO2 corrosion of steel, Int. J. Adv. Eng. Technol. 7 (2014) 790.
[3] B. Mishra, S. Al-Hassan, D. Olson, M. Salama, Development of a predictive model for activation-controlled corrosion of steel in solutions containing carbon dioxide, Corrosion 53 (1997) 852–859.
[4] Y. Zhu, M.L. Free, R. Woollam, W. Durnie, A review of surfactants as corrosion inhibitors and associated modeling, Progr. Mater. Sci. 90 (2017) 159–223.
[5] M.A. Kappes, Evaluation of thiosulfate as a substitute for hydrogen sulfide in sour corrosion fatigue studies, The Ohio State University, 2011, p. 280.
[6] M. Koteeswaran, CO2 and H2S corrosion in oil pipelines, University of Stavanger, Norway, 2010.
[7] W. Sun, S. Nešić, D. Young, R.C. Woollam, Equilibrium expressions related to the solubility of the sour corrosion product mackinawite, Ind. Eng. Chem. Res. 47 (2008) 1738–1742.
[8] W. Sun, D.V. Pugh, S.N. Smith, S. Ling, J.L. Pacheco, R.J. Franco, A parametric study of sour corrosion of carbon steel, National Association of Corrosion Engineers, Houston, TX, 2010.

CHAPTER 10

Weight loss method of corrosion assessment

Weight loss (WL) which is also called mass loss, gravimetric, coupon test, or immersion testing is one of the easiest, cost-effective, and extensively used methods of corrosion testing and corrosion rate determination. The WL method of corrosion testing has been widely used by researchers. Procedure of the WL method of corrosion testing is given in ASTM G31 standard. A typical WL experiment includes the following steps.

(i) *Preparation of test electrolyte*: In WL experiments, metallic coupons or specimens are allowed to immerse and corrode for a specific time. An electrolyte is any chemical species that undergoes ionization when dissolved in any ionizing solvents such as water. Some of the commonly used electrolytes are soluble acids, bases, and salts. Most of the useful electrolytic solutions are prepared through the dilution of analytic grade electrolyte in water. The presence of the electrolytes triggers the anodic and cathodic reactions therefore they play a very crucial role in the corrosion process. Generally, electrolytes serve to remove the oxidized metals at the anode and provide species that can reduce at the cathode. However, corrosion in oil and gas industries is attributed to the presence of dissolved gases and moisture. All these solutions should be freshly prepared (usually 24 h before use).

(ii) *Preparation of coupons*: before immersing the metallic specimens into the electrolyte, they should be properly cleaned, washed, degreased, and dried. Cleaning of the specimens is carried out by rubbing their surface through different grades of emery papers (abrasive paper or sandpaper). Generally, emery papers of 600–1200 grit size are used for cleaning the metallic surface. This type of cleaning is carried out to clean and remove the surface impurities such as rusts and scales. It is important to mention that low grit size is consistent with high surface roughness. Therefore, cleaning should be successively carried out from less grit size to high grit size. After abrasive cleaning, metallic specimens are washed with water to remove the residual solid deposits. For washing purposes, distilled or deionized water is recommended to use in order to avoid any possibility of corrosion. Degreasing of the metallic surfaces is generally carried out by washing them with ethanol and finally from acetone. The metallic specimens after abrasive cleaning, washing, degreasing, and drying can be used for immersion testing. Cleaning of the metallic surface before WL experiment can be termed as initial cleaning.

(iii) *Coupons immersion*: the cleaned and dried specimens are weighted accurately and immersed in an electrolyte for a specific time. The immersion time of the specimens depends upon the nature of metal and electrolyte. It ranges from a few minutes to several hours.

(iv) *Final cleaning*: after a specific immersion time, metallic specimens are taken out. They can be termed as corroded specimens. The corroded specimens are washed and degreased again to remove any corrosion products collected over the metallic surface. Cleaning of metallic specimens after immersion test can be termed as final cleaning.

(v) *Calculation of WL and corrosion rate (C_R)*: WL can be calculated by subtracting the weight of corroded specimen from the weight of noncorroded specimen. Generally, WL is presented as ΔW.

$$\text{Weight loss (DW)} = \text{Initial weight}(W_{\text{initial}}) - \text{Final weight}(W_{\text{final}}) \quad (10.1)$$

The value of WL (ΔW) can be used to determine corrosion rate. Corrosion rate can be expressed in various ways. The grams per square inch per hour (g sq^{-1} in h^{-1}) and milligrams per square centimeter per day (mg cm^{-2} d^{-1}) are the commonly used expressions of corrosion rate. However, mil per year (mpy) is one of the most desirable ways of corrosion rate expression. Corrosion rate can be calculated by using the following equation [1,2]:

$$C_R(\text{mpy}) = \frac{\Delta W K}{DAt} \quad (10.2)$$

where ΔW is weight (mg), D is the density of metal (g cm^{-3}), A is the area of coupon (sq. in), and t is the immersion time (h). K is the conversion factor that depends upon desired units. For mpy (mils per year) and mm y^{-1} (millimeter year^{-1}), K, attains the values of 534 and 8.76×10^4, respectively. The percentage inhibition efficiency ($\eta\%$) and surface coverage (θ) of a corrosion inhibitor (Inh) can be calculated using the values of corrosion rate (C_R) as per the following relationship [3]:

$$\eta\% = \frac{C_R - C_{R(\text{inh})}}{C_R} \times 100 \quad (10.3)$$

$$\theta = \frac{C_R - C_{R(\text{inh})}}{C_R} \quad (10.4)$$

where C_R and $C_{R(\text{Inh})}$ are the corrosion rates in the absence and presence of a corrosion inhibitor (Inh), respectively.

10.1 Advantages and disadvantages of weight loss study

10.1.1 Advantages of weight loss technique

WL is a very versatile technique of corrosion monitoring as it can be fabricated from any commercially accessible metal or alloy. More so, through suitable designs, various corrosion phenomena including (but not limited) galvanic (bimetallic) attack, differential aeration corrosion, stress assisted corrosion, and heat-affected zones may be studied. Some major advantages of WL technique are:

(i) WL is one of the best and simplest methods of corrosion monitoring.
(ii) It is easy to perform.
(iii) High accuracy and reproducibility of experimental data.
(iv) It does not require any expensive and sophisticated instruments except a high-accuracy weighing machine.
(v) It provides short-term corrosion rate data.
(vi) Using WL, localized corrosion can be recognized and measured.
(vii) WL technique can be used to all environments, e.g., liquids, gases, and solids.
(viii) Inhibition performance of an inhibitor can be easily accessed with high accuracy.
(ix) WL and corrosion rate can be easily and readily measured.
(x) Through WL, corrosion deposits such as rusts and scales can also be easily inspected and analyzed.

10.1.2 Disadvantages of weight loss technique

Though WL is one of the simplest and extremely versatile techniques of corrosion testing, it also has some limitations. Some of the common limitations of WL technique are:

(i) WL technique does not provide any information about the mechanism of corrosion.
(ii) It also does not provide any clue about the mechanism of interaction between a corrosion inhibitor and the metal surface.
(iii) It also suffers from either too high or too low immersion time.
(iv) Extremely tedious as it requires cleaning of the entire coupon surface.
(v) It is a destructive method of corrosion monitoring.

10.2 WL technique of corrosion monitoring: case studies

WL experiment is widely used to demonstrate the corrosion inhibition performance of the organic corrosion inhibitors. WL studies showed that in most of cases inhibition efficiency of corrosion inhibitors increases on increasing their concentration. This is attributed due to the fact that increase in the inhibitor concentration causes subsequent increase in the surface coverage. Inhibitor molecules get adsorbed over the metallic surface and form inhibitive films that avoid the aggressive attack of electrolyte. However, after certain concentrations (called optimum concentration) further increase in inhibitor concentration does not cause any significant change in the protection effectiveness of the inhibitor. This observation shows that at optimum concentration all active sites on the metallic surface, responsible for the corrosion are covered with the inhibitor molecules. It is also reported that below optimum concentration inhibitor molecules adsorb using their flat orientations because of the intermolecular force of attraction between inhibitor molecules and metallic surface. However, at higher inhibitor concentration, the intermolecular force of repulsion between inhibitor molecules gets dominated that forces inhibitor molecules to attain the vertical orientation. In some cases, it is also reported that increase in corrosion inhibitors concentration beyond optimum concentration reduces their protection. At high inhibitor concentration, reduction in the inhibition effectiveness of the corrosion inhibitors is attributed due to the following effects:

(i) Due to reduced solubility of organic inhibitors.
(ii) At high concentration, intermolecular force of attraction among the inhibitor molecules may dominate over the intermolecular force of attraction between metal surface and inhibitor molecules. This will lead to desorption of adsorbed inhibitor molecules.
(iii) At high concentration, intermolecular force of repulsion among the inhibitor molecules forces inhibitor molecules to gain the vertical orientation.

Because of its simplicity, high accuracy, and reproducibility of the data, WL experiment is widely used to study the corrosion inhibition effect of various organic molecules for different metal-electrolyte systems. Zhao and Mu [4] studied the corrosion inhibition effect of three anionic surfactants namely, sodium dodecyl sulfate, dodecylbenzene sulfonic acid sodium salt, and dodecyl sulfonic acid sodium salt, abbreviated as SDS, DBSASS, and DSASS, respectively, for aluminium in acidic solution of hydrochloric acid using WL method. Results showed that optimum concentration of SDS, DBSASS, and DSASS was 1×10^{-3} M. Further increase in their concentration does not cause any significant change in the inhibition performance. Studies showed that SDS, DBSASS, and DSASS become effective by adsorbing on the metallic surface and their adsorption on aluminium surface obeyed the Langmuir adsorption isotherm model.

In another study, corrosion inhibition power of three 5-(Phenylthio)-3H-pyrrole-4-carbonitriles (PPCs), designated as PPC I, PPC II, and PPC III, differing in the nature of substituents is demonstrated for mild steel in 1M HCl using WL and other methods [5]. WL results showed that inhibition performance of PPCs increases on increasing their concentration from 10 to 50 mg L^{-1} concentration. Inhibition efficiencies of PPCs followed the sequence: PPC III (98.69%) > PPC II (96.99%) > PPC I (94.75%). Analyses showed that inhibition efficiency of the PPCs greatly depends upon the nature of substituent(s) present in their molecular structures. Presence of the electron donating hydroxyl (–OH) substituent(s) increases the inhibition efficiency. WL study showed that adsorption of the PPC I, PPC II, and PPC III on metallic surface obeyed the Langmuir adsorption isotherm model. Inhibition performance of PPCs was also determined at various temperatures using WL method and results showed that increase in the inhibitor's concentration causes subsequent reduction in their inhibition performance. Results derived from WL were further supported by electrochemical impedance spectroscopy, potentiodynamic polarization, scanning electron microscope, atomic force microscope, and Monte Carlo simulations. Results of these studies were in good agreement to the results derived through WL measurement. Variation in the inhibition efficiencies of 5-(Phenylthio)-3H-pyrrole-4-carbonitriles (PPCs) at different concentrations and temperatures is presented in Fig. 10.1.

Ansari et al. [6] investigated the corrosion inhibition effect of three pyridyl substituted triazole-based Schiff's bases (SBs), designated as SB-1, SB-2, and SB-3, for mild steel in acidic solution of 1M HCl using WL and other various methods. WL study revealed that protection effectiveness of the SBs increases on increasing their concentrations.

Among the tested SBs, SB-1 showed the maximum inhibition efficiency of 96.6% at 150 mg L^{-1} concentration. Study showed that difference in the protection effectiveness of the SBs is attributed due to the presence of different substituents in their molecular structures. Results showed that both electron-donating hydroxyl (–OH) at *ortho*-position (SB-2) and electron-withdrawing nitro (–NO$_2$) substituent at *para*-position (SB-3), decrease their

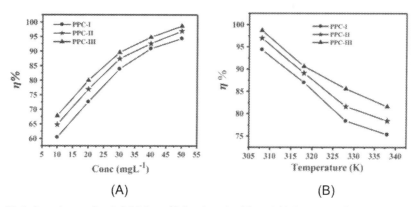

FIG. 10.1 Variation of corrosion inhibition efficiencies of 5-(Phenylthio)-3H-pyrrole-4-carbonitriles (*PPCs*) at different concentrations (A) and temperatures (B) [5]. (Reproduced with permission@ Copyright Elsevier).

inhibition efficiency as compared to the nonsubstituted compound (SB-1). However, this decrease in the inhibition efficiency was more pronounced in the presence of electron withdrawing nitro ($-NO_2$) substituent. WL analysis showed that adsorption of studied SBs on metallic surface obeyed the Langmuir adsorption isotherm model. Increase in the temperature caused subsequent decrease in the inhibition performance of the tested SBs. Results derived through WL were supported by open circuit potential, electrochemical impedance spectroscopy, potentiodynamic polarization, scanning electron microscope, energy dispersive X-ray, and density functional theory (DFT) methods. Results of these studies well supported the results derived from WL analyses. Variation in the corrosion inhibition efficiencies at different concentrations and temperatures is presented in Fig. 10.2. Table 10.1 represents the summary of results derived for some important class of compounds from WL.

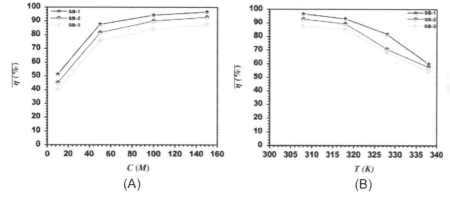

FIG. 10.2 Variation of corrosion inhibition efficiencies of pyridyl substituted triazole based Schiff's bases (*SBs*) at different concentrations (A) and temperatures (B) [5,6]. (Reproduced with permission@ Copyright Elsevier).

TABLE 10.1 Weight loss results for some important class of organic corrosion inhibitors.

S. No.	Corrosion inhibitor(s)	Metal & electrolyte	Efficiency % Opt. conc.	Adsorption behavior	References
1	5-(Phenylthio)-3H-pyrrole-4-carbonitrile (PPCs)	Mild steel/1M HCl	98.69% at 50 mg L^{-1} (PPC III)	Langmuir adsorption	[5]
2	pyridyl substituted triazole based Schiff's bases	Mild steel/1M HCl	96.6% at 50 mg L^{-1} (SB-1)	Langmuir adsorption	[6]
3	Triazine derivatives (HTs)	Mild steel/1M HCl	98.6% at 80 mg L^{-1} (HT-1)	Langmuir adsorption	[7]
4	Glucosamine-Based, Pyrimidine-Fused Heterocycles (CARBs)	Mild steel/1M HCl	96.52% at 7.41 × 10^{-5} M (CARB-4)	Langmuir adsorption	[8]
5	2-Aminobenzene-1,3-dicarbonitriles (ABDNs)	Mild steel/1M HCl	97.83% at 100 mg L^{-1} (ABDN-3)	Langmuir adsorption	[9]
6	Hydrazinobenzothiazole compounds	N80 steel/15% HCl	97.3% at 500 mg L^{-1} (BTHP)	Langmuir adsorption	[10]
7	Spiropyrimidinethione derivatives	Mild steel/15% HCl	89.4% at 250 mg L^{-1} (MPTS)	Langmuir adsorption	[11]
8	Carbohydrate-based compounds	N80 steel/15% HCl	95.9% at 400 mg L^{-1} (BIHT)	Langmuir adsorption	[12]
9	2-Amino-5-nitro-4,6-diarylcyclohex-1-ene-1,3,3-tricarbonitriles (ANDTs)	Mild steel/1M HCl	98.96% at 25 mg L^{-1} (ANDT-3)	Langmuir adsorption	[13]
10	Imidazolium based zwitterions (AIZs)	Mild steel/1M HCl	96.08% at 0.55 mM (AIZ-3)	Langmuir adsorption	[14]
11	2-amino-4-arylquinoline-3-carbonitriles (AACs)	Mild steel/1M HCl	96.95% at 50 mg L^{-1} (AAC-3)	Langmuir adsorption	[15]
12	2,4-Diamino-5-(phenylthio)-5H-chromeno [2,3-b] pyridine-3-carbonitriles (DHPCs)	Mild steel/1M HCl	97.91% at 12.70 × 10^{-5} M (DHPC-3)	Langmuir adsorption	[16]
13	2-amino-3-((4-((S)-2-amino-2-carboxyethyl)-1H-imidazol-2-yl)thio) propionic acid (AIPA)	Mild steel/1M HCl	96.52% at 0.456 mM	Langmuir adsorption	[17]
14	5-Arylpyrimido-[4,5-b]quinoline-diones (APQD)	Mild steel/1M HCl	96.52% at 20 mg L^{-1} (APQD-4)	Langmuir adsorption	[18]
15	Functionalized chitosan macromolecules	Mild steel/1M HCl	92% at 200 mg L^{-1} (CS-TCH)	Langmuir adsorption	[19]
16	Pyrimidine derivatives (PPs)	N80 steel/15% HCl	89.1% at 250 mg L^{-1} (PP-1)	Langmuir adsorption	[20]
17	Isatin-β-thiosemicarbzone derivatives (TZs)	Mild steel/20% H$_2$SO$_4$	99.3% at 300 mg L^{-1} (TZ-2)	Langmuir adsorption	[21]
18	Pyridine derivatives	N80 steel/15% HCl	90.24% at 200 mg L^{-1} (ADP)	Langmuir adsorption	[22]
19	Pyran derivatives (APs)	N80 steel/15% HCl	97.64% at 300 mg L^{-1} (AP-1)	Langmuir adsorption	[23]
20	Pyranpyrazole derivatives (EPPs)	Mild steel/1M HCl	98.8% at 100 mg L^{-1} (EPP-1)	Langmuir adsorption	[24]

10.3 Summary

WL or gravimetric is one of the best and simplest methods of corrosion monitoring. It is associated with various advantages such as ease to perform and high accuracy. It does not need to use any expensive and complicated instruments. It provides short-term corrosion rate data. Several phenomena related to corrosion such as localized corrosion can be recognized and measured using WL in various environments. However, the WL method of corrosion monitoring is also associated with various disadvantages. Though, WL method provides short-term corrosion rate data, it does not furnish any information about the mechanism of corrosion and the nature of interaction between a metal surface and inhibitor molecules. Nevertheless, WL is one of the most preliminary and most extensively utilized methods of corrosion testing. Literature study showed that WL method is widely used in investigating the corrosion inhibition effect of organic compounds. Generally, WL method provides two vital parameters namely corrosion rate (C_R) and surface coverage (θ). Corrosion rate can be expressed in various ways. Mostly, corrosion rate is presented in the grams per square inch per hour (g sq^{-1} in h^{-1}) and milligrams per square centimeter per day (mg cm^{-2} d^{-1}). However, mpy is one of the most desirable ways of corrosion rate expression. A lower C_R value and higher θ value are associated with high inhibition efficiency and vice versa.

10.4 Useful websites

https://www.sciencedirect.com/topics/engineering/weight-loss-method
https://link.springer.com/chapter/10.1007%2F978-1-85233-845-9_9
https://www.achrnews.com/articles/91492-the-benefits-and-limitations-of-corrosion-coupons

References

[1] V. Thangarasu, R. Anand, Comparative evaluation of corrosion behavior of Aegle Marmelos Correa diesel, biodiesel, and their blends on aluminum and mild steel metals. In: Advanced Biofuels, Woodhead Publishing, Elsevier, UK, 2019, pp. 443–471.

[2] P. Pearson, A. Cousins, Assessment of Corrosion in Amine-based Post-Combustion Capture of Carbon Dioxide Systems. In: Absorption-Based Post-Combustion Capture of Carbon Dioxide, Woodhead Publishing, Elsevier, UK, 2016, pp. 439–463.

[3] C. Verma, M. Quraishi, E. Ebenso, Electrochemical studies of 2-amino-1, 9-dihydro-9-((2-hydroxyethoxy) methyl)-6H-purin-6-one as green corrosion inhibitor for mild steel in 1.0 M hydrochloric acid solution, Int. J. Electrochem. Sci. 8 (2013) 7401–7413.

[4] T. Zhao, G. Mu, The adsorption and corrosion inhibition of anion surfactants on aluminium surface in hydrochloric acid, Corrosion Sci. 41 (1999) 1937–1944.

[5] C. Verma, E. Ebenso, I. Bahadur, I. Obot, M. Quraishi, 5-(Phenylthio)-3H-pyrrole-4-carbonitriles as effective corrosion inhibitors for mild steel in 1 M HCl: experimental and theoretical investigation, J. Mol. Liq. 212 (2015) 209–218.

[6] K. Ansari, M. Quraishi, A. Singh, Schiff's base of pyridyl substituted triazoles as new and effective corrosion inhibitors for mild steel in hydrochloric acid solution, Corrosion Sci. 79 (2014) 5–15.

[7] A. Singh, K. Ansari, J. Haque, P. Dohare, H. Lgaz, R. Salghi, M. Quraishi, Effect of electron donating functional groups on corrosion inhibition of mild steel in hydrochloric acid: experimental and quantum chemical study, J. Taiwan Inst. Chem. Eng. 82 (2018) 233–251.

[8] C. Verma, L.O. Olasunkanmi, E.E. Ebenso, M.A. Quraishi, I.B. Obot, Adsorption behavior of glucosamine-based, pyrimidine-fused heterocycles as green corrosion inhibitors for mild steel: experimental and theoretical studies, The Journal of Physical Chemistry C 120 (2016) 11598–11611.

[9] C.B. Verma, M. Quraishi, A. Singh, 2-Aminobenzene-1, 3-dicarbonitriles as green corrosion inhibitor for mild steel in 1 M HCl: electrochemical, thermodynamic, surface and quantum chemical investigation, J. Taiwan Inst. Chem. Eng. 49 (2015) 229–239.

[10] P. Shaw, I. Obot, M. Yadav, Functionalized 2-hydrazinobenzothiazole with carbohydrates as a corrosion inhibitor: electrochemical, XPS, DFT and Monte Carlo simulation studies, Mater. Chem. Front. 3 (2019) 931–940.

[11] M. Yadav, R. Sinha, S. Kumar, T. Sarkar, Corrosion inhibition effect of spiropyrimidinethiones on mild steel in 15% HCl solution: insight from electrochemical and quantum studies, RSC Adv. 5 (2015) 70832–70848.

[12] M. Yadav, T. Sarkar, I. Obot, Carbohydrate compounds as green corrosion inhibitors: electrochemical, XPS, DFT and molecular dynamics simulation studies, RSC Adv. 6 (2016) 110053–110069.

[13] C. Verma, M. Quraishi, A. Singh, 2-Amino-5-nitro-4, 6-diarylcyclohex-1-ene-1, 3, 3-tricarbonitriles as new and effective corrosion inhibitors for mild steel in 1 M HCl: Experimental and theoretical studies, J. Mol. Liq. 212 (2015) 804–812.

[14] V. Srivastava, J. Haque, C. Verma, P. Singh, H. Lgaz, R. Salghi, M. Quraishi, Amino acid based imidazolium zwitterions as novel and green corrosion inhibitors for mild steel: experimental, DFT and MD studies, J. Mol. Liq. 244 (2017) 340–352.

[15] C. Verma, M. Quraishi, L. Olasunkanmi, E.E. Ebenso, L-Proline-promoted synthesis of 2-amino-4-arylquinoline-3-carbonitriles as sustainable corrosion inhibitors for mild steel in 1 M HCl: experimental and computational studies, RSC Adv. 5 (2015) 85417–85430.

[16] C. Verma, L.O. Olasunkanmi, I. Obot, E.E. Ebenso, M. Quraishi, 2, 4-Diamino-5-(phenylthio)-5 H-chromeno [2, 3-b] pyridine-3-carbonitriles as green and effective corrosion inhibitors: gravimetric, electrochemical, surface morphology and theoretical studies, RSC Adv. 6 (2016) 53933–53948.

[17] J. Haque, V. Srivastava, C. Verma, M. Quraishi, Experimental and quantum chemical analysis of 2-amino-3-((4-((S)-2-amino-2-carboxyethyl)-1H-imidazol-2-yl) thio) propionic acid as new and green corrosion inhibitor for mild steel in 1 M hydrochloric acid solution, J. Mol. Liq. 225 (2017) 848–855.

[18] C. Verma, L. Olasunkanmi, I. Obot, E.E. Ebenso, M. Quraishi, 5-Arylpyrimido-[4, 5-b] quinoline-diones as new and sustainable corrosion inhibitors for mild steel in 1 M HCl: a combined experimental and theoretical approach, RSC Adv. 6 (2016) 15639–15654.

[19] D.S. Chauhan, K. Ansari, A. Sorour, M. Quraishi, H. Lgaz, R. Salghi, Thiosemicarbazide and thiocarbohydrazide functionalized chitosan as ecofriendly corrosion inhibitors for carbon steel in hydrochloric acid solution, Int. J. Biol. Macromol. 107 (2018) 1747–1757.

[20] J. Haque, K. Ansari, V. Srivastava, M. Quraishi, I. Obot, Pyrimidine derivatives as novel acidizing corrosion inhibitors for N80 steel useful for petroleum industry: a combined experimental and theoretical approach, J. Ind. Eng. Chem. 49 (2017) 176–188.

[21] K. Ansari, M. Quraishi, A. Singh, Isatin derivatives as a non-toxic corrosion inhibitor for mild steel in 20% H2SO4, Corrosion Sci. 95 (2015) 62–70.

[22] K. Ansari, M. Quraishi, A. Singh, Pyridine derivatives as corrosion inhibitors for N80 steel in 15% HCl: electrochemical, surface and quantum chemical studies, Measurement 76 (2015) 136–147.

[23] A. Singh, K. Ansari, M. Quraishi, H. Lgaz, Y. Lin, Synthesis and investigation of pyran derivatives as acidizing corrosion inhibitors for N80 steel in hydrochloric acid: theoretical and experimental approaches, J. Alloys Comp. 762 (2018) 347–362.

[24] P. Dohare, K. Ansari, M. Quraishi, I. Obot, Pyranpyrazole derivatives as novel corrosion inhibitors for mild steel useful for industrial pickling process: experimental and quantum chemical study, J. Ind. Eng. Chem. 52 (2017) 197–210.

CHAPTER 11

Electrochemical methods of corrosion assessment

Corrosion is an electrochemical process that involves oxidation and reduction half-cell reactions. As electrons are discharged as anode (oxidation) and consumed at cathode (reduction), there is a flow of current (electrons) from anode to cathode. This current can be determined electronically. Consequently, electrochemical techniques can be used to demonstrate the corrosion-related properties of metals and their components in association with numerous electrolyte solutions. It is important to mention that corrosion-related properties are unique for every metal/electrolyte system.

In a typical electrochemical testing practice, a polarization assembly is set up that consists of a reference electrode, an auxiliary (counter) electrode, a working (test) electrode, and an electrolytic solution. The working electrode is nothing but a piece of metal to be studied. These electrodes are immersed in the electrolyte solution and are coupled with an electronic device called a potentiostat. In the electrolyte solution, a voltage, which is known as electrochemical potential (E_{corr}) or corrosion potential, is developed between the three electrodes. The corrosion potential can be determined through potentiostat in the form of energy (potential) difference between the reference electrode and the working electrode. Various electrochemical experiments can be designed to measure the potential and current of the electrode oxidation/reduction reactions. There are numerous electrochemical techniques that are being used to demonstrate the corrosion inhibition effect of the corrosion inhibitors. However, potentiodynamic (Tafel) polarization (PDP) and electrochemical frequency modulation are the most advanced and useful electrochemical techniques.

11.1 Electrochemical impedance spectroscopy

Electrochemical impedance spectroscopy (EIS) is one of the most advanced and useful electrochemical techniques of corrosion monitoring. It has become a standardized tool for applied and fundamental corrosion research. EIS can also be successfully used for high impedance (resistance) systems such as lining and coating along with low conducting systems such as pure water. Procedure and setup for EIS method of corrosion monitoring is given in ASTM G106-89 standard. Similar to most of electrochemical techniques, the EIS technique utilizes three-electrode cells, controlled by a potentiostat. In general, for EIS study, an AC (potential or voltage) signal of different frequencies, typically ranging from 10^5 to 10^{-2} or 10^{-3} of ±10 mV amplitude is applied at various desecrate frequencies, generally 5–10 frequency/decade, alternating current response, $i\ (\omega)$, is measured at every frequency (ω). Obviously, full frequency sweeps furnish information about phase-shift that can be used in the combination with a suitable equivalent circuit to derive significant information from the complex interface of the corrosion system. The frequency-dependent impedance $(Z_{(\omega)})$ of such systems can be calculated using the following relationship [1]:

$$Z_{(\omega)} = \frac{\text{Voltage}, V}{\text{Current}, I} \tag{11.1}$$

where ω $(=2\pi f)$ is the angular frequency of the applied AC signal. Typically, an electrochemical equivalent circuit consists of various elements such as resistance (R), inductance (L), and capacitance (C). Explanations of EIS data are based on the equivalent circuit employed. Now, various software packages and programs are available for fitting and analyzing the impedance data/spectra. Randles is the most common equivalent circuit that is used to describe the electrochemical properties of various metal/electrolyte interfaces. A typical Randles' equivalent circuit is presented in Fig. 11.1, which consists of a solution resistance (R_s), charge transfer resistance (R_{ct}), and a double layer capacitance (C_{dl}).

It is important to notice that charge transfer resistance can be regarded as equivalent to linear polarization resistance (R_p). The nature and properties of the interface can be presented as follows [1]:

$$Z_{(\omega)} = R_s + \frac{R_p}{1+(j\omega R_p C_{dl})^\beta} \tag{11.2}$$

R_p or R_{ct} can be derived through various ways. One of the most common convenient methods of determination of R_p or R_{ct} is to use the Nyquist diagram. A typical Nyquist diagram fitted in Randles equivalent circuit is presented in Fig. 11.2 that represents a perfect semicircle.

The high-frequency response is utilized to calculate the component of R_s participated in the measurement. R_s can be directly interpreted from the abscissa when the angular frequency (ω) approaches infinite (f_{max} or $f\rightarrow 0$). On the other hand, the total resistance, R_{total} ($R_s + R_p$) can be read from the abscissa when the angular frequency tends to be zero (f_{min} or $f\rightarrow 0$). Therefore, R_p or R_{ct} can be calculated from subtracting the solution resistance from the lower frequency measurement. The value of double-layer capacitance (C_{dl}) can be calculated using the following relation [1]:

$$C_{dl} = \frac{1}{2\pi f_{max} R_p} \tag{11.3}$$

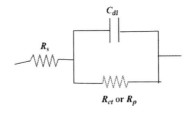

FIG. 11.1 A typical Randles, equivalent circuit.

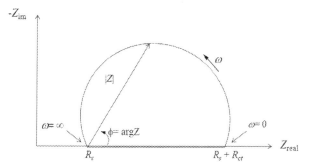

FIG. 11.2 Nyquist plot fitted in a simple Randle equivalent circuit.

The values of R_p (or R_{ct}) and C_{dl} are widely used to demonstrate the corrosion inhibition efficiency of various organic corrosion inhibitors. Using the values of R_p (or R_{ct}), percentage of corrosion inhibition efficiency ($\eta\%$) of an organic compound can be derived using following equation [2]:

$$\eta\% = \frac{R_{p(inh)} - R_p}{R_{p(inh)}} \times 100 \tag{11.4}$$

where, R_p and $R_{p(inh)}$ are the polarization resistance for metallic corrosion in the absence and presence of corrosion inhibitor, respectively. Generally, a high value of R_p and low value of C_{dl} consist with high inhibition efficiency and vice versa.

11.2 Potentiodynamic polarization

Julius Tafel (1905) proposed a relationship between the current (I) and the overpotential (η), while studying the electrocatalytic reduction protons (to form hydrogen molecules) on various metallic electrodes including Sn, Bi, Hg, Au, Cu, Hg, etc. [1].

$$\eta = a + b \log I \tag{11.5}$$

where η is the overpotential which is defined as difference in the potential of working (test) electrode and the equilibrium potential. The linear relationship has been reported between electrode potential and log I when the working electrode is polarized at sufficient large potentials and far away from the corrosion potential (E_{corr}), both in cathodic and anodic directions (Fig. 11.3). The portions in polarization curves in which such relationships prevail are known as Tafel regions or Tafel portions. Mathematically, it can be presented as follows [1]:

$$\begin{aligned} I &= I_{corr}\left[\exp\left(\frac{2.303\eta}{\beta_a}\right) - \exp\left(-\frac{2.303\eta}{\beta_c}\right)\right] \\ &= I_{corr}\left[\exp\left(\frac{2.303(E - E_{corr})}{\beta_a}\right) - \exp\left(-\frac{2.303(E - E_{corr})}{\beta_c}\right)\right] \end{aligned} \tag{11.6}$$

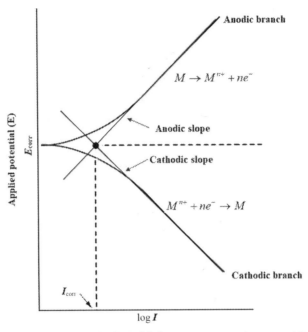

FIG. 11.3 Extrapolation of anodic and cathodic Tafel slopes to get corrosion current (I_{corr}).

where E is the applied potential, E_{corr} is the corrosion potential, η ($E - E_{corr}$) is the overpotential, I is the applied current, I_{corr} is the corrosion current, and β_a and β_c are the anodic and cathodic Tafel parameters or constants. These constants are derived through E-log I as the cathodic and anodic slope in the Tafel portions.

Extrapolation of the linear segment or portion of either anodic or cathodic or both, on intersection point is observed at corrosion potential (E_{corr}) from which corrosion current density (I_{corr}) is readily available from the log I axes. Therefore, extrapolation of polarization curves gives some vital parameters including corrosion current (I_{corr}), anodic (β_a), and cathodic (β_c) Tafel constants along with the corrosion potential (E_{corr}). To obtain Tafel regions anodic and cathodic curves have to be polarized away from (e.g., ±250 mV) from the corrosion potential. A sufficiently large value of η, in the anodic ($\eta=\eta_a$) and cathodic ($\eta=\eta_c$) direction can be illustrated as follows [1]:

$$\eta_a = \beta_a \log \frac{I}{I_{corr}} \qquad (11.7)$$

$$\eta_c = \beta_c \log \frac{I}{I_{corr}} \qquad (11.8)$$

The Tafel polarization can be measured statically or dynamically (either in the current-controlled mode or in the potential-controlled mode). Both the methods have their own advantages and drawbacks. For example, static polarization techniques generally provide better Tafel parameters but they are very time-consuming. On the other hand, dynamic polarization techniques are relatively fast but less accurate. Tafel polarization is widely used to describe the corrosion inhibition effect of various organic compounds. These measurements provide some vital information about the nature of corrosion inhibitors and their mechanism of action. Percentage of inhibition efficiency of an organic compound can be derived using the following relation [2]:

$$\eta\% = \frac{I_{corr}^0 - I_{corr}^{inh}}{I_{corr}^0} \times 100 \qquad (11.9)$$

where, I_{corr}^0 and I_{corr}^{inh} are the corrosion current densities in the absence and presence of corrosion inhibitor, respectively. Obviously, a lower value of I_{corr} is consistent with high protection effectiveness. It is important to mention that most of the organic compounds become effective corrosion inhibitors by retarding the rate of anodic and cathodic reactions. They may be classified as anodic or cathodic or mixed type. Generally, a shift or displacement in the E_{corr} value of an inhibited Tafel curve with respect to the E_{corr} of value of uninhibited Tafel curve is used to define the nature of corrosion inhibitor. The corrosion inhibitors can be classified as anodic or cathodic-type, if displacement in the E_{corr} value is more than −85 mV. The corrosion inhibitors can be classified as mixed type if this displacement is less than −85 mV. The literature study suggested that most of the organic compounds tested as corrosion inhibitors are mixed type.

11.3 Electrochemical impedance spectroscopy and potentiodynamic polarization studies of corrosion inhibitors: case studies

Electrochemical techniques especially, EIS and PDP are widely used to study the inhibition effect of organic compounds. These techniques provide the vital parameters in the term of which relative corrosion inhibition effectiveness and mechanism of corrosion inhibition can be explained. Table 11.1 represents the collection of information derived from EIS and PDP studies for some important organic corrosion inhibitors. Haque et al. [3] studied the corrosion inhibition effect of two functionalized tetrahydropyridines (THPs) for mild steel in 1M HCl using experimental and computational methods. EIS- and PDP-based electrochemical studies were carried out to study the nature of corrosion inhibitors and mechanism of corrosion. EIS study revealed that both investigated THPs acted as effective corrosion inhibitors and their inhibition effect increases on increasing their concentration. Presence of the THPs in the electrolyte increases the values of charge transfer resistance thereby they acted as interface-type corrosion inhibitors. It was also observed that diameters of the Nyquist curves were much higher in the presence of THPs as compared to in their absence. This increase in the diameter in the Nyquist curves was consistent with the THPs concentration. This observation suggested that studied molecules become effective by adsorbing the interface of metal and electrolyte. PDP studies showed that both investigated functionalized THPs retards anodic as well as cathodic reactions and behaved as mixed-type corrosion inhibitors. PDP analysis suggested that a significant reduction in the values of corrosion current (I_{corr}) was observed in the presence of THPs. This decrease in I_{corr} values was consistent with their concentration. These observations suggest that THPs become effective by adsorbing the blocking of the active sites present over the metallic surface.

TABLE 11.1 Summary of the results derived for some corrosion inhibitors with their electrochemical mode of action.

S. No.	Corrosion inhibitor(s)	Metal & electrolyte	Efficiency % Opt. conc.	Nature of inhibitors	References
1	Functionalized tetrahydropyridines	Mild steel/1M HCl	94.30% at 7.95×10^{-5} M (THP-2)	Mixed-type inhibitors	[3]
2	Condensed Uracils	Mild steel/1M HCl	96.70% at 12.8×10^{-4} M (CU-3)	Mixed-type inhibitors	[4]
3	Phthalazine derivatives	Mild steel/1M HCl	82% at 2 mM (PTO)	Mainly anodic-type inhibitors	[5]
4	Triazolotriazepine derivatives	Carbon steel/1M HCl	97.8% at 1×10^{-3} M (TTY5)	Mixed-type inhibitors	[6]
5	Chalcone derivatives (AEs)	Mild steel/1M HCl	95% at 5×10^{-3} M (AE-3)	Mixed-type inhibitors	[7]
6	Benzimidazole derivatives	Mild steel/1M HCl	97.6% at 1×10^{-3} M (BB1C-3)	Mixed-type inhibitors	[8]
7	5-((2-ethyl-1Hbenzo[d]imidazol-1-yl)methyl)-1,3,4-oxadiazole-2-thiol	Mild steel/0.5M HCl	93.34% at 760 μM	Mainly cathodic-type inhibitor	[9]
8	Tetrahydropyrimido-Triazepine derivatives (TRCs)	Mild steel/1M HCl	91% at 1×10^{-3} M (TRC-Cl)	Mixed-type inhibitors	[10]
9	Salicylaldeyde-Chitosan Schiff Base	J55 steel/3.5% NaCl + CO_2	95.4% at 150 mg L^{-1}	Mixed-type inhibitor.	[11]
10	Pyrimido[2,1-B] benzothiazoles	Mild steel/1M HCl	94.88% at 42.8×10^{-5} M (PBT-IV)	Mainly cathodic-type inhibitors	[12]
11	2-thioureidobenz-heteroazoles	Mild steel/2M HCl	96.1% at 2 mM (BTT)	Mainly cathodic-type inhibitors	[13]
12	Quinoxaline derivatives	Mild steel/1M HCl	89.27% at 100 ppm (Me-4-PQPB)	Mixed-type inhibitors	[14]
13	2-Amino-5-nitro-4,6-diarylcyclohex-1-ene-1,3,3-tricarbonitriles	Mild steel/1M HCl	98.96% at 25 mg L^{-1} (ANDT-3)	Mixed-type inhibitors	[15]
14	Glucosamine-Based, Pyrimidine-Fused Heterocycles (CARBs)	Mild steel/1M HCl	96.52% at 7.41×10^{-5} M (CARB-4)	Mainly cathodic-type inhibitors	[2]
15	5-(Phenylthio)-3H-pyrrole-4-carbonitrile	Mild steel/1M HCl	98.69% at 50 mg L^{-1} (PPC III)	Mainly anodic-type inhibitors	[16]
16	2-amino-4-arylquinoline-3-carbonitriles	Mild steel/1M HCl	96.95% at 50 mg L^{-1} (AAC-3)	Mainly cathodic-type inhibitors	[17]
17	2,4-Diamino-5-(phenylthio)-5H-chromeno [2,3-b] pyridine-3-carbonitriles	Mild steel/1M HCl	97.91% at 12.70×10^{-5} M (DHPC-3)	Mainly cathodic-type inhibitors	[18]
18	Pyrimidine derivatives (PPs)	N80 steel/15% HCl	89.1% at 250 mg L^{-1} (PP-1)	Mainly cathodic-type inhibitors	[19]
19	Isatin-β-thiosemicarbzone derivatives (TZs)	Mild steel/20% H_2SO_4	99.3% at 300 mg L^{-1} (TZ-2)	Mainly cathodic-type inhibitors	[20]

Yadav and Quraishi [4] reported the corrosion inhibition effect of four condensed uracils (CUs) differing in the nature of substituents for mild steel corrosion in 1M HCl using experimental and DFT methods. Experimental results revealed that presence of electron-donating –OMe and –Me substituents in CU-3 and CU-4 increase their corrosion inhibition efficiency as compared to the non-substituted CU (CU-1). More so, electron-withdrawing –NO_2 substituent decreases the protection effectiveness. The inhibition efficiencies of investigated CUs followed the sequence: CU-3 (–OMe) > CU-4 (–Me) > CU-1 ((–H) > CU-2 (–NO_2). EIS study suggested that investigated CU become effective increasing the values of charge transfer resistance for mild steel corrosion process. Increase in the values of charge transfer resistance indicated that investigated CUs behaved as interface-type inhibitors, i.e., they become effective by adsorbing at the interface of metal and electrolyte. EIS study also suggested that diameters of Nyquist curves in the presence of CUs were much higher as compared to in their absence. This observation further suggested that CUs become effective by adsorbing the interface of metal and electrolyte. DPD study conducted to study the nature of corrosion inhibitors. Analyses showed that investigated CUs adversely affect the rate of anodic and cathodic reactions. PDP results showed that the displacement in E_{corr} values of inhibited Tafel curves with respect to the uninhibited Tafel curves was less than –85 mV, indicating that all investigated CUs behaved as mixed-type corrosion inhibitors.

In another study, phthalazine derivatives, corrosion inhibition of mild steel in 1M HCl was reported using electrochemical and computational methods [5]. Electrochemical studies showed that inhibition efficiencies of the investigated phthalazine derivatives followed the sequence: PTD < PT < PTO. EIS study showed that values of charge transfer resistance and diameter of the Nyquist curves significantly increased in the presence of investigated compounds. This increase in the diameter of the Nyquist curves and charge transfer resistance was consistent with the concentration of investigated compounds. PDP study revealed that investigated compounds acted as mixed-type corrosion inhibitors as their presence adversely affected the nature of both anodic and cathodic Tafel reactions. Electrochemical results and the order of inhibition efficiency were well supported by computational studies.

Three new triazolotriazepine derivatives were tested as corrosion inhibitors for carbon steel in acidic solution of 1M HCl using electrochemical and computational methods [6]. All these compounds differ in the length of carbon (hydrocarbon) chains. It was observed that increase in the hydrophobic chain length causes subsequent increase in the inhibition performance of the studied compounds. Inhibition efficiencies of studied compounds followed the order: TTY (–CH_2–CH_3) < TTY4 (–$(CH_2)_9$–CH_3) < TTY4 (–$(CH_2)_{13}$–CH_3). EIS studies revealed that presence of TTYs increase the values of charge transfer resistance and the diameter of the Nyquist curves. This observation revealed that investigated compounds become effective by adsorbing at the interface of metal and electrolyte. The Nyquist curves for carbon steel corrosion in 1M HCl with and without TTY, TTY4, and TTY5 are presented in Fig. 11.4. Careful observation of this figure reveals that Nyquist curves represent the slightly depressed semicircle instead of the perfect semicircle. The deviation from the perfect semicircle is attributed due to inhomogeneities and roughness of the carbon steel surface attributed to a phenomenon called the "dispersing effect."

PDP curves for carbon steel corrosion in 1M HCl with and without TTY, TTY4, and TTY5 are presented in Fig. 11.5. It can be clearly seen that the nature of both anodic and cathodic curves are greatly changed in the presence of investigated compounds. More so, it can also

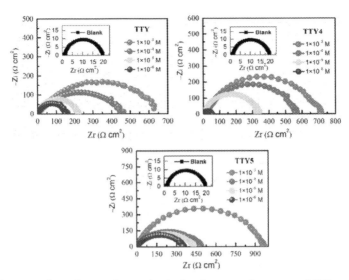

FIG. 11.4 Nyquist curves for carbon steel corrosion in the absence and presence of different concentrations of TTY, TTY4 and TTY5 [6]. (Reproduced with permission@ Copyright Elsevier).

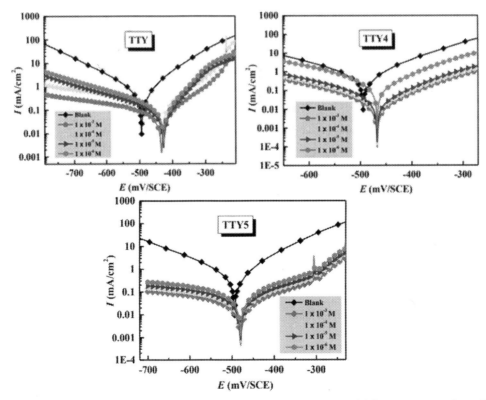

FIG. 11.5 Tafel curves for carbon steel corrosion in the absence and presence of different concentrations of TTY, TTY4, and TTY5 [6]. (Reproduced with permission@ Copyright Elsevier).

be seen that corrosion current values for carbon steel corrosion are greatly reduced in the presence of TTY, TTY4, and TTY5. This observation suggests that TTY, TTY4, and TTY5 become effective by adsorbing and blocking the active sites present over the metallic surface. The displacement in the values of E_{corr} was less than −85 mV indicating that all studied compounds behaved as mixed-type corrosion inhibitors.

11.4 Summary

Electrochemical techniques especially EIS and PDP are widely used in studying the corrosion inhibition property of organic corrosion inhibitors. Using the EIS method, interfacial behavior of corrosion inhibition can be derived. Most of the organic corrosion inhibitors become effective by adsorbing at the interface of metal and electrolyte. These inhibitors are called interface-type corrosion inhibitors. Generally, adsorption of organic inhibitors at the metal–electrolyte interface increases the value of charge transfer resistance and diameter of the Nyquist curves. Imperfection in the semicircle of Nyquist curves with and without inhibitor is attributed due to the surface roughness which resulted through the structural or interfacial origin. Using PDP method, organic corrosion inhibitors may be classified as anodic or cathodic or mixed type. Generally, presence of organic inhibitors shifts the corrosion potential (E_{corr}) in either anodic or cathodic direction with respect to the E_{corr} of uninhibited Tafel curve. An inhibitor can be classified as anodic- or cathodic-type, if the displacement in the E_{corr} values of inhibited and uninhibited Tafel curves is more than −85 mV. However, if this displacement is less than −85 mV then the inhibitor can be classified as mixed type. Most of the organic corrosion inhibitors become effective by retarding the both anodic as well as cathodic reactions and act as mixed-type corrosion inhibitors. Both EIS and PDP studies are widely used in investigating the corrosion inhibition property of organic compounds.

11.5 Useful websites

https://www.gamry.com/application-notes/EIS/basics-of-electrochemical-impedance-spectroscopy/
https://www.ameteksi.com/-/media/ameteksi/download_links/documentations/library/princetonappliedresearch/application_note_corr-1.pdf?dmc=1
https://www.palmsenscorrosion.com/knowledgebase/tafel-plot-and-evans-diagram/

References

[1] A. Berradja, Electrochemical techniques for corrosion and tribocorrosion monitoring: methods for the assessment of corrosion rates. In: Corrosion Inhibitors, IntechOpen, London, UK, 2019.
[2] C. Verma, L.O. Olasunkanmi, E.E. Ebenso, M.A. Quraishi, I.B. Obot, Adsorption behavior of glucosamine-based, pyrimidine-fused heterocycles as green corrosion inhibitors for mild steel: experimental and theoretical studies, J. Phys. Chem. C 120 (2016) 11598–11611.
[3] J. Haque, C. Verma, V. Srivastava, M. Quraishi, E.E. Ebenso, Experimental and quantum chemical studies of functionalized tetrahydropyridines as corrosion inhibitors for mild steel in 1 M hydrochloric acid, Results Phys. 9 (2018) 1481–1493.

[4] D.K. Yadav, M.A. Quraishi, Application of some condensed uracils as corrosion inhibitors for mild steel: gravimetric, electrochemical, surface morphological, UV–visible, and theoretical investigations, Ind. Eng. Chem. Res. 51 (2012) 14966–14979.

[5] A.Y. Musa, R.T. Jalgham, A.B. Mohamad, Molecular dynamic and quantum chemical calculations for phthalazine derivatives as corrosion inhibitors of mild steel in 1 M HCl, Corrosion Sci. 56 (2012) 176–183.

[6] Y. El Bakri, L. Guo, E.M. Essassi, Electrochemical, DFT and MD simulation of newly synthesized triazolotriazepine derivatives as corrosion inhibitors for carbon steel in 1 M HCl, J. Mol. Liq. 274 (2019) 759–769.

[7] H. Lgaz, K.S. Bhat, R. Salghi, S. Jodeh, M. Algarra, B. Hammouti, I.H. Ali, A. Essamri, Insights into corrosion inhibition behavior of three chalcone derivatives for mild steel in hydrochloric acid solution, J. Mol. Liq. 238 (2017) 71–83.

[8] E. Ech-Chihbi, A. Nahlé, R. Salim, F. Benhiba, A. Moussaif, F. El-Hajjaji, H. Oudda, A. Guenbour, M. Taleb, I. Warad, Computational, MD simulation, SEM/EDX and experimental studies for understanding adsorption of benzimidazole derivatives as corrosion inhibitors in 1.0 M HCl solution, J. Alloys Comp. 844 (2020) 155842.

[9] P.R. Ammal, M. Prajila, A. Joseph, Effective inhibition of mild steel corrosion in hydrochloric acid using EBIMOT, a 1, 3, 4-oxadiazole derivative bearing a 2-ethylbenzimidazole moiety: electro analytical, computational and kinetic studies, Egypt. J. Petrol. 27 (2018) 823–833.

[10] F. Benhiba, H. Serrar, R. Hsissou, A. Guenbour, A. Bellaouchou, M. Tabyaoui, S. Boukhris, H. Oudda, I. Warad, A. Zarrouk, Tetrahydropyrimido-Triazepine derivatives as anti-corrosion additives for acid corrosion: chemical, electrochemical, surface and theoretical studies, Chem. Phys. Lett. 743 (2020) 137181.

[11] K. Ansari, D.S. Chauhan, M. Quraishi, M.A. Mazumder, A. Singh, Chitosan Schiff base: an environmentally benign biological macromolecule as a new corrosion inhibitor for oil & gas industries, Int. J. Biol. Macromol. 144 (2020) 305–315.

[12] C. Verma, M. Quraishi, I. Obot, E.E. Ebenso, Effect of substituent dependent molecular structure on anti-corrosive behavior of one-pot multicomponent synthesized pyrimido [2, 1-B] benzothiazoles: computer modelling supported experimental studies, J. Mol. Liq. 287 (2019) 110972.

[13] R. Murthy, P. Gupta, C. Sundaresan, Theoretical and electrochemical evaluation of 2-thioureidobenzheteroazoles as potent corrosion inhibitors for mild steel in 2 M HCl solution, J. Mol. Liq. 319 (2020) 114081.

[14] L.O. Olasunkanmi, M.M. Kabanda, E.E. Ebenso, Quinoxaline derivatives as corrosion inhibitors for mild steel in hydrochloric acid medium: electrochemical and quantum chemical studies, Physica E: Low-Dimensional Syst. Nanostruct. 76 (2016) 109–126.

[15] C. Verma, M. Quraishi, A. Singh, 2-Amino-5-nitro-4, 6-diarylcyclohex-1-ene-1, 3, 3-tricarbonitriles as new and effective corrosion inhibitors for mild steel in 1 M HCl: Experimental and theoretical studies, J. Mol. Liq. 212 (2015) 804–812.

[16] C. Verma, E. Ebenso, I. Bahadur, I. Obot, M. Quraishi, 5-(Phenylthio)-3H-pyrrole-4-carbonitriles as effective corrosion inhibitors for mild steel in 1 M HCl: experimental and theoretical investigation, J. Mol. Liq. 212 (2015) 209–218.

[17] C. Verma, M. Quraishi, L. Olasunkanmi, E.E. Ebenso, L-Proline-promoted synthesis of 2-amino-4-arylquinoline-3-carbonitriles as sustainable corrosion inhibitors for mild steel in 1 M HCl: experimental and computational studies, RSC Adv. 5 (2015) 85417–85430.

[18] C. Verma, L.O. Olasunkanmi, I. Obot, E.E. Ebenso, M. Quraishi, 2, 4-Diamino-5-(phenylthio)-5 H-chromeno [2, 3-b] pyridine-3-carbonitriles as green and effective corrosion inhibitors: gravimetric, electrochemical, surface morphology and theoretical studies, RSC Adv. 6 (2016) 53933–53948.

[19] J. Haque, K. Ansari, V. Srivastava, M. Quraishi, I. Obot, Pyrimidine derivatives as novel acidizing corrosion inhibitors for N80 steel useful for petroleum industry: a combined experimental and theoretical approach, J. Ind. Eng. Chem. 49 (2017) 176–188.

[20] K. Ansari, M. Quraishi, A. Singh, Isatin derivatives as a non-toxic corrosion inhibitor for mild steel in 20% H2SO4, Corrosion Sci. 95 (2015) 62–70.

CHAPTER 12

Computational methods of corrosion assessment

Molecular modelings through computational simulations have gained particular attention toward the testing of corrosion inhibition effectiveness of the organic corrosion inhibitors. These techniques can be regarded as time-saving, cost-effective, and environmental friendly protocols of corrosion monitoring. These simulations do not involve the use of any toxic and expensive chemicals that will help in reducing the environmental and safety risks associated with using traditional methods of corrosion monitoring. Using computational modeling corrosion inhibition effectiveness of the organic compounds to be used as corrosion inhibitors can be theoretically predicted even before their syntheses. Various computational modelings are successfully used in evaluating the corrosion inhibition properties of organic corrosion inhibitors. However, density functional theory (DFT) and Monte Carlo (MC) and molecular dynamics (MD) simulations are the most advanced and most frequently used computational modelings.

12.1 Density functional theory

The use of DFT-based computational modeling has been significantly enhanced in the designing and validating the inhibition efficiency of organic corrosion inhibitors derived experimentally. The inhibition efficiency of organic corrosion inhibitors can be predicted with the help of various reactivity parameters and electronic or molecular properties. According to Hohenberg–Kohn theorems, DFT modeling focuses on the electron density, $\rho(r)$, which is a carrier of the information related to the molecule or atom in the ground state. The electron density provides information on the combined contribution of all electrons of the system. Therefore, according to the Hohenberg–Kohn theorems, the ground state of an electronic system (atom or molecule) is just a function of electron density. Obviously, the knowledge of the electron density is needed to derive all the properties of an atom or molecule (electronic system).

In the form of various parameters, DFT study provides information about the interactions between metallic surface and inhibitor molecules. Effect of substituents and molecular/electronic structure of organic corrosion inhibitors on their corrosion inhibition effect can also be successfully described with the aid of DFT. Generally, inhibition effectiveness of the organic inhibitors can be correlated to their values of energies of frontier molecular orbitals (FMOs), i.e., energy of highest occupied molecular orbital, E_{HOMO}, and energy of lowest unoccupied molecular orbital, E_{LUMO}. Various other global and local parameters such as energy bandgap ($\Delta E = E_{LUMO} - E_{HOMO}$), electronegativity ($\chi$), hardness ($\eta$), softness ($\rho$) dipole moment (DM) ($\mu$), and fraction of electron transfer (ΔN), etc., can be calculated using the energies of FMOs [1,2]. Detail description and effect of some major DFT-based reactivity parameters is given below.

12.1.1 Frontier molecular orbitals

FMOs (HOMO and LUMO) are the most important DFT-based parameters. Energies of the E_{HOMO} and E_{LUMO} can be directly correlated to the order of corrosion inhibition efficiency of organic compounds. It is important to mention that organic inhibitors interact with metallic surfaces using donor–acceptor (coordination) interactions in which organic inhibitors are assumed as charge (electrons) donor and metallic surface can be assumed as charge acceptor. This process of change sharing is known as donation. However, extra charges present in metal can also be transferred to organic inhibitors via a process known as retro-donation. It is well-documented in literature that an organic compound with strong ability of donation as well as retro-donation would be an effective corrosion inhibitor. Obviously, a compound with high value of E_{HOMO} would be a better corrosion inhibitor than the compound with relatively less value of E_{HOMO}. Conversely, a low value of E_{LUMO} is consistent with high protection effectiveness. Another important reactivity parameter can be derived from subtracting the E_{HOMO} from E_{LUMO}. This is called the energy bandgap, ΔE ($E_{LUMO} - E_{HOMO}$). A low value of ΔE is associated with high chemical reactivity and therefore high inhibition efficiency.

According to Koopman's theorem, the negative of the E_{HOMO} and E_{LUMO} can be approximated as the ionization potential (I) and electron affinity (A), respectively [1].

$$I = -E_{HOMO} \tag{12.1}$$

$$A = -E_{LUMO} \tag{12.2}$$

Literature study suggests that values of E_{HOMO} (–I), E_{LUMO} (–EA) and energy bandgap ($\Delta E = E_{LUMO} - E_{HOMO}$) are widely used to manifest the corrosion inhibition effect of organic corrosion inhibitors.

Significance of FMOs: information derived from DFT studies

As described above, the corrosion inhibition effect of organic corrosion inhibitors can be derived using several DFT-based theoretical parameters. Along with various DFT parameters, DFT analyses also provide pictorial presentation of the FMOs in the term of which corrosion inhibition efficiency of a series of compounds can be described. An organic corrosion inhibitor with higher HOMO and/or LUMO contribution would be a better corrosion inhibitor as compared to the inhibitor molecule with lower HOMO and/or LUMO contribution.

Generally, electron-donating substituents such as –OH, –CH$_3$, and –OCH$_3$, etc., increase the FMOs contributions and converse is true for electron-withdrawing substituents (–NO$_2$ and –CN). For example, while studying the corrosion inhibition three 2-amino-4-arylquinoline-3-carbonitriles (AACs) for mild steel in 1M HCl, we observed that electron-withdrawing

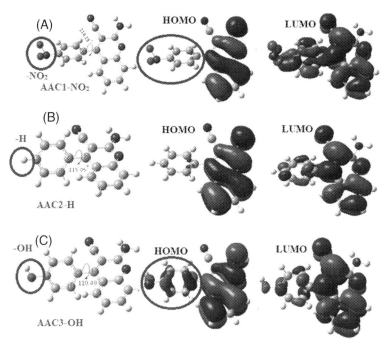

FIG. 12.1 The optimized, HOMO and LUMO frontier molecular orbital pictures of AACs [3]. (Author original work; Permission not required@ Copyright the Royal Society of Chemistry). *AACs*, 2-amino-4-arylquinoline-3-carbonitriles; *HOMO*, highest occupied molecular orbital; *LUMO*, lowest unoccupied molecular orbital.

–NO2 substituent decreases the contribution of HOMO as compared to the nonsubstituted and electron donating hydroxyl (–OH) substituted molecules [3]. The optimized, HOMO and LUMO FMO pictures of AACs are presented in Fig. 12.1.

We reported a similar observation, while studying the corrosion inhibition effectiveness of four 5-Arylpyrimido-[4,5-b]quinoline-diones (APQDs) for mild steel in acidic medium in 1M HCl [4]. DFT studies showed that electron-withdrawing nitro (–NO_2) substituent decreases HOMO's contribution as compared to APQDs molecules without any substituent and with electron-donating hydroxyl (–OH) substituent(s). The optimized, HOMO and LUMO FMO pictures of APQDs are presented in Fig. 12.2.

In another study, during studying the corrosion inhibition effect of four glucosamine-based, pyrimidine-fused heterocycles (CARBs), it was observed that presence of electron-withdrawing –NO_2 substituent decreased the contribution of both HOMO and LUMO with respect to HOMO and LUMO contribution of nonsubstituted and substituted by –CH_3 and –OH substituents. The optimized, HOMO and LUMO FMO pictures of CARBs are presented in Fig. 12.3.

It is important to mention that organic compounds with higher localized areas of HOMO and/or LUMO act as superior corrosion inhibitors than that of the compounds having lesser localized areas of HOMO and/or LUMO. DFT also provides vital information about the effect of substituents. Generally, electron-donating substituents increase inhibition efficiency of organic compounds because of their ability to increase the electron density at the donor

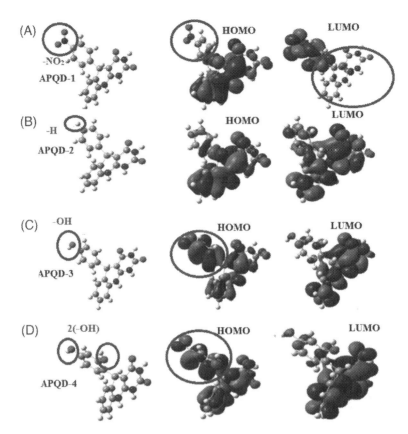

FIG. 12.2 The optimized, HOMO and LUMO frontier molecular orbital pictures of APQDs [4]. (Author original work; Permission not required@ Copyright the Royal Society of Chemistry). *APQDs*, 5-Arylpyrimido-[4,5-b]quinoline-diones; *HOMO*, highest occupied molecular orbital; *LUMO*, lowest unoccupied molecular orbital.

sites. By DFT analysis, it can be observed that electron-donating substituents generally increase HOMO's and/or LUMO's contribution and converse is true for electron-withdrawing substituents.

12.1.2 Dipole moment

In general terms, DM is used to describe the polarity of a molecule and used to measure the polarity of a polar covalent bond. The DM is defined as the product of charges present over the two covalently atoms and the distance between them [1].

$$\mu = qR \tag{12.3}$$

where q is the charge and R is the distance. Though the SI unit of the DM is coulomb meter (Cm), it is most commonly reported in non-SI, Debye (D). Both units are related as follows:

$$1D = 3.33564 \times 10^{-30} Cm \tag{12.4}$$

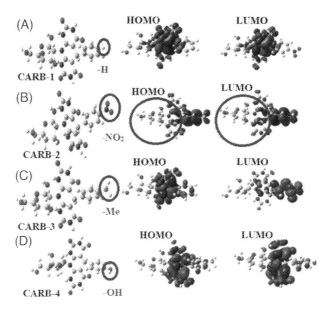

FIG. 12.3 The optimized, HOMO and LUMO frontier molecular orbital pictures of CARBs [5]. (Reproduced with permission@ Copyright the American Chemical Society). *HOMO,* highest occupied molecular orbital; *LUMO,* lowest unoccupied molecular orbital.

The DM is a global parameter, i.e., describes the overall polarity of organic compounds instead of a single covalent bond. Literature survey shows that both positive and negative trend of corrosion inhibition efficiency is reported with the value of DM.

12.1.3 Chemical potential and electronegativity

DFT modeling also provides useful insight about chemical selectivity and reactivity in the terms of global molecular properties including chemical potential (μ) and electronegativity (χ). For N-electron system with an external potential, $v(r)$ and total energy, E, the chemical potential which is the negative of electronegativity (χ) can be presented as follows [1]:

$$\chi = -\mu = -\left(\frac{\partial E}{\partial N}\right)_{v(r)} \tag{12.5}$$

According to Iczkowski and Margrave, the above equation can be simplified in the terms of ionization potential (I) and electron affinity (A) as follows [1]:

$$\chi = -\mu = -\left(\frac{I+A}{2}\right) \tag{12.6}$$

$$\chi = -\mu = -\left(\frac{E_{HOMO} + E_{LUMO}}{2}\right) \tag{12.7}$$

According to Sanderson's principle of electronegativity equalization, when two or more atoms combine together to form a molecule, their electronegativities turn out to be adjusted to the equal intermediate value.

12.1.4 Global hardness and softness

The hardness (η) is a global parameter and it defines the overall electronegativity of the molecule rather than the electronegativity of an isolated atom. For an electronic system, η can be presented as follows [1]:

$$\eta = \frac{1}{2}\left(\frac{\partial \mu}{\partial N}\right)_{v(r)} = \left(\frac{\partial^2 E}{\partial N^2}\right)_{v(r)} \tag{12.8}$$

According to the valence state parabola model and Janak's theorem, hardness can also be approximated in the terms of E_{HOMO} and E_{LUMO} as follows [1]:

$$\eta = \frac{I-A}{2} = -\left(\frac{E_{LUMO} - E_{HOMO}}{2}\right) \tag{12.9}$$

The global hardness is also called absolute hardness. Obviously, the global softness (σ) can be defined as inverse of the global hardness [1]:

$$\sigma = \frac{1}{\eta} = \left(\frac{\partial N}{\partial \mu}\right)_{v(r)} = \left(\frac{\partial N^2}{\partial^2 E}\right)_{v(r)} \tag{12.10}$$

12.1.5 Fraction of electrons (charge) transferred

According to the Pearson, fraction of electron transferred from an organic corrosion inhibitor to the metal surface is given by [1]:

$$\Delta N = \frac{\chi_M - \chi_{inh}}{[2(\eta_M + \eta_{inh})]} \tag{12.11}$$

In the above equation, χ_M and χ_{inh} represent the electronegativity of bulk metal and the inhibitor molecule. The interactions between metal surface and inhibitor molecule can be regarded as a charge-sharing process between two chemical species of different electronegativities. In such interaction, charge flow occurs from a species of less electronegativity (inhibitor molecule) to the species of more electronegativity (metal surface), until their chemical potentials become the same. For the calculation of the fraction of electrons transferred, many researchers used an electronegativity value of 7.0 eV for iron. The hardness of metal surface can be assumed as zero (e.g., $\eta_{Fe} = 0$), by considering that for a bulk metal, ionization potential is equal to electron affinity, i.e., $I = A$. It is important to mention that ΔN does not give any information about the number of electrons leaving the inhibitor molecule and entering into the acceptor (metal). Therefore, the expression "electron-donation ability" has been preferred to use over the fraction of electron transfer. Lukovits proposed that if ΔN < 3.6 then inhibition efficiency of the organic corrosion inhibitors increases on increasing electron-donating ability on the metal surface.

Kokaji proposed that use of the work function of metal (ϕ) would be preferred at the place of electronegativity. In the terms of wave function, Eq. (12.11) can be written as [1]:

$$\Delta N = \frac{\phi - \chi_{inh}}{[2(\eta_M + \eta_{inh})]} \qquad (12.12)$$

This equation can be further simplified as follows by assuming η_M to zero:

$$\Delta N = \frac{\phi - \chi_{inh}}{2\eta_{inh}} \qquad (12.13)$$

12.2 Molecular dynamics and Monte Carlo simulations

MD and MC simulations are established as vital computational modelings for studying the interaction of inhibitor molecules with metallic surfaces. Whether the interaction between an inhibitor molecule and a metallic surface is a spontaneous or nonspontaneous process can be derived through these simulations. It is important to mention that orientation of inhibitor molecules on metallic surfaces largely affects their corrosion inhibition effectiveness. In general, an organic compound with flat (or horizontal with metallic surface) orientation will cover the large metal surface and would be a more effective corrosion inhibitor as compared to the organic compound having vertical orientation. MD and MC simulations provide information about the orientation of the organic inhibitor molecules in the form of picture (top- and side-view). Most of the organic inhibitors become effective by adsorbing on the metallic surface. Generally, adsorption of organic corrosion inhibitors on metallic surfaces is accompanied with the discharge of energy, i.e., adsorption of inhibitor molecules on metallic surface is a spontaneous and exothermic process. In the simulation process, total energy (E_{total}) of an inhibitor molecule (adsorbate) is the sum of inhibitor energy ($E_{inhibitor}$), rigid adsorption energy (E_{rigid}) and deformation energy (E_{def}) [6]:

$$E_{total} = E_{inhibitor} + E_{rigid} + E_{def} \qquad (12.14)$$

where E_{rigid} is the energy absorbed or released when an unrelaxed inhibitor molecule adsorbed on the metallic surface before its geometric optimization and E_{def} is the energy released or absorbed when an adsorbed inhibitor molecule is relaxed on the metallic surface. The magnitude of energy released or absorbed by the adsorption of one mole of organic molecules is defined as adsorption energy (E_{ads}) or interaction energy (E_{int}). The interaction energy (E_{int}) can be derived using following equation [6]:

$$E_{int} = E_{total} - (E_{surface+solution} + E_{inhibitor}) \qquad (12.15)$$

In the above equation, $E_{surface+solution}$ represents the total energy of metal surface and solution (electrolyte) with and without inhibitor molecules. Generally, negative of the adsorption or interaction energy is called as binding energy ($E_{binding}$) [6]:

$$E_{int} = -E_{binding} \tag{12.16}$$

In the aqueous solution, adsorption of inhibitor molecules on metallic surfaces depends upon various types of interactions including metal–water, inhibitor–water, metal–inhibitor, and water–water. In aqueous electrolytes, organic inhibitors may get solvated that can adversely affect the adsorption tendency of the inhibitor molecules. The solvation energy ($E_{solvation}$) for metal–inhibitor interaction in aqueous electrolytes can be derived as follows [6]:

$$E_{solvation} = E_{inhibitor+water} - (E_{water} + E_{inhibitor}) \tag{12.17}$$

where $E_{inhibitor+water}$ is the sum of inhibitor and water energies. The $E_{inhibitor}$ and E_{water} are the total energy of inhibitor molecules and water, respectively.

The MD and MC simulations are widely used in identifying the corrosion inhibition nature and adsorption behavior of the organic compounds over the metallic surface. In most of the studies, magnitude of interaction energy (E_{int}) acquires the negative sign indicating that metal–inhibitor interaction is a spontaneous process. Obviously, a higher negative value of E_{ads} is consistent with the high adsorption tendency (or corrosion inhibition effectiveness). For example, while studying the corrosion inhibition property of four glucosamine-based, pyrimidine-fused heterocycles (CARBs), differing in the nature of substituents, Verma et al. [5] observed that presence of both electron-withdrawing (–NO$_2$) and donating (–CH$_3$ & –OH) substituents increase the inhibition efficiency as compared to the nonsubstituted inhibitor molecule. This increase in the inhibition efficiency was more pronounced in the presence of electron-donating substituents. Using MC simulations study, it was observed that the inhibitor molecule exhibiting the best inhibition efficiency (CARB-4) was associated with the highest value of adsorption energy. Adsorption energies (in negative) of the tested inhibitor molecules follow the sequence: CARB-4 (–234.898 kcal mol^{-1}) > CARB-3 (–222.651 kcal mol^{-1}) > CARB-2 (–213.372 kcal mol^{-1}) > CARB-1 (–210.235 kcal mol^{-1}). The order of adsorption energy was well consistent with the order of inhibition efficiency derived through experimental studies, i.e., a more effective inhibitor was associated with high protection efficiency.

The top and side views of the CARB molecules on the metallic surface are shown in Fig. 12.4. It can be clearly seen (marked by yellow circles) that in the absence of any substituent ((–H; CARB-1) and presence of electron-withdrawing nitro (–NO$_2$; CARB-2) substituent, a large portion of inhibitor molecules acquire the vertical orientation and only a small portion of the inhibitor molecules participate in adsorption process. However, in the presence of electron-donating methyl (–CH$_3$; CARB-3) and hydroxyl (–OH; CARB-4), almost entire molecules participate in the adsorption process.

Similar observations have been extensively reported in the literature [4]. Our research team, during studying the anticorrosive effect of four 5-Arylpyrimido-[4,5-b]

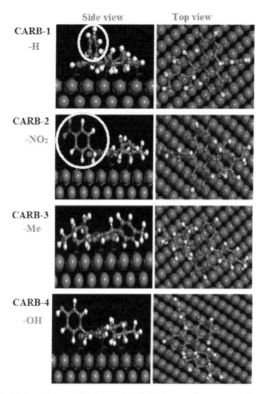

FIG. 12.4 Side (left-hand side) and top (right-hand side) views of most stable equilibrium configuration of CARBs on Fe (110) surface derived using MC simulations [5]. (Reproduced with permission@ Copyright the American Chemical Society). *MC*, Monte Carlo.

quinoline-diones (APQDs) for mild steel corrosion in acidic solution, reported that inhibition efficiency of the inhibitor molecules get enhanced in the presence containing electron-donating hydroxyl (–OH) substituent(s) as compared to the nonsubstituted inhibitor molecule. On the other hand, inhibition efficiency of the inhibitor molecule get reduced in the presence of electron-withdrawing nitro (–NO$_2$) substituent [4].

Using MD simulations, a more negative value of adsorption energy (E_{ads}) was associated with high inhibition efficiency. The values of E_{ads} followed the order: APQD-4 (–115.78 kJ/mol) > APQD-3 (–107.30 kJ/mol) > APQD-2 (–93.62 kJ/mol) > APQD-1 (–83.44 kJ/mol). Fig. 12.5 represents the orientation (top and side views) of APQDs on a metallic surface. It can be clearly seen that electron releasing –OH substituent(s) help inhibitor molecules in getting them adsorbed using flatter orientation while converse is true in the presence of electron-withdrawing –NO$_2$ substituent. Corrosion inhibition effects of various organic corrosion inhibitors and their adsorption behavior as well as orientation on metallic surfaces are extensively reported in the literature.

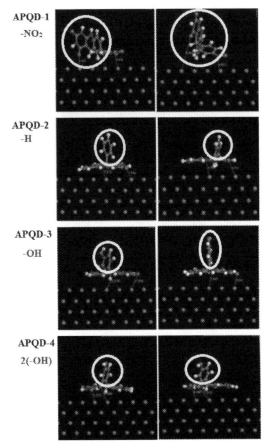

FIG. 12.5 Side (left-hand side) and top (right-hand side) views of most stable equilibrium configuration of APQDs [4]. (Author original work; Permission not required@ Copyright the Royal Society of Chemistry). *APQDs*, 5-Arylpyrimido-[4,5-b]quinoline-diones.

12.3 Summary

Recently, use of computational modelings especially DFT, MD, and MC simulations has acquired particular attention in describing the corrosion inhibition effect of the organic corrosion inhibitors. These computer-based modelings can be regarded as environmental friendly approaches as unlike traditional corrosion monitoring methods, the computational modelings don't require the use of any toxic and expensive chemicals and instruments. More so, using these modelings corrosion inhibition effect of various organic compounds can be determined before their syntheses. These modelings provide some vital parameters in the term of which corrosion inhibition effect and adsorption behavior of organic compounds on the metallic surface can be explained. Generally, localized FMOs give information about molecular sites involved in electron sharing. The molecular sites localized in HOMO are responsible for the electron donation whereas localized sites in LUMO are responsible for electron acceptance. MD and MC simulations provide information about the nature and effectiveness of metal–inhibitor interaction. An inhibitor with high value of adsorption energy (E_{ads}) would act as better adsorbate and corrosion inhibitor as compared to the corrosion inhibitor with lower value of E_{ads}. MD and MC simulations studies also provide pictorial orientation on the metallic surface. An inhibitor molecule with flat orientation will act as a better corrosion inhibitor as compared to the inhibitor with a relatively vertical orientation.

12.4 Useful links

https://www.hindawi.com/journals/isrn/2013/175910/
http://newton.ex.ac.uk/research/qsystems/people/coomer/dft_intro.html
https://dft.uci.edu/teaching/lausanne/B.pdf
http://people.virginia.edu/~lz2n/mse627/notes/MD-basics.pdf
https://udel.edu/~arthij/MD.pdf

References

[1] I. Obot, D. Macdonald, Z. Gasem, Density functional theory (DFT) as a powerful tool for designing new organic corrosion inhibitors. Part 1: an overview, Corrosion Sci. 99 (2015) 1–30.

[2] D.K. Verma, Density functional theory (DFT) as a powerful tool for designing corrosion inhibitors in aqueous phase, Adv. Eng. Test. (2018) 87–105.

[3] C. Verma, M. Quraishi, L. Olasunkanmi, E.E. Ebenso, L-Proline-promoted synthesis of 2-amino-4-arylquinoline-3-carbonitriles as sustainable corrosion inhibitors for mild steel in 1 M HCl: experimental and computational studies, RSC Adv. 5 (2015) 85417–85430.

[4] C. Verma, L. Olasunkanmi, I. Obot, E.E. Ebenso, M. Quraishi, 5-Arylpyrimido-[4, 5-b] quinoline-diones as new and sustainable corrosion inhibitors for mild steel in 1 M HCl: a combined experimental and theoretical approach, RSC Adv. 6 (2016) 15639–15654.

[5] C. Verma, L.O. Olasunkanmi, E.E. Ebenso, M.A. Quraishi, I.B. Obot, Adsorption behavior of glucosamine-based, pyrimidine-fused heterocycles as green corrosion inhibitors for mild steel: experimental and theoretical studies, J. Phys. Chem. C 120 (2016) 11598–11611.

[6] C. Verma, H. Lgaz, D. Verma, E.E. Ebenso, I. Bahadur, M. Quraishi, Molecular dynamics and Monte Carlo simulations as powerful tools for study of interfacial adsorption behavior of corrosion inhibitors in aqueous phase: a review, J. Mol. Liq. 260 (2018) 99–120.

CHAPTER 13

Ionic liquids as green corrosion inhibitors

Ionic liquids (ILs) are salts that exist in a liquid state below the boiling point of water. Because of their various advantages such as low toxicity, low vapor pressure, extremely high solubility, polarity, and stability, ILs are widely used for various biological and industrial applications. ILs are also widely used as corrosion inhibitors for different metals and alloys in various electrolytes. ILs are highly soluble in the aqueous electrolytes because of their high polarity that comes out in the forms of their high inhibition efficiency. ILs acquire several advantages over the traditional organic corrosion inhibitors. One of such advantages is their environmental benign. Different classes of ILs are extensively used as green corrosion inhibitors. The present chapter describes the various aspects of corrosion inhibition using ILs.

13.1 Ionic liquids: property and application

ILs are commonly known as ionic fluids, ionic glasses, fused salts, ionic melts, and ionic electrolytes. Generally, they exist in a liquid state below 100°C (boiling point of water). However, 3-methylimidazolium dicyanamide ([EMIM] [N $(CN)_2$]$^-$) and ethylammonium nitrate (EAN), etc., are liquid below room temperature. The 3-methylimidazolium dicyanamide ([EMIM] [N $(CN)_2$]$^-$) and EAN acquire the melting points of -21°C C and 12°C, respectively. EAN was established as the first IL and it was reported by Paul Walden in 1914. However, after the establishment of EAN as the first IL, various classes of ILs including pyridinium, tetra-ammonium, phosphonium, imidazolium, etc., are extensively developed especially in the 1970s–80s.

ILs are widely used for different purposes (Fig. 13.1). ILs acquire various useful properties that make them suitable candidates for various applications. Though ILs consist of cations and anions, they are poor conductors of electricity. ILs can be categorized into different classes based on their chemical nature. Fig. 13.2 represents the brief classification of ILs. Acidic ILs are those that are able to donate the proton (Arrhenius acid) and basic ILs are those that are able to accept the proton (Arrhenius base). Both acidic and basic ILs are widely used as corrosion inhibitors. Nevertheless, functionalized and supported ILs are also utilized as metallic corrosion inhibitors.

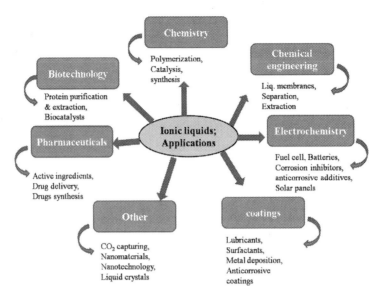

FIG. 13.1 Industrial and biological applications of ionic liquids in the fields of science and engineering.

ILs can be further classified based on the cations present in their molecular structure. On this basis, ILs may be of imidazolium, tetra-ammonium, pyridinium, phosphonium, etc., based. Literature studies showed that imidazolium-based ILs are most extensively developed and investigated. During the last three decades, synthesis and consumption of ILs as green corrosion inhibitors has gained immense attention. It is important to mention that

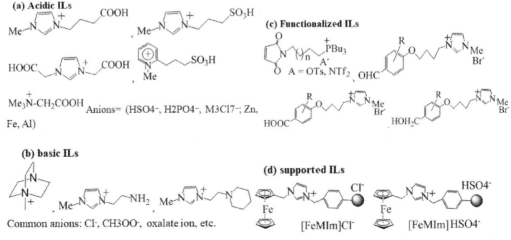

FIG. 13.2 Brief classification of ionic liquids. (A) acidic, (B) basic, (C) functionalized, and (D) supported ionic liquids.

properties of ILs including their corrosion inhibition effectiveness can be suitably modified by chaining the nature of cations and/or anions. Because of this property, ILs are called as designer chemicals, i.e., designer corrosion inhibitors. Some common properties of ILs are given below:

(i) Negligible vapor pressure and negligible ability of evaporation. Therefore, they can be recovered and reused.
(ii) Nonflammable and high stability to thermal and chemical treatment.
(iii) Wider liquid range compared to water and other common organic solvents/chemicals.
(iv) Soluble in organic and aqueous solvents/electrolytes.
(v) Thermally stable up to 200°C.
(vi) Ability to exhibit Lewis, Bronsted, and other types of acidity.
(vii) Tendency to act as solvents for inorganic, organic, and polymeric compounds.
(viii) Ease to prepare and use.
(ix) Environmental benign nature.

13.2 Mechanism of corrosion inhibition using ionic liquids

It is well established that ILs become effective by adsorbing on the metallic surface using their electron-rich sites called as adsorption centers. After getting adsorbed, ILs inhibit both anodic oxidation as well as cathodic hydrogen evolution Tafel reactions. The nature of anodic and cathodic Tafel reaction with and without ILs in sulfate (H_2SO_4) and chloride (HCl) based electrolytes can be presented as follows [1]:

(a) *Anodic reactions without ILs*

$$M + nH_2O \leftrightarrow M(H_2O)_{n(ads)} \tag{13.1}$$

$$M(H_2O)_{n(ads)} + 2Cl^-/SO_4^{2-} \leftrightarrow M\left[(H_2O)_n 2Cl^-/SO_4^{2-}\right]_{ads} \tag{13.2}$$

$$M\left[(H_2O)_n 2Cl^-/SO_4^{2-}\right]_{ads} \leftrightarrow M[(H_2O)_n 2Cl/SO_4]_{ads} + 2e^- \tag{13.3}$$

$$M[(H_2O)_n 2Cl/SO_4]_{ads} + 2e^- \leftrightarrow M^{2+} + H^+ + OH^- + 2Cl^-/SO_4^{2-} \tag{13.4}$$

It can be clearly seen from Eqs. 13.2–13.4 that the presence of chloride and sulfate ions accelerates the rate of anodic dissolution reaction. However, in the presence of ILs, the rate of anodic reaction gets reduced due to the adsorption of organic cations. Obviously, organic cations preferably get adsorbed and inhibit anodic reaction.

(b) *Anodic reactions with ILs*

$$M + nH_2O \leftrightarrow M(H_2O)_{n(ads)} \tag{13.5}$$

$$M(H_2O)_{n(ads)} + 2Cl^-/SO_4^{2-} \leftrightarrow M\left[(H_2O)_n 2Cl^-/SO_4^{2-}\right]_{ads} \tag{13.6}$$

$$M\left[(H_2O)_n\, 2Cl^-/SO_4^{2-}\right]_{ads} + ILs^+ \rightarrow M(H_2O)_n\, 2Cl/SO_4 - ILs\big]_{ads}^{-} \quad (13.7)$$

$$M(H_2O)_n\, 2Cl/SO_4 - ILs\big]_{ads}^{-} + 2IL^+ + 2Cl^-/SO_4^{2-} \rightarrow$$

$$M(H_2O)_n\, 2Cl/SO_4 - ILs\big]_{ads}^{-} + [2IL]^+ 2Cl^-/SO_4^{2-} \quad (13.8a)$$

The above reactions can be summarized as:

$$M + X^- \leftrightarrow (MX)_{ads}^- \quad (13.8)$$

$$(MX)_{ads}^- + IL^+ \leftrightarrow (MX^-IL^+)_{ads} \quad (13.9)$$

(c) *Cathodic reactions without ILs*

Cathodic hydrogen evolution reaction, in aqueous electrolytes without ILs proceeds as follows:

$$M + H_3O^+ \leftrightarrow M(H_3O)_{ads}^+ \quad (13.10)$$

$$M(H_3O)_{ads}^+ \leftrightarrow M(H_3O)_{ads} + e^- \quad (13.11)$$

$$M(H_3O)_{ads} + e^- + H^+ \leftrightarrow M + H_2O + H_2 \quad (13.12)$$

(d) *Cathodic reactions with ILs*

$$M + IL^+ \leftrightarrow M(IL)_{ads}^+ \quad (13.13)$$

$$M(IL)_{ads}^+ + e^- \leftrightarrow M(IL)_{ads} \quad (13.14)$$

It is important to mention that in aqueous electrolytes, metallic surfaces become charged due to the rapid oxidation. Therefore, during the initial stage of interaction with a metallic surface, ILs interact through electrostatic forces. However, later they may get adsorbed through charge (electron sharing). Anodic and cathodic reactions with and without a choline-based IL are presented in Fig. 13.3.

13.3 Ionic liquids as corrosion inhibitors: case studies

13.3.1 Ionic liquids as corrosion inhibitors for iron alloys in HCl

Hydrochloric acid solutions are extensively employed as electrolytes. It is important to mention that lower concentrations (3%–8%) of HCl solution are used for the removal of the surface rusts and scales. Removal of surface scales using acidic solutions is called descaling. On the other hand, highly concentrated HCl solutions (14%–16%) are widely employed in industrial acid pickling and acid cleaning processes. ILs are extensively utilized as corrosion inhibitors for steel alloys (mild steel and carbon steel) in HCl based electrolytes. Out of various classes of ILs, imidazolium-derived ILs are most widely used. Imidazolium-based ILs containing hydrophobic hydrocarbon chain(s) act as effective corrosion inhibitors for steel alloys. It can be believed that in such molecules, polar hydrophilic imidazolium ring interacts

Anodic sites

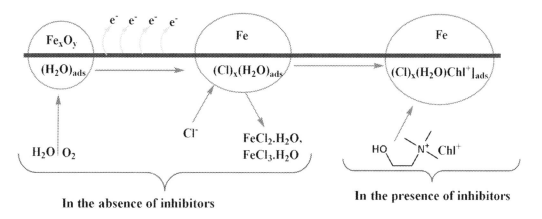

Cathodic sites

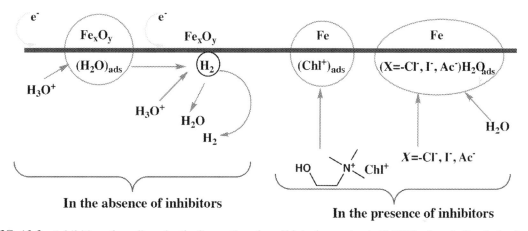

FIG. 13.3 Inhibition of anodic and cathodic reactions for mild steel corrosion in 1M HCl using choline derived ionic liquids [2].

and adsorbs on metallic surface, and hydrophobic hydrocarbon chain(s) orient toward the solution side.

Deyab and coworkers [3] reported the corrosion inhibition effect of 1-methyl imidazolium (IL1 and IL2) and 1,2-dimethyl imidazolium (IL3 and IL4) based four ILs for carbon steel in hydrochloric acid solution. It was observed that 1,2-dimethyl imidazolium-based ILs (IL3 and IL4) were relatively more effective corrosion inhibitors than that of the 1-methyl imidazolium based ILs (IL1 and IL2). Potentiodynamic polarization study showed that all studied ILs become effective by retarding the anodic and well as cathodic half-cell reactions and thereby they behaved as mixed-type corrosion inhibitors. EIS study revealed that all studied ILs adsorb at the interface of metal and electrolyte and increase resistance for the charge transfer process. Therefore, IL1-IL4 act as interface type of corrosion inhibitors. Adsorption of the investigated ILs on carbon steel surface was further supported by Fourier-Transformed Infra-Red (FT-IR) spectroscopic analyses.

Subasree and Selvi [4] studied the corrosion inhibition property of two ILs namely 3-hexadecyl-1-methyl-1H-imidazol-3-ium bromide [$C_{16}M_1Im$][Br] and 3-hexadecyl-1,2-dimethyl-1H-imidazol-3-ium bromide [$C_{16}M_2Im$][Br] for mild steel in 1M hydrochloric acid solution. Both investigated ILs were synthesized using previously developed methods. Chemical structure of the [$C_{16}M_1Im$][Br] and [$C_{16}M_2Im$][Br] was characterized using FT-IR and 1H NMR spectroscopic studies. Corrosion inhibition performance of [$C_{16}M_1Im$][Br] and [$C_{16}M_2Im$][Br] was tested using weight loss, PDP, EIS, SEM, EDX, AFM, and UV-vis studies. Results showed that [$C_{16}M_1Im$][Br] having two methyl (–CH_3) substituent showed better corrosion inhibition effectiveness as compared to the [$C_{16}M_2Im$][Br] containing only one methyl substituent. Potentiodynamic polarization study revealed that [$C_{16}M_1Im$][Br] and [$C_{16}M_2Im$][Br] acted as mixed-type but slightly anodic-type corrosion inhibitors. EIS study suggested that [$C_{16}M_1Im$][Br] and [$C_{16}M_2Im$][Br] acted as interface-type corrosion inhibitors as their presence increases the value of charge transfer resistance. Adsorption of [$C_{16}M_1Im$][Br] and [$C_{16}M_2Im$][Br] on metallic surface was studied by surface investigation. SEM and AFM studies showed that the metallic surfaces protected by [$C_{16}M_1Im$][Br] and [$C_{16}M_2Im$][Br] were much more smoother as compared to the non-protected metallic surface. EDX and UV-vis spectral analyses showed that [$C_{16}M_1Im$][Br] and [$C_{16}M_2Im$][Br] effectively adsorb on metallic surface.

In another study, Azeez et al. [5] studied the corrosion inhibition effect of three imidazolium ILs namely, 1-methyl-3-propylimidazolium iodide (MPIMI), 1-butyl-3-methylimidazolium iodide (BMIMI), and 1-hexyl-3-methylimidazolium iodide (HMIMI) for mid steel in 1M hydrochloric acid solution. The investigated ILs differ in the size of the hydrocarbon chain. Study showed that HMIMI, BMIMI & PMIMI become effective by adsorbing on the metallic surface and their adsorption follows the Langmuir adsorption isotherm. Potentiodynamic polarization study suggested that HMIMI, BMIMI, & PMIMI acted as mixed-type corrosion inhibitors as they adversely affect anodic and cathodic half-cell reactions. EIS study suggested that HMIMI, BMIMI, & PMIMI become effective by adsorbing on the interface of metal and electrolyte, and thereby they acted as interface-type corrosion inhibitors. Adsorption of HMIMI, BMIMI, & PMIMI on the metal-electrolyte interface increases the values of charge transfer resistance which was manifested by increase in the diameter of Nyquist plots inhibited by HMIMI, BMIMI, & PMIMI. Adsorption of studied ILs on the metallic surface was further studied by SEM, AFM, and FT-IR measurements. The presence of studied ILs significantly improved the surface morphology of the SEM and AFM images. Experimental studies were supported by DFT-based quantum chemical calculations.

A brief summary of some recent studies on ILs as corrosion inhibitors for steel alloys in hydrochloric acid solutions is given in Table 13.1. Literature study shows that most of

TABLE 13.1 Chemical structures, abbreviations, nature of metals, and electrolytes of some major works reported on anticorrosive effect of ionic liquids for iron alloys in HCl electrolytes.

Chemical str. of ILs	%IE/ Conc.	System	References	Chemical str. of ILs	%IE/ Conc.	System	References
IL1: n=7; IL2: n=9; IL3: n=7; IL4: n=9	83.9%/200 ppm (IL4)	Carbon steel/1 M HCl	[3]	$[C_{16}M_1Im][Br]$, $[C_{16}M_2Im][Br]$	71.83%/200 ppm ($[(C_{16}M_2Im)][Br]$)	Mild steel/1 M HCl	[4]
MPIMI: n=1; BMIMI: n=2; HMIMI: n=3	93.1%/ 5×10^{-3} M (HMIMI)	Mild steel/1 M HCl	[5]	A: n=1; B: n=2; C: n=3	93.5%/ 15 mM	Mild steel/0.5 M HCl	[6]
(CPEPB)	91.67%/ 1×10^{-3} M	Carbon steel/1 M HCl	[7]	(G2IL): n = 2; (G3IL): n = 3; (G6IL): n = 6	93.0%/ 5×10^{-3} M (G6IL)	Carbon steel/1 M HCl	[8]
(DBImL), (DBImA)	88%/ 100 ppm (DBImL)	API 5LX52 steel /1 M HCl & H_2SO_4	[9]	(CTAB), (SDS)	90% (CTAB/90+ SDS/10)	Mil steel/2 M HCl	[10]
(IBMMBj[+Br-)	97.9%/ 6.84 $\times 10^{-4}$ M	Carbon steel/1 M HCl	[11]	$[C_4C_1im][FeCl4]$	74.89% (0.2ml water content)	Fe/1 M HCl	[12]

(continued)

TABLE 13.1 (Cont'd)

Chemical str. of ILs	%IE/ Conc.	System	References	Chemical str. of ILs	%IE/ Conc.	System	References
[EMIM] + [EtSO4]−, [EMIM] + [Ac]−, [BMIM] + [SCN]−, [BMIM] + [Ac]−, [BMIM] + [DCA]−	90%/ 500 ppm ([BMIM][DCA])	Mild steel /1 M HCl	[13]	[EMIM]+[BF4]−, [BDMIM]+[BF4]−, [C10MIM]+[BF4]−	98.08%/ 500 ppm ([C10MIM][BF4])	Mild steel/ 1 M HCl	[14, 15]
	86.7%/ 500 ppm [PDMIM][NTf2]	Carbon steel/1 M HCl	[16]	(EMIm Cl), (BMIm Cl), (BMIm PF6), (BMIm BF4), (BMIm Br), (HMIm Cl)	80%/ 3wt% (HMIm Cl)	Carbon steel/2 M HCl	[17]
(I), (II), (III), (IV)	80.9%/ 5 × 10⁻³ M (Comp. II)	Carbon steel/1 M HCl	[18]	[HMIM][TfO] [HMIM][BF4], [HMIM][PF6] & [HMIM][I]	81.16%/ 500 ppm ([HMIM][TfO])	Mild steel /1 M HCl	[19]
(ODA-TS), (OA-TS)	98.0%/ 0.341 mM (ODA-TS)	Mild steel/1 M HCl	[20]	(PPIB1), (PPIB4)	94.2%/ 1 × 10⁻² M (PPIB4)	C38 steel /1 M HCl	[21]
(MA1), (MA2)	97.6%/ 1 × 10⁻² M (MA2)	Mild steel/1 M HCl	[22]		—	Mild steel/1 M HCl & H2SO4	[23]

Structure	Efficiency / Conc.	Material / Medium	Ref.	Structure	Efficiency / Conc.	Material / Medium	Ref.
[Cbil][Cl] [Cbil][I] [Cbil][Ac]	95.1% / 1.59×10^{-3} M	Mild steel / 1 M HCl	[24]		83.43% / 1×10^{-3} M (s1)	Carbon steel / 1 M HCl	[25]
	96.59% / 17.91×10^{-4} M ([Chl][Ac])	Mild steel / 1 M HCl	[2]	(TDTB), (TDTM), (EMIM)(ESO4)	92.16% / 8×10^{-2} M (TDTM)	Mild steel / 1 M HCl	[26]
C_2-IMIC$_4$-S: R=C_2H_5 C_{10}-IMIC$_4$-S: R=$C_{10}H_{21}$	97.9% / 15 mM (C_{10}IMIC$_4$-S)	Carbon steel / 0.5 M HCl	[27]	[VAIMJ]: R= Me, [VPIMJ]: R= C_3H_7 [VBIMJ]: R= C_4H_9	99.4% / 5 mM ([VBIM]I)	X70 steel / 0.5 M HCl	[28]
[VAIM][PF6]: X= PF$_6$ [VAIM][BF4]: X= BF$_4$	86.94% / 0.8 mM ([VAIM][PF6])	Mild steel / 1 M HCl	[29]	[BMEB]+BF4	97.4% / 250 ppm	Mild steel / 1N HCl	[30]
(P1) R=CH_2CH_2–OH (P2) R=CH_2CH_3 (P3) R=CH_3	91.4% / 250 ppm (P1)	X-65 steel / 1 M HCl	[31]	IM-CDs	92.6% (1 M HCl) / 200 ppm	Q235 steel / 1 M HCl	[32]
[(HOC2)MIm]PF6: X= PF6 [(HOC2)MIm]NTf$_2$: X= NTF$_2$	75.13% / 16 mM [(HOC$_2$)MIm]PF$_6$	Mild steel / 1 M HCl	[33]	[bmim][Cl]: X= Cl [bmim][Ac]: X= CH_3OO [bmim][CF$_3$SO$_3$]: CF$_3$SO$_3$	97.15% / 8.67×10^{-4} M / [bmim][Ac]	Mild steel / 1 M HCl	[34]
TTA	85% / 100 ppm (TMA)	API-X52 Fe / 1 M HCl	[35]	[BsMIM][HSO4], [BsMIM][OTs]	NA	Fe, Ni, and 304 SS / water	[36]

(continued)

TABLE 13.1 (Cont'd)

Chemical str. of ILs	%IE/ Conc.	System	References	Chemical str. of ILs	%IE/ Conc.	System	References
IL	94.8%/ 1mM	carbon/ 0.5 M HCl	[37]	[BTBA]+[AlCl4]−	92.1%/ 400 ppm	Carbon steel/2N HCl	[38]
TDTB, TDTM (EMIM)(ESO4)	92.16%/ 8 × 10⁻² M (TDTB)	Mild steel/1 M HCl	[39]	VIPS, PVIPS	93.3%/ 500 ppm (PVIPS)	Carbon steel/0.5 M HCl	[40]
[PheME][Sac], [LeuME][Sac], [AlaME][Sac]	79.90%/100 ppm ([PheME][SacI])	Mild steel/1 M HCl	[41]	n=1: IPyr-C₂H₅; n=3: IPyr-C₄H₉	92.4%/1 × 10⁻³ M (IPyr-C₄H₇)	Mild steel/1 M HCl	[42]
Br⁻	92.36%/ 5 × 10⁻³ M	Cast iron/0.5M HCl	[43]		92%/ 400 ppm	Mild steel/1 M HCl	[44]
[NH₂emim]Br	94.16/10 mM ([NH₂emim]Br)	Mild steel/H₂S and HCl	[45]	[BTBA]+[FeCl4]−	99.5%/ 400 ppm	Carbon steel/1N HCl	[46]

the investigated ILs inhibit steel alloys corrosion in acidic solution of HCl by adsorbing on the metallic surface that follows the Langmuir adsorption isotherm model. Potentiodynamic polarization study showed that most of the ILs act as mixed-type corrosion inhibitors. However, some cathodic or anodic predominance has also been reported in few studies. EIS studies revealed that mostly ILs behave as interface-type corrosion inhibitors as their presence increases the resistance for the charge transfer process. The adsorption mechanism of corrosion inhibition was further supported by SEM, AFM, and FT-IR analyses.

13.4 Ionic liquids as corrosion inhibitors for iron alloys in H_2SO_4

Sulphuric acid based electrolytic media are extensively utilized. Generally, less concentrated H_2SO_4 solutions are used for academic research purposes and highly concentrated H_2SO_4 solutions are utilized for industrial purposes. Recently, various reports dealing with the anticorrosive effect of ILs, especially imidazolium-based ILs for steel alloys in sulfuric acid solutions are reported. Corrales-Luna and coworkers [47] described the anticorrosive effect of 1-ethyl 3-methylimidazolium thiocyanate, designated as (EMIM)+(SCN)−, for API X52 iron alloy in 0.5 M HCl and 0.5 M H_2SO_4 media. The inhibition effect of (EMIM)+(SCN)− was investigated using experimental and computational methods. The (EMIM)+(SCN)− exhibited the highest protection efficiency of 74.4% (at 100 ppm) and 82.9% (at 75 ppm) in 0.5 M HCl and 0.5M H_2SO_4 solutions, respectively. Experimental analyses revealed that inhibition efficiency of (EMIM)+(SCN)− enhances with its concentration whereas temperature exerts an inverse effect. The (EMIM)+(SCN)− becomes effective by adsorbing on the metallic surface that obeyes the Langmuir adsorption isotherm. Various surface investigation techniques such as XPS, AFM, SEM, and EDX were used to describe the adsorption behavior of the (EMIM)+(SCN)−. Experimental results were supported by computation studies carried out through MD simulations. Most of the ILs act as mixed-type corrosion inhibitors for steel alloys in sulfuric acid media. They become effective by adsorbing on the metallic surface that followed the Langmuir adsorption isotherm model. A brief summary of some recent studies on ILs as corrosion inhibitors for steel alloys in sulfuric acid solutions is given in Table 13.2.

13.5 Miscellaneous

ILs have also been used as corrosion inhibitors for other metallic alloys in the electrolytes other than HCl and H_2SO_4. One of such common electrolytes is NaCl solutions. The ILs are successfully used as corrosion inhibitors for aluminum, magnesium, zinc, copper, and other metals. A brief summary of some such recent reports is given in Table 13.3. It is important to mention that most of the evaluated ILs act as mixed-type and interface-type corrosion inhibitors. They become effective by adsorbing on the metallic surface following the Langmuir adsorption isotherm model.

TABLE 13.2 Chemical structures, abbreviations, nature of metals, and electrolytes of some major works reported on anticorrosive effect of ionic liquids for iron alloys in H_2SO_4 electrolytes.

Chemical str. of ILs	%IE/Conc.	System	References	Chemical str. of ILs	%IE/Conc.	System	References
$(EMIM)^+(SCN)^-$	77.4%/75 ppm (H_2SO_4)	API 5L X52 0.5 M/ H_2SO_4 & HCl	[47]	[DBIM+]I⁻: R=C_4H_9; [DPIM+]I⁻: R=C_3H_7	96%/100 ppm	API 5L X52/1 M H_2SO_4	[48]
[FBMIm]Br	99.48%/0.01 M	Mild steel/0.5 M H_2SO_4	[49]	LB104; R=C_4H_9 P8, R=$C_{10}H_{21}$ P10, R=$C_{14}H_{29}$ P14	99.8%/0.5 M (P8)	Mild steel/0.5 M H_2SO_4	[50]
CPA6, ESA8	83%/100 ppm (ESA8)	API-X60 Fe/1 M H_2SO_4	[51]	IL1: R=C_8H_{17}; IL2: R=$C_{10}H_{21}$; IL3: R=$C_{12}H_{25}$	98.4%/100 ppm (IL2)	304 SS Fe/0.5 M H_2SO_4	[52]
1-NpMe-TPC: Naphthyl; 4-MeOBz-TPB: -Ph-p-oMe	99.81%/10^{-2} M (1-NpMe-TPC)	Mild steel/0.5 M H_2SO_4	[53]	IPTPPB	99.52%/10^{-2} M	Mild steel/0.5 M H_2SO_4	[54]
[C_6]: n=1; [C_9]: n=3; [C_{12}]: n=6	83%/100 ppm ([C_{12}])	API X60 Fe/1 M H_2SO_4	[55]		88%/800 ppm	Steel/1 M H_2SO_4	[56]
(MTABr), (TOMABr)	85.7%/5 × 10^{-4} M	AISI 431 steel/2 M H_2SO_4	[57]	R= IL1:C_4H_9; IL2:C_8H_{17}; IL3:$C_{12}H_{25}$; IL4:$C_{18}H_{37}$; IL5: $C_{22}H_{45}$	94%/100 ppm (IL3)	Carbon steel/1 M H_2SO_4	[58]

Structure	Efficiency / Conc.	Material / Medium	Ref.	Structure	Efficiency / Conc.	Material / Medium	Ref.
([BsMIM]-[HSO4]), ([BsMIM][BF4])	86% / 4mM (2-MeHImn 4-OHCin)	Mild steel / 1 M H_2SO_4	[59]	(DDI), (TMA) (TML)	95% / 100 ppm (DDI)	API 5L X52 / 1 M H_2SO_4	[60]
(VImC4PF6) (VImC8PF6) (VImC12PF6) (VImC18PF6) (VImC22PF6)	97% / 100 ppm (VIm-$C_{12}PF_6$)	Carbon steel / 1 M H_2SO_4	[61]	(EOPC)	90.99% / 10^{-3} M	Carbon steel / 0.5 M H_2SO_4	[62]
IL	86% / 600 ppm	Mild steel / 1 M H_2SO_4	[63]		74.8% / 10 mM	Mild steel / 0.5 M H_2SO_4	[64]
[AEIM]Br: C_2H_5; [ABIM]Br: C_4H_9; [AHIM]Br: C_6H_{13}; [AOIM]Br: C_8H_{17}	97.5% / 5 mM ([AOIM]Br)	Carbon steel / 0.5 M H_2SO_4	[65]	ABTPPB	99.08% / 10^{-2} M	Mild steel / 0.5 M H_2SO_4	[66]
	83.9% / 75 ppm	API 5L X52 0.5 M / H_2SO_4	[67]	IL1: R=C_8H_{17}; IL2: R=$C_{10}H_{21}$; IL3: R=$C_{12}H_{25}$	95.9% / 100 ppm (IL2)	304 SS / Fe / 0.5 M H_2SO_4	[68]
(EMIM)+(SCN)-							

TABLE 13.3 Chemical structures, abbreviations, nature of metals, and electrolytes of some major works reported on anticorrosive effect of ionic liquids for other alloys in different electrolytes.

Chemical str. of ILs	%IE/Conc.	System	References	Chemical str. of ILs	%IE/Conc.	System	References
(pyrrolidinium triflate structure)	80.2%/800 ppm	Mild steel/3.5% NaCl	[69]	(imidazolinium structure with CH3, HO-phenyl)	86%/4mM	Mild steel/0.01 M NaCl	[59]
(DMICL) imidazolium Cl	98.5%/2 mM	Mild steel/NaCl (pH 3.8 & pH 6.8)	[70]	(long alkyl chain, Br; SDS structure with Na) (CTAB) (SDS)	84%/2.5 mM	Mild steel/3.5% NaCl	[71]
(BMIM-Cl)	Coating	Mild steel/3.5% NaCl	[72]	[EMIM][BF4], [BMIM][Otf], [EMIM][Otf]	NA	Carbon steel 1020/CO2	[73]
Coating (amine carboxylate structures)	87%/100 ppm	Carbon steel/production water	[74]	(TSIL) imidazolium triflate Cl	78.7%/100 ppm	Carbon steel/1 M HCl/open	[75]
TESFI (H3C-S+-CH3, F3C-S-N-S-CF3)	97.8%/120 ppm	304 SS Fe/5% sulfamic acid	[76]	[emim]-[Otf], [emim]-[DCA], [emim][acetate], [emim][tosylate]	88%/30 wt/wt	Fe/CO2 Capture system	[77]
AA: L-Histidine; L-Glutamic acid; L-Aspartic acid	NA	SS Fe/36% HCl + 5% Sb2O + 4% SnCl2	[78]	[beim]Br	91.29%/100 ppm	Copper/0.1 M Na2SO4	[79]
[DTP][NTf2] (n=7), [ITP][NTf2] (n=11), and [OTP][NTf2] (n=15).	86.4%/0.30 Mm ([OTP][NTf2])	AZ31B Mg/0.05 wt.% NaCl	[80]	MOBB (bis-methoxyphenyl imidazolium Br)	98.8%/0.5 mM (H2SO4)	6061 Al-15 vol% SiC (P)/0.1 M HCl&H2SO4	[81]

13.6 Summary and outlook

ILs are effective and relatively new class of environmental friendly class of corrosion inhibitors. They are associated with various advantageous properties such as high thermal and chemical stability, negligible vapor pressure, high solubility in polar and nonpolar media, etc. Different series of ILs including pyridinium, tetra-ammonium, imidazolium, and phosphonium based ILs are extensively used as corrosion inhibitors for different metals and alloys in various electrolytes. Literature investigation showed that imidazolium based ILs is most extensively investigated. Potentiodynamic polarization study showed that ILs become effective by blocking or adversely affecting both anodic as well as cathodic reactions and thereby act as mixed-type corrosion inhibitors. However, in few reports anodic or cathodic predominance of ILs has also been reported. EIS investigation showed that ILs adsorb at the interface of electrode (metal) and electrolyte and act as interface-type corrosion inhibitors. Such type adsorption of ILs on metallic surfaces results in the increase in the charge transfer resistance for the corrosion process. Adsorption behavior of corrosion inhibition using ILs is widely supported by SEM, AFM, EDX, XPS, UV-vis, and FT-IR methods. Adsorption of most of the ILs on metallic surface obeyed the Langmuir adsorption isotherm. Computational simulations especially using DFT, MD, and MC simulations are widely used to describe the adsorption behavior of ILs with the metallic surface.

13.7 Useful websites

https://en.wikipedia.org/wiki/Ionic_liquid
https://www.organic-chemistry.org/topics/ionic-liquids.shtm
https://www.sigmaaldrich.com/technical-documents/articles/chemfiles/ionic-liquids0.html

References

[1] N.V. Likhanova, M.A. Domínguez-Aguilar, O. Olivares-Xometl, N. Nava-Entzana, E. Arce, H. Dorantes, The effect of ionic liquids with imidazolium and pyridinium cations on the corrosion inhibition of mild steel in acidic environment, Corros. Sci. 52 (2010) 2088–2097.

[2] C. Verma, I. Obot, I. Bahadur, E.-S.M. Sherif, E.E. Ebenso, Choline based ionic liquids as sustainable corrosion inhibitors on mild steel surface in acidic medium: gravimetric, electrochemical, surface morphology, DFT and Monte Carlo simulation studies, Appl. Surf. Sci. 457 (2018) 134–149.

[3] M. Deyab, M. Zaky, M. Nessim, Inhibition of acid corrosion of carbon steel using four imidazolium tetrafluoroborates ionic liquids, J. Mol. Liq. 229 (2017) 396–404.

[4] N. Subasree, J.A. Selvi, Imidazolium based ionic liquid derivatives; synthesis and evaluation of inhibitory effect on mild steel corrosion in hydrochloric acid solution, Heliyon 6 (2020) e03498.

[5] F.A. Azeez, O.A. Al-Rashed, A.A. Nazeer, Controlling of mild-steel corrosion in acidic solution using environmentally friendly ionic liquid inhibitors: effect of alkyl chain, J. Mol. Liq. 265 (2018) 654–663.

[6] S. Cao, D. Liu, H. Ding, H. Lu, J. Gui, Towards understanding corrosion inhibition of sulfonate/carboxylate functionalized ionic liquids: an experimental and theoretical study, J. Colloid Interface Sci. 579 (2020) 315–329.

[7] A. Bousskri, A. Anejjar, M. Messali, R. Salghi, O. Benali, Y. Karzazi, S. Jodeh, M. Zougagh, E.E. Ebenso, B. Hammouti, Corrosion inhibition of carbon steel in aggressive acidic media with 1-(2-(4-chlorophenyl)-2-oxoethyl) pyridazinium bromide, J. Mol. Liq. 211 (2015) 1000–1008.

[8] S.M. Tawfik, Ionic liquids based gemini cationic surfactants as corrosion inhibitors for carbon steel in hydrochloric acid solution, J. Mol. Liq. 216 (2016) 624–635.

[9] I. Lozano, E. Mazario, C. Olivares-Xometl, N. Likhanova, P. Herrasti, Corrosion behaviour of API 5LX52 steel in HCl and H2SO4 media in the presence of 1, 3-dibencilimidazolio acetate and 1, 3-dibencilimidazolio dodecanoate ionic liquids as inhibitors, Mater. Chem. Phys. 147 (2014) 191–197.

[10] A. Yousefi, S. Javadian, J. Neshati, A new approach to study the synergistic inhibition effect of cationic and anionic surfactants on the corrosion of mild steel in HCl solution, Ind. Eng. Chem. Res. 53 (2014) 5475–5489.

[11] P. Kannan, J. Karthikeyan, P. Murugan, T.S. Rao, N. Rajendran, Corrosion inhibition effect of novel methyl benzimidazolium ionic liquid for carbon steel in HCl medium, J. Mol. Liq. 221 (2016) 368–380.

[12] S. Ullah, M.A. Bustam, A.M. Shariff, G. Gonfa, K. Izzat, Experimental and quantum study of corrosion of A36 mild steel towards 1-butyl-3-methylimidazolium tetrachloroferrate ionic liquid, Appl. Surf. Sci. 365 (2016) 76–83.

[13] S. Yesudass, L.O. Olasunkanmi, I. Bahadur, M.M. Kabanda, I. Obot, E.E. Ebenso, Experimental and theoretical studies on some selected ionic liquids with different cations/anions as corrosion inhibitors for mild steel in acidic medium, J. Taiwan Inst. Chem. Eng. 64 (2016) 252–268.

[14] Y. Sasikumar, A. Adekunle, L.O. Olasunkanmi, I. Bahadur, R. Baskar, M.M. Kabanda, I. Obot, E.E. Ebenso, Experimental, quantum chemical and Monte Carlo simulation studies on the corrosion inhibition of some alkyl imidazolium ionic liquids containing tetrafluoroborate anion on mild steel in acidic medium, J. Mol. Liq. 211 (2015) 105–118.

[15] X. Zhou, H. Yang, F. Wang, [BMIM] BF4 ionic liquids as effective inhibitor for carbon steel in alkaline chloride solution, Electrochim. Acta 56 (2011) 4268–4275.

[16] L.C. Murulana, A.K. Singh, S.K. Shukla, M.M. Kabanda, E.E. Ebenso, Experimental and quantum chemical studies of some bis (trifluoromethyl-sulfonyl) imide imidazolium-based ionic liquids as corrosion inhibitors for mild steel in hydrochloric acid solution, Ind. Eng. Chem. Res. 51 (2012) 13282–13299.

[17] A. Yousefi, S. Javadian, N. Dalir, J. Kakemam, J. Akbari, Imidazolium-based ionic liquids as modulators of corrosion inhibition of SDS on mild steel in hydrochloric acid solutions: experimental and theoretical studies, RSC Adv. 5 (2015) 11697–11713.

[18] B. Hammouti, S. Alamry, A. Alzahrani, Z. Moussa, M. Messali, M. Ibrahim, Corrosion inhibition of carbon steel by imidazolium and pyridinium cations ionic liquids in acidic environment, Port. Electrochim. Acta 29 (2011) 375–389.

[19] M.E. Mashuga, L.O. Olasunkanmi, A.S. Adekunle, S. Yesudass, M.M. Kabanda, E.E. Ebenso, Adsorption, thermodynamic and quantum chemical studies of 1-hexyl-3-methylimidazolium based ionic liquids as corrosion inhibitors for mild steel in HCl, Materials 8 (2015) 3607–3632.

[20] A.M. Atta, G.A. El-Mahdy, H.A. Al-Lohedan, A.R.O. Ezzat, A new green ionic liquid-based corrosion inhibitor for steel in acidic environments, Molecules 20 (2015) 11131–11153.

[21] A. Zarrouk, M. Messali, H. Zarrok, R. Salghi, A. Ali, B. Hammouti, S. Al-Deyab, F. Bentiss, Synthesis, characterization and comparative study of new functionalized imidazolium-based ionic liquids derivatives towards corrosion of C38 steel in molar hydrochloric acid, Int. J. Electrochem. Sci 7 (2012) 6998–7015.

[22] M. Messali, M. Asiri, A green ultrasound-assisted access to some new 1-benzyl-3-(4-phenoxybutyl) imidazolium-based ionic liquids derivatives—potential corrosion inhibitors of mild steel in acidic environment, J. Mater. Environ. Sci 5 (2013) 770–785.

[23] M. Ezhilarasi, B. Prabha, T. Santhi, Novel pyrazole based ionic liquid as a corrosion inhibitor for mild steel in acidic media, Chem. Sci. 4 (2015) 758–767.

[24] E. Kowsari, S. Arman, M. Shahini, H. Zandi, A. Ehsani, R. Naderi, A. PourghasemiHanza, M. Mehdipour, In situ synthesis, electrochemical and quantum chemical analysis of an amino acid-derived ionic liquid inhibitor for corrosion protection of mild steel in 1M HCl solution, Corros. Sci. 112 (2016) 73–85.

[25] F. El-Hajjaji, M. Messali, A. Aljuhani, M. Aouad, B. Hammouti, M. Belghiti, D. Chauhan, M. Quraishi, Pyridazinium-based ionic liquids as novel and green corrosion inhibitors of carbon steel in acid medium: electrochemical and molecular dynamics simulation studies, J. Mol. Liq. 249 (2018) 997–1008.

[26] A.A. Nkuna, E.D. Akpan, I. Obot, C. Verma, E.E. Ebenso, L.C. Murulana, Impact of selected ionic liquids on corrosion protection of mild steel in acidic medium: experimental and computational studies, J. Mol. Liq. 314 (2020) 113609.

[27] S. Cao, D. Liu, H. Ding, J. Wang, H. Lu, J. Gui, Task-specific ionic liquids as corrosion inhibitors on carbon steel in 0.5 M HCl solution: an experimental and theoretical study, Corros. Sci. 153 (2019) 301–313.

[28] L. Feng, S. Zhang, Y. Qiang, S. Xu, B. Tan, S. Chen, The synergistic corrosion inhibition study of different chain lengths ionic liquids as green inhibitors for X70 steel in acidic medium, Mater. Chem. Phys. 215 (2018) 229–241.

[29] Y. Guo, B. Xu, Y. Liu, W. Yang, X. Yin, Y. Chen, J. Le, Z. Chen, Corrosion inhibition properties of two imidazolium ionic liquids with hydrophilic tetrafluoroborate and hydrophobic hexafluorophosphate anions in acid medium, J. Ind. Eng. Chem. 56 (2017) 234–247.

[30] P. Kannan, T.S. Rao, N. Rajendran, Anti-corrosion behavior of benzimidazoliumtetrafluroborate ionic liquid in acid medium using electrochemical noise technique, J. Mol. Liq. 222 (2016) 586–595.

[31] H. El Sayed, S. Elsaeed, H. Ashour, E. Zaki, H. El Nagy, Novel acrylamide ionic liquids as anti-corrosion for X-65 steel dissolution in acid medium: adsorption, hydrogen evolution and mechanism, J. Mol. Struct. 1168 (2018) 106–114.

[32] D. Yang, Y. Ye, Y. Su, S. Liu, D. Gong, H. Zhao, Functionalization of citric acid-based carbon dots by imidazole toward novel green corrosion inhibitor for carbon steel, J. Cleaner Prod. 229 (2019) 180–192.

[33] Y. Guo, Z. Chen, Y. Zuo, Y. Chen, W. Yang, B. Xu, Ionic liquids with two typical hydrophobic anions as acidic corrosion inhibitors, J. Mol. Liq. 269 (2018) 886–895.

[34] C. Verma, L.O. Olasunkanmi, I. Bahadur, H. Lgaz, M. Quraishi, J. Haque, E.-S.M. Sherif, E.E. Ebenso, Experimental, density functional theory and molecular dynamics supported adsorption behavior of environmental benign imidazolium based ionic liquids on mild steel surface in acidic medium, J. Mol. Liq. 273 (2019) 1–15.

[35] P. Arellanes-Lozada, O. Olivares-Xometl, N.V. Likhanova, I.V. Lijanova, J.R. Vargas-García, R.E. Hernández-Ramírez, Adsorption and performance of ammonium-based ionic liquids as corrosion inhibitors of steel, J. Mol. Liq. 265 (2018) 151–163.

[36] Y. Ma, F. Han, Z. Li, C. Xia, Corrosion behavior of metallic materials in acidic-functionalized ionic liquids, ACS Sustainable Chemistry & Engineering 4 (2016) 633–639.

[37] S. Cao, D. Liu, H. Ding, J. Wang, H. Lu, J. Gui, Corrosion inhibition effects of a novel ionic liquid with and without potassium iodide for carbon steel in 0.5 M HCl solution: an experimental study and theoretical calculation, J. Mol. Liq. 275 (2019) 729–740.

[38] P. Kannan, A. Varghese, K. Palanisamy, A.S. Abousalem, Evaluating prolonged corrosion inhibition performance of benzyltributylammonium tetrachloroaluminate ionic liquid using electrochemical analysis and Monte Carlo simulation, J. Mol. Liq. 297 (2020) 111855.

[39] A.A. Nkuna, E.D. Akpan, I. Obot, C. Verma, E.E. Ebenso, L.C. Murulana, Impact of selected ionic liquids on corrosion protection of mild steel in acidic medium: Experimental and computational studies, J. Mol. Liq. 314 (2020) 113609.

[40] J. Wang, D. Liu, S. Cao, S. Pan, H. Luo, T. Wang, H. Ding, B.B. Mamba, J. Gui, Inhibition effect of monomeric/polymerized imidazole zwitterions as corrosion inhibitors for carbon steel in acid medium, J. Mol. Liq. 312 (2020) 113436.

[41] R. Aslam, M. Mobin, I.B. Obot, A.H. Alamri, Ionic liquids derived from α-amino acid ester salts as potent green corrosion inhibitors for mild steel in 1M HCl, J. Mol. Liq. 318 (2020) 113982.

[42] F. El-Hajjaji, E. Ech-chihbi, N. Rezki, F. Benhiba, M. Taleb, D.S. Chauhan, M. Quraishi, Electrochemical and theoretical insights on the adsorption and corrosion inhibition of novel pyridinium-derived ionic liquids for mild steel in 1 M HCl, J. Mol. Liq. 314 (2020) 113737.

[43] S.M. Ali, K.M. Emran, M. Messali, Improved protection performance of modified sol-gel coatings with pyridinium-based ionic liquid for cast iron corrosion in 0.5 M HCL solution, Prog. Org. Coat. 130 (2019) 226–234.

[44] E.K. Ardakani, E. Kowsari, A. Ehsani, Imidazolium-derived polymeric ionic liquid as a green inhibitor for corrosion inhibition of mild steel in 1.0 M HCl: Experimental and computational study, Colloids Surf. A 586 (2020) 124195.

[45] F. Cui, Y. Ni, J. Jiang, L. Ni, Z. Wang, Experimental and theoretical studies of five imidazolium-based ionic liquids as corrosion inhibitors for mild steel in H2S and HCl solutions, Chem. Eng. Commun. (2020) 1–14, doi:https://doi.org/10.1080/00986445.2020.1802257.

[46] P. Kannan, A. Varghese, A.T. Mathew, K. Palanisamy, A.S. Abousalem, M. Kalaiyarasan, N. Rajendran, Exploring the inhibition performance of tetrachloroferrate ionic liquid in acid environment using scanning electrochemical microscope and theoretical approaches, Surfaces and Interfaces (2020) 100594.

[47] M. Corrales-Luna, T.Le Manh, M. Romero-Romo, M. Palomar-Pardavé, E.M. Arce-Estrada, 1-Ethyl 3-methylimidazolium thiocyanate ionic liquid as corrosion inhibitor of API 5L X52 steel in H2SO4 and HCl media, Corros. Sci. 153 (2019) 85–99.

[48] P. Arellanes-Lozada, V. Díaz-Jiménez, H. Hernández-Cocoletzi, N. Nava, O. Olivares-Xometl, N.V. Likhanova, Corrosion inhibition properties of iodide ionic liquids for API 5L X52 steel in acid medium, Corros. Sci. 175 (2020) 108888.

[49] P.D. Pancharatna, S. Lata, G. Singh, Imidazolium based ionic liquid as an efficient and green corrosion constraint for mild steel at acidic pH levels, J. Mol. Liq. 278 (2019) 467–476.

[50] Y. Li, S. Zhang, Q. Ding, B. Qin, L. Hu, Versatile 4, 6-dimethyl-2-mercaptopyrimidine based ionic liquids as high-performance corrosion inhibitors and lubricants, J. Mol. Liq. 284 (2019) 577–585.

[51] N.V. Likhanova, P. Arellanes-Lozada, O. Olivares-Xometl, H. Hernández-Cocoletzi, I.V. Lijanova, J. Arriola-Morales, J. Castellanos-Aguila, Effect of organic anions on ionic liquids as corrosion inhibitors of steel in sulfuric acid solution, J. Mol. Liq. 279 (2019) 267–278.

[52] M. Zaky, M. Nessim, M. Deyab, Synthesis of new ionic liquids based on dicationic imidazolium and their anti-corrosion performances, J. Mol. Liq. 290 (2019) 111230.

[53] M. Goyal, S. Kumar, I. Bahadur, E.E. Ebenso, H. Lgaz, I.-M. Chung, Interfacial adsorption behavior of quaternary phosphonium based ionic liquids on metal-electrolyte interface: electrochemical, surface characterization and computational approaches, J. Mol. Liq. 298 (2020) 111995.

[54] M. Goyal, H. Vashisht, A. Kumar, S. Kumar, I. Bahadur, F. Benhiba, A. Zarrouk, Isopentyltriphenylphosphonium bromideionic liquid as a newly effective corrosion inhibitor on metal-electrolyte interface in acidic medium: Experimental, surface morphological (SEM-EDX & AFM) and computational analysis, J. Mol. Liq. 316 (2020) 113838.

[55] O. Olivares-Xometl, I.V. Lijanova, N.V. Likhanova, P. Arellanes-Lozada, H. Hernández-Cocoletzi, J. Arriola-Morales, Theoretical and experimental study of the anion carboxylate in quaternary-ammonium-derived ionic liquids for inhibiting the corrosion of API X60 steel in 1 M H_2SO_4, J. Mol. Liq. 318 (2020) 114075.

[56] A. Jannat, R. Naderi, E. Kowsari, H. Zandi, M. Saybani, R. Safari, A. Ehsani, Electrochemical techniques and quantum chemical analysis as tools to study effect of a dicationic ionic liquid on steel behavior in H_2SO_4, J. Taiwan Inst. Chem. Eng. 99 (2019) 18–28.

[57] R. Fuchs-Godec, The erosion–corrosion inhibition of AISI 431 martensitic stainless steel in 2.0 M H_2SO_4 solution using N-alkyl quaternary ammonium salts as inhibitors, Ind. Eng. Chem. Res. 49 (2010) 6407–6415.

[58] D. Guzmán-Lucero, O. Olivares-Xometl, R. Martínez-Palou, N.V. Likhanova, M.A. Domínguez-Aguilar, V. Garibay-Febles, Synthesis of selected vinylimidazolium ionic liquids and their effectiveness as corrosion inhibitors for carbon steel in aqueous sulfuric acid, Ind. Eng. Chem. Res. 50 (2011) 7129–7140.

[59] A.L. Chong, J.I. Mardel, D.R. MacFarlane, M. Forsyth, A.E. Somers, Synergistic corrosion inhibition of mild steel in aqueous chloride solutions by an imidazolinium carboxylate salt, ACS Sustainable Chemistry & Engineering 4 (2016) 1746–1755.

[60] O. Olivares-Xometl, C. López-Aguilar, P. Herrastí-González, N.V. Likhanova, I. Lijanova, R. Martínez-Palou, J.A. Rivera-Márquez, Adsorption and corrosion inhibition performance by three new ionic liquids on API 5L X52 steel surface in acid media, Ind. Eng. Chem. Res. 53 (2014) 9534–9543.

[61] N.V. Likhanova, O. Olivares-Xometl, D. Guzmán-Lucero, M.A. Domínguez-Aguilar, N. Nava, M. Corrales-Luna, M.C. Mendoza, Corrosion inhibition of carbon steel in acid environment by imidazolium ionic liquids containing vinyl-hexafluorophosphate as anion, Int. J. Electrochem. Sci 6 (2011) 4514–4536.

[62] O. Benali, O. Cherkaoui, A. Lallam, Adsorption and corrosion inhibition of new synthesized pyridazinium-based ionic liquid on carbon steel in 0.5 MH_2so_4, J. Mater. Environ. Sci 6 (2015) 598–606.

[63] A.P. Hanza, R. Naderi, E. Kowsari, M. Sayebani, Corrosion behavior of mild steel in H_2SO_4 solution with 1, 4-di [1′-methylene-3′-methyl imidazolium bromide]-benzene as an ionic liquid, Corros. Sci. 107 (2016) 96–106.

[64] X. Zheng, S. Zhang, M. Gong, W. Li, Experimental and theoretical study on the corrosion inhibition of mild steel by 1-octyl-3-methylimidazolium L-prolinate in sulfuric acid solution, Ind. Eng. Chem. Res. 53 (2014) 16349–16358.

[65] Y. Qiang, S. Zhang, L. Guo, X. Zheng, B. Xiang, S. Chen, Experimental and theoretical studies of four allyl imidazolium-based ionic liquids as green inhibitors for copper corrosion in sulfuric acid, Corros. Sci. 119 (2017) 68–78.

[66] M. Goyal, H. Vashisht, S. Kumar, I. Bahadur, Anti-corrosion performance of eco-friendly inhibitor (2-aminobenzyl) triphenylphosphonium bromide ionic liquid on mild steel in 0.5 M sulfuric acid, J. Mol. Liq. 261 (2018) 162–173.

[67] M.C. Luna, T.Le Manh, R.C. Sierra, J.M. Flores, L.L. Rojas, E.A. Estrada, Study of corrosion behavior of API 5L X52 steel in sulfuric acid in the presence of ionic liquid 1-ethyl 3-methylimidazolium thiocyanate as corrosion inhibitor, J. Mol. Liq. 289 (2019) 111106.

[68] M. Nessim, M. Zaky, M. Deyab, Three new gemini ionic liquids: synthesis, characterizations and anticorrosion applications, J. Mol. Liq. 266 (2018) 703–710.

[69] A. El-Shamy, K. Zakaria, M. Abbas, S.Z. El Abedin, Anti-bacterial and anti-corrosion effects of the ionic liquid 1-butyl-1-methylpyrrolidinium trifluoromethylsulfonate, J. Mol. Liq. 211 (2015) 363–369.

[70] D. Yang, M. Zhang, J. Zheng, H. Castaneda, Corrosion inhibition of mild steel by an imidazolium ionic liquid compound: the effect of pH and surface pre-corrosion, RSC Adv. 5 (2015) 95160–95170.

[71] A. Yousefi, S. Javadian, M. Sharifi, N. Dalir, A. Motaee, An experimental and theoretical study of biodegradable gemini surfactants and surfactant/carbon nanotubes (CNTs) mixtures as new corrosion inhibitor, Journal of Bio-and Tribo-Corrosion 5 (2019) 82.

[72] A.K. Dermani, E. Kowsari, B. Ramezanzadeh, R. Amini, Utilizing imidazole based ionic liquid as an environmentally friendly process for enhancement of the epoxy coating/graphene oxide composite corrosion resistance, J. Ind. Eng. Chem. 79 (2019) 353–363.

[73] A. Acidi, M. Hasib-ur-Rahman, F. Larachi, A. Abbaci, Ionic liquids [EMIM][BF 4], [EMIM][Otf] and [BMIM][Otf] as corrosion inhibitors for CO 2 capture applications, Korean J. Chem. Eng. 31 (2014) 1043–1048.

[74] M. Ontiveros-Rosales, O. Olivares-Xometl, N.V. Likhanova, I.V. Lijanova, D. Guzman-Lucero, M.D.C. Mendoza-Herrera, Use of the ionic liquid trioctylmethyl ammonium dodecanedioate as a corrosion inhibitor of steel in production water, Res. Chem. Intermed. 43 (2017) 641–660.

[75] E. Kowsari, M. Payami, R. Amini, B. Ramezanzadeh, M. Javanbakht, Task-specific ionic liquid as a new green inhibitor of mild steel corrosion, Appl. Surf. Sci. 289 (2014) 478–486.

[76] M. Deyab, Sulfonium-based ionic liquid as an anticorrosive agent for thermal desalination units, J. Mol. Liq. 296 (2019) 111742.

[77] M. Hasib-ur-Rahman, F. Larachi, Prospects of using room-temperature ionic liquids as corrosion inhibitors in aqueous ethanolamine-based CO2 capture solvents, Ind. Eng. Chem. Res. 52 (2013) 17682–17685.

[78] A.M. Sadanandan, P.K. Khatri, R. Saxena, S.L. Jain, Guanidine based amino acid derived task specific ionic liquids as noncorrosive lubricant additives for tribological performance, J. Mol. Liq. 313 (2020) 113527.

[79] G. Vastag, A. Shaban, M. Vraneš, A. Tot, S. Belić, S. Gadžurić, Influence of the N-3 alkyl chain length on improving inhibition properties of imidazolium-based ionic liquids on copper corrosion, J. Mol. Liq. 264 (2018) 526–533.

[80] H. Su, L. Wang, Y. Wu, Y. Zhang, J. Zhang, Insight into inhibition behavior of novel ionic liquids for magnesium alloy in NaCl solution: Experimental and theoretical investigation, Corros. Sci. 165 (2020) 108410.

[81] S.K. Shetty, A.N. Shetty, Eco-friendly benzimidazolium based ionic liquid as a corrosion inhibitor for aluminum alloy composite in acidic media, J. Mol. Liq. 225 (2017) 426–438.

Green corrosion inhibitors from one step multicomponent reactions

14.1 Multicomponent reactions as a green synthetic strategy

It is well-established that synthetic organic compounds are the most effective and economic corrosion inhibitors. Most of the traditional corrosion inhibitors are synthesized using multistep reactions that are usually associated with several disadvantages including being expensive, tedious, and time-consuming. More so, because of their multistep, traditional syntheses are connected with the use and discharge of huge amounts of environmentally malignant and expensive solvents, chemicals, and catalysts in the surrounding environment. These undesirable species adversely affect the aquatic and soil life along with polluting the environment. Because of the increasing ecological awareness and strict environmental regulations, use of compounds derived through traditional multistep reactions is highly restricted. In view of this one step multicomponent reaction (MCR) has emerged as one of the powerful environmental friendly strategies for the synthesis of organic compounds useful for various industrial and biological applications including as corrosion inhibitors.

MCRs define a chemical transformation in which three or more reactant molecules react/combine together to form a product [1–3]. The MCRs have been recognized for over 150 years. Synthesis of α-aminocyanide by the reaction of benzaldehyde, ammonia, and hydrogen cyanide was the first document of MCR. This reaction was reported by Laurent and Gerhardt in 1838. After that in 1950, Strecker proposed a common synthetic scheme for α-aminocyanides using aldehyde, ammonia, and hydrogen cyanide. Later on, different series of organic compounds, especially heterocyclic compounds such pyridine, imidazole, quinoline, isoquinoline, indole, etc., are effectively synthesized using MCRs. Some of the recently used MCRs include organiboran, free radical, isocyanides, and metal-mediated three or more component reactions. Various named-reactions including alkyne trimerisation and polymerization, Biginelli, Gewald, Bucherer–Bergs, Grieco three-component, Hantzsch, Kabachnik–Fields, Mannich, Passerini, Pauson–Khand, Petasis, Strecker, Ugi, Asinger, etc., are the common MCRs.

The MCRs have several advantages over the traditional multistep reactions. Some of the major advantages of MCRs include:

(i) Ease and rapid access to large libraries of organic/heterocyclic compounds.
(ii) Low waste production.

(iii) Excellent atom economy and experimental simplicity.
(iv) Lower number of works-up steps and simplified purification.
(v) Excellent regio- and chemo-selectivity.
(vi) Require typical solvents like water, alcohols, amines, etc.
(vii) Useful for the synthesis of simple to highly complex molecules.
(viii) Utilize simple, cheaply, and readily available starting materials/chemicals.
(ix) Facile automation.
(x) Low adverse impact on the environment.
(xi) Encouraging economic factors: cost-effective.
(xii) Resource and time-saving.

14.2 Corrosion inhibitors derived from multicomponent reactions

Literature study shows that a large number of organic compounds derived through MCRs serve as effective corrosion inhibitors for different metals and alloys in various electrolytic solutions. Generally, these compounds become effective by adsorbing on the metallic surface using their electron-rich polar functional groups such as hydroxyl (–OH), 1°-amine (–NH$_2$), 2°-amine (>NH), 3°-amine (>N–), methoxy (–OH), nitro (–NO$_2$), nitrile (–CN), ester (–COOR), acid halide (–COX), amide (–CONH$_2$), ether (–O–), carbonyl (>C=O), thio-carbonyl (>C=S), imine (>C=N–), etc., and multiple (double and triple) bonds. Generally, polar substituents enhance inhibition efficiency of their parent compounds by acting as adsorption centers as well as enhancing their solubility in the polar electrolytes. In aromatic substituted compounds effect of substituents on the corrosion inhibition efficiency of organic compounds can be derived through their Hammett substituent constant (σ) values. On the other hand, the effect of substituent on the corrosion inhibition effect of aliphatic substituted compounds (cyclic or acyclic) can be determined using their Taft constant (σ^*) values [4]. Hammett equation can be presented as [4–6]:

$$\log = \frac{K_R}{K_H} = \rho\sigma \tag{14.1}$$

$$\log = \frac{1-\eta\%_R}{1-\eta\%_H} = \rho\sigma \tag{14.2}$$

$$\log = \frac{\eta\%_R}{\eta\%_H} = \log\frac{C_{rH}}{C_{rR}} = \rho\sigma - \log\frac{\theta_R}{\theta_H} \tag{14.3}$$

where K, $\eta\%$, and θ are the equilibrium constant, inhibition efficiency and surface coverage, respectively. The R represents the substituent present in the molecular structure of the inhibitor molecule. σ is commonly known as Hammett constant and it describes the total electronic effect of the substituents on the metal-inhibitor bondings. ρ is the reaction parameter which magnitude mainly depends on the nature of metal-inhibitor interactions. The effect of substituents on the corrosion inhibition effect of organic compounds is widely investigated. Generally, electron-donating substituents including methyl (–CH$_3$), methoxy (–OCH$_3$), 1°-amine/amino (–NH$_2$), 2°-amine (>NH), 3°-amine (>N–) and hydroxyl (–OH) substituents increase electron density at the donor sites (active centers). Therefore, electron-donating substituents increase the inhibition efficiency and converse is true for the electron-withdrawing substituents. However, few exceptions have also been reported.

Various experimental methods including weight loss, electrochemical (OCP, LPR, EIS, and PDP) surface scanning electron microscopy (SEM), energy dispersive X-ray (EDX), atomic force microscopy (AFM), X-Ray Diffraction (XRD), X-ray photoelectron spectroscopy (XPS), etc. are widely used to describe the nature of organic corrosion inhibitors on the metallic surface. These techniques are also useful to describe the adsorption behavior of organic corrosion inhibitors over the metallic surface. Computational methods such as DFT and MD/MC simulations are used to describe the adsorption behavior of organic corrosion inhibitors with the metallic surface. Using computational modelings, adsorption behavior of the organic compounds can be described through various theoretical parameters including energies of frontier molecular orbitals (E_{LUMO} and E_{HOMO}), energy band gap (ΔE), softness (σ), fraction of electron transfer (ΔN), harness (η), electronegativity (χ), nucleophilic (ω), and electrophilic (ψ) sites, dipole moment (μ), binding energy (E_{bin}), interaction energy (E_{int}), and adsorption energy (E_{ads}) [7,8]. The spontaneous or non-spontaneous nature of metal and inhibitor molecule interaction can be experimentally derived using the value of standard Gibb's free energy of adsorption (G^0_{ads}). Obviously, negative value of G^0_{ads} is consistent with the spontaneous interaction/adsorption whereas converse is true for its positive value. It is also important to mention that G^0_{ads} value of -20 kJ mol^{-1} or less negative is consistent with the physisorption (electrostatic interactions) and its value of -40 kJ mol^{-1} or more negative is associated with chemisorption (charge sharing). The literature study revealed that in most of the studies values of G^0_{ads} range in between -20 kJ mol^{-1} and -40 kJ mol^{-1} that indicate that interaction of organic corrosion inhibitors with metallic surface mostly involve the physiochemisorption, i.e., mixed-type adsorption mechanism.

14.3 Case studies: green corrosion inhibitors derived from multicomponent reactions

Organic compounds, especially heterocyclic compounds derived from one step MCRs, are extensively used as corrosion inhibitors for different metals and alloys in various electrolytic solutions. Synthetic schemes, nature of metal and electrolyte, highest efficiency and optimum concentration of some organic corrosion inhibitors derived through MCRs are presented in Table 14.1. It is important to mention that most of the organic compounds derived from MCRs behave as mixed-type corrosion inhibitors, i.e., they suppress both anodic and cathodic half-cell reactions. However, in few reports some cathodic or anodic predominance have also been reported. Through EIS analysis, it has been extensively reported that organic compounds derived from MCRs act as interface-type corrosion inhibitors as they become effective by adsorbing at the interface of metal and electrolyte. Mostly, adsorption of organic corrosion inhibitors over the metallic surface obeys the Langmuir adsorption isotherm. Adsorption behavior of organic compounds is extensively reported by surface analyses especially through scanning electron microscope (SEM), EDX, AFM, XRD, and XPS analyses. Different authors employed computational analyses, especially DFT and MD/MC simulations to support their experimental results.

Haque et al. [9] reported the synthesis and corrosion inhibition effect of two pyrimidine derivatives differing in the nature of substituents for N80 steel in 15% HCl solution. Both the pyrimidine derivatives (PPs) were derived through MCRs in high yield and showed

TABLE 14.1 Synthetic schemes, nature of metal and electrolyte, highest efficiency, and optimum concentration of some organic corrosion inhibitors derived through multicomponent reactions.

Synthetic scheme(s)	%EI & Conc.	Metal/electrolyte	References
[scheme with thiobarbituric acid + Ar-CHO + thiourea → pyrimidine product; PP-1= -CH=CH-Ph; PP-2= -C$_6$H$_5$OH; C_2H_5OH, HCl]	89.1%/ (PP-1)	N80/15% HCl	[9]
[scheme with amino sugar + p-R-benzaldehyde + barbituric acid → fused pyrimidine product; CARB-1: R=-H, CARB-2: R=-NO$_2$, CARB-3: R=-CH$_3$, CARB-4: R=-OH; PTSA (0.1g), 80°C, 12hrs]	96.52%/7.41 × 10^{-5} M (CARB-4)	Mild steel/1M HCl	[10]
[scheme with R-C$_6$H$_4$-CHO + ethyl acetoacetate + ethyl cyanoacetate + H$_2$N-NH$_2$ → pyrano-pyrazole; EPP-1=-CH$_3$, EPP-2=-H, EPP-3=-NO$_2$; TEA, H$_2$O]	98.8%/100 mg L^{-1} (EPP-1)	Mild steel/1M HCl	[11]
[scheme with malononitrile + R-C$_6$H$_4$-CHO + NaN$_3$ → tetrazole-styryl product; PTR: R=-H, NTR: R=-NO$_2$, MTR: R=-OH; H$_2$O, 500C]	98.69%/40 mg L^{-1} (HTA)	Mild steel/1M HCl	[12]
[scheme with resorcinol + R-C$_6$H$_4$-CHO + malononitrile → quinoline product; Q-1: R=-H, Q-2: R=-CH$_3$, Q-3: R=-OCH$_3$, Q-4: R=-NMe$_2$; H$_2$O, NH$_4$OAc, MW, 6-10 min]	98.09%/150 mg L^{-1}	Mild steel/1M HCl	[13]
[scheme with succinimide + 2-aminopyridine + 4-pyridinecarboxaldehyde → SAP; C$_2$H$_5$OH, stirr, 50°C, 2hrs]	94.5%/200 ppm	Mild steel/1M HCl	[14]
[scheme with 2,5-dihydroxyacetophenone + R-C$_6$H$_4$-CHO + malononitrile → fused chromene product; N-1: R=-H, N-2: R=-CH$_3$, N-3: R=-OCH$_3$; Silica Gel, H$_2$O, 80°C]	98.09%/6.54 × 10^{-5} M (N-3)	Mild steel/1M HCl	[15]

Synthetic scheme(s)	%EI & Conc.	Metal/electrolyte	References
(scheme with glucose, aniline derivative, and barbituric acid; PTSA (0.1g), 80°C, 12 hrs; GPH-1: R=-H, GPH-2: R=-CH3, GPH-3: R=-OCH3)	97.82% 10.15×10^{-5} M (GPH-3)	Mild steel/1M HCl	[16]
(scheme with R-C6H4-CHO, malononitrile, aniline; L-proline (20%), H2O, Reflux; AAC-1: R=-NO2, AAC-2: R=-H, AAC-3: R=-OH)	96.95%/50 ppm (AAC-3)	Mild steel/1M HCl	[17]
(scheme with substituted benzaldehyde, barbituric acid, aniline; L-proline (20%), H2O, Reflux; APQD-1: R1=-H, R2=-NO2, APQD-2: R1=-H, R2=-H, APQD-3: R1=-H, R2=-OH, APQD-4: R1=-OH, R2=-OH)	98.30% at 20 mg L^{-1} (APQD-4)	Mild steel/1M HCl	[18]
(scheme with thiophenol, malononitrile, salicylaldehyde derivative; Et3N, C2H5OH, Reflux; DHPC-1: R=-NO2, DHPC-2: R=-H, DHPC-3: R=-OH)	95.69%/10.22×10^{-5} M (DHPC-3)	Mild steel/1M HCl	[19]
(scheme with R-CHO, 4-chloroaniline, triethyl phosphite; US, neat, RT, 20-45s; APCl-1: R=-Ph, APCl-2: R=-C6H4-p-OCH3, APCl-3: R=-CH=CH-Ph)	96.90/564×10^{-6} M (APCl-3)	Mild steel/1M HCl	[20]
(scheme with benzaldehyde, barbituric acid, malononitrile; L-Proline (5 mol%), EtOH, RT; CU-1: R=-H, CU-2: R=-NO2, CU-3: R=-OCH3, CU-4: R=-CH3)	96.1%/450 ppm/14.4×10^{-4} M (CU-3)	Mild steel/1M HCl	[21]
(scheme with benzaldehyde, thiophenol, malononitrile; H3BO3, CTAB, H2O, US; ADPT-I: R=-OCH3, ADPT-II: R=-H, ADPT-III: R=-NO2)	97.6%/1.22 mmol L^{-1} (ADTP-1)	Mild steel/1M HCl	[22]

(continued)

TABLE 14.1 (Cont'd)

Synthetic scheme(s)	%EI & Conc.	Metal/electrolyte	References
NaN₃ + R₂-CH₂-X + R₁-CH=CH-H → triazole, Cu(I), R1=R, Ph; R2=-R, Ph; X=-Cl, -Br, -I	95.9%/10 ppm	API 5 L X52 steel/1M HCl	[23]
2-naphthol + R-C₆H₄-CHO + H₂N-CO-NH₂ → CPHU-type product, H₃PW₁₂O₄₀, 100°C, Et₄NCl; CPHU: R=-Cl; HNPU: R=-H; HNMU: R=-OMe	90%/8 ppm (CPHU)	Mild steel/0.5M HCl	[24]
Substituted benzaldehyde + nitroalkene + malononitrile → ANTD product, Piperidine, RT; ANTD-1: R₁=R₂=-H; ANTD-2: R₁=-H, R₂=-OMe; ANTD-3: R₁=-OH, R₂=-OMe	98.96%/25 mg L⁻¹ (ANTD-3)	Mild steel/1M HCl	[25]
Isatin + HCHO + thiosemicarbazide + morpholine → 5-BMOH, EtOH; 1a: R=-H; 1b: R=-NO₂; 1c: -Br	94.67%/200 ppm (5-BMOH)	Mild steel/1M HCl	[26]
Pyridine-4-CHO + urea + R-H → UPP-type, EtOH, 10-12h, 65°C; UPyP R= pyrrolidinyl; UPP R= piperidinyl; UMP R= morpholinyl	86%/200 ppm (UPP)	Mild steel/1M HCl	[27]
2-aminopyridine + R-C₆H₄-CHO + CH₂(CN)₂ → ANC product, VB1/H₂O/90°C; ANC-1: R=4-OCH₃; ANC-2: R=4-CH₃; ANC-3: R=4-NO₂	93%/200 ppm (ANC-1)	N80 steel/15% HCl	[28]
4-R-C₆H₄-CHO + C₆H₅-COCH₃ + CH₂(CN)₂ → biphenyl product, NH₄OAc, C₂H₅OH; ATN: R=-CH₃; AMN: R=-OCH₃	97.14%/0.33 mM (AMN)	Mild steel/1M HCl	[29]

Synthetic scheme(s)	%EI & Conc.	Metal/electrolyte	References
ArCHO + RCOCH$_3$ + CH$_2$(CN)$_2$ → (NH$_4$OAc, MW) aminocyanopyridine; AMP: R=-C$_6$H$_5$; ADP: R=-2,4-OH-C$_6$H$_3$; Ar=4-OMe-C$_6$H$_4$-	90.3%/200 ppm (ADP)	N80/15% HCl	[30]
4-R-C$_6$H$_4$-CHO + CH$_2$(CN)$_2$ → (NH(CH$_2$CH$_3$)$_2$, CH$_3$OH, H$_2$O, r.t.) pyridine derivative; PC-1: R=-H; PC-2: R=-CH$_3$; PC-3: R=-OCH$_3$	97.4%/400 mg L^{-1} (PC-1)	Mild steel/1M HCl	[31]
HCHO + R-CH(NH$_2$)-COOH + OHC-CHO → (HOAc, H$_2$O) AIZs; AIZ-1: R=-H; AIZ-2: R=-CH$_3$; AIZ-3: R=-CH$_2$-C$_6$H$_5$	96.08%/0.55 mM/200 ppm (AIZ-3)	Mild steel/1M HCl	[32]
barbituric/thiobarbituric acid + 4-R-C$_6$H$_4$-CHO + H$_2$N-C(X)-NH$_2$ → (H$_2$O, MW) pyrimidopyrimidine; PPD-1: R=-OCH$_3$, X=-S; PPD-2: R=-H, X=-S; PPD-3: R=-OCH$_3$, X=-O; PPD-4: R=-H, X=-O	97.1%/400 ppm (PDP-1)	Mild steel/1M HCl	[33]
C$_6$H$_5$-XH + NC-CH$_2$-CN + salicylaldehyde → (Et$_3$N, C$_2$H$_5$OH, Reflux); PCC-1: X=-S; PCC-2: X=-O	92.4%/200 mg/L (PPC-1)	N-80 steel/15% HCl	[34]
NH$_2$-NH-C(S)-NH$_2$ + RNCS → (50% EtOH/D) RNH-C(S)-NH-NH-C(S)-NH$_2$; Inh I: R=-Ph; Inh II: R=-C$_6$H$_4$-p-OCH$_3$; Inh III: R=-C6H4-p-Cl	95.6%/200 ppm (BMBD)	Cu/3.5% NaCl & 1M HCl	[35]
4-OCH$_3$-C$_6$H$_4$-CHO + H$_2$N-C(S)-NH$_2$ + CH$_3$COCH$_2$COOC$_2$H$_5$ → (TBAB, Solvent free, MWI, 3-4 min) dihydropyrimidinethione	83.33%/500 ppm	Al/1M HCl	[36]

reasonably good corrosion inhibition efficiencies at relatively low concentration. Gravimetric analysis showed that inhibition efficiency of PPs increases on increasing the concentration. Analyses showed that PP-1 and PP-2 exhibited highest protection efficiency (at 250 mg L^{-1}) of 89.1% and 73.1%, respectively. Experimental analyses further showed that PPs become effective by adsorbing on the metallic surface and their adsorption mechanism followed the Langmuir adsorption isotherm. Potentiodynamic polarization study revealed that PPs suppressed both anodic and cathodic half-cell reactions thereby behaving as mixed-type corrosion inhibitors. However, some cathodic predominance was also observed. EIS study showed that PPs behave as interface-type corrosion inhibitors and they become effective by adsorbing at the interface of metal and electrolyte. Adsorption behavior of PPs was further investigated using surface and computational studies. A good agreement in the results of experimental and computational analyses was observed.

Verma and coworkers [10] described the effect of electron-donating hydroxyl (–OH) and methyl (–Me) and electron-withdrawing nitro (–NO$_2$) on the corrosion inhibition effectiveness of four glucosamine-based substituted pyrimidine-fused heterocycles (CARBs) for mild steel corrosion in 1M HCl. All the CARBs were derived through MCR of glucosamine, barbituric acid and substituted aromatic aldehydes using PTSA as a catalyst. All investigated CARBs were derived in good yield. Experimental analyses showed that both electron withdrawing as well as electron-donating substituents increases in the inhibition effectiveness of CARBs with respect to the inhibition efficiency of un-substituted CARB (CARB-1). However, increase in the inhibition efficiency was more pronounced in the presence of electron-donating substituents. Weight loss study showed that all tested compounds act as good corrosion inhibitors and their inhibition effectiveness was concentration and temperature-dependent. Obviously, inhibition efficiency of CARBs increased with their concentration and decreased with the rise in temperature.

Potentiodynamic polarization studies showed that CARBs adversely affected the anodic oxidation and cathodic hydrogen evolution reactions and behaved as mixed-type corrosion inhibitors. EIS analyses revealed that CARBs act as interface-type corrosion inhibitors as they increase the values of charge transfer resistance through their adsorption at the interface of metal and electrolyte.

Adsorption of the CARBs on metallic surface followed the Langmuir adsorption isotherm model. Adsorption behavior of CARBs on metallic surfaces was supported by SEM and AFM analyses. SEM and AFM studies showed that presence of CARBs in the corrosive electrolyte causes significant smoothness in the metallic surface morphologies. It was also observed that smoothness in the metallic surface morphologies was consistent with their order of inhibition efficiency. SEM and AFM images of mild steel surface corroded in 1M HCl solution with and without CARBs are presented in Figs 14.1 and 14.2, respectively. It can be clearly seen that surface morphology of the metallic specimens remarkably improved in the presence of CARBs.

In AFM analysis, authors observed that average surface roughness of uninhibited metallic surface was 388 nm. The high metallic surface roughness was attributed to aggressive attack of metal surface through electrolyte molecules. However, in the presence of inhibitor molecules average surface roughness values were greatly reduced due to adsorption and formation of the inhibitive film over the metallic surface. The average surface roughnesses of metal in the presence of CARB-1, CARB-2, CARB-3, and CARB-4 were 171, 143, 117, and 84 nm, respectively.

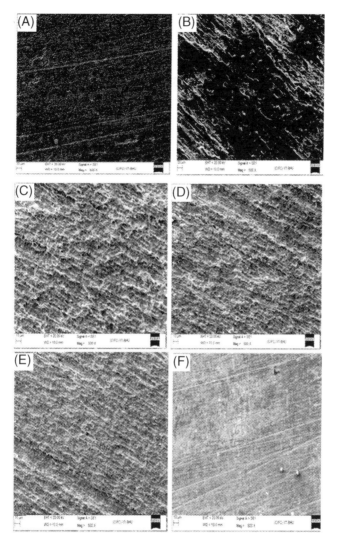

FIG. 14.1 SEM images of (A) polished, (B) corroded in 1M HCl without CARBs, and presence of CARB-1 (C), CARB-2 (D), CARB-3 (E), and CARB-4 (F) [10].

14.4 Summary

From ongoing discussion, it is clear that compounds derived through one-step MCRs are extensively used as corrosion inhibitors for different metals and alloys in various electrolytes. Organic compounds derived through MCRs are treated as environmental friendly because of the association of MCRs with several advantages including high atom economy, ease to perform, facile automation, cost-effectivity, lower number of work-IP, purification steps, etc. Chemical compounds derived through MCRs containing polar substituents and multiple

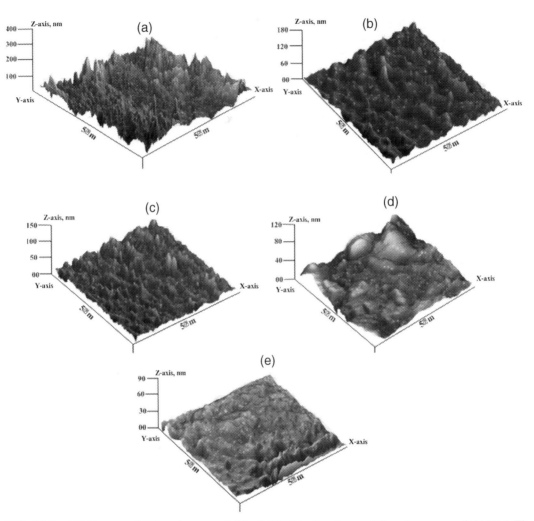

FIG. 14.2   AFM images of MS surface corroded in 1M HCl in the absence (A), and presence of CARB-1 (B), CARB-2 (C), CARB-3 (D), and CARB-4 (E) [10].

bonds act as most effective corrosion inhibitors. The effect of polar substituents on the corrosion inhibition effectiveness of organic compounds can be accessed through Hammett substituents constant. Generally, presence of electron-donating substituents increases inhibition efficiency while converse is true for electron-withdrawing substituents. Most of the corrosion inhibitors derived through MCRs become effective by adsorbing on the metallic surface. These corrosion inhibitors inhibit both anodic and cathodic Tafel reactions and mostly act as mixed-type inhibitors. Using EIS study, it is observed that organic compounds derived through MCRs act as interface-type corrosion inhibitors as they increase the value of charge transfer resistance for the corrosion process. Adsorption of these corrosion inhibitors on metallic surface followed the Langmuir adsorption isotherm. Adsorption behavior of these corrosion inhibitors can be described with the help of surface analyses and computational modeling and simulation.

14.5 Useful websites

https://www.sciencedirect.com/topics/chemistry/multicomponent-reaction#:~:text=Multicomponent%20reactions%20(MCRs)%20are%20one,all%20the%20reactants%20%5B82%5D.
https://www.organic-chemistry.org/topics/multicomponent-reactions.shtm
https://en.wikipedia.org/wiki/Multi-component_reaction
https://www.frontiersin.org/articles/10.3389/fchem.2018.00502/full

References

[1] J.R. Donald, R.R. Wood, S.F. Martin, Application of a sequential multicomponent assembly process/Huisgen cycloaddition strategy to the preparation of libraries of 1, 2, 3-triazole-fused 1, 4-benzodiazepines, ACS Comb. Sci. 14 (2012) 135–143.

[2] L.K. Ransborg, M. Overgaard, J. Hejmanowska, S. Barfüsser, K.A. Jørgensen, Ł. Albrecht, Asymmetric formation of bridged benzoxazocines through an organocatalytic multicomponent dienamine-mediated one-pot cascade, Organ. Lett. 16 (2014) 4182–4185.

[3] B.A. Granger, Development of Multicomponent Assembly Processes and Their Application to the Synthesis of Novel Heterocyclic Scaffolds and the Total Synthesis of Actinophyllic Acid: Application of an Iminium Ion Mediated Cascade, UT Electronic Theses and Dissertations, The University of Texas at Austin, August 2013.

[4] C. Verma, L. Olasunkanmi, E.E. Ebenso, M. Quraishi, Substituents effect on corrosion inhibition performance of organic compounds in aggressive ionic solutions: a review, J. Mol. Liq. 251 (2018) 100–118.

[5] E. Grunwald, J.E. Leffler, Rates and Equilibria of Organic Reactions, Wiley, New York, USA, 1963.

[6] V.S. Sastri, Green Corrosion Inhibitors: Theory and Practice, John Wiley & Sons, New York, USA, 2012.

[7] I. Obot, D. Macdonald, Z. Gasem, Density functional theory (DFT) as a powerful tool for designing new organic corrosion inhibitors. Part 1: an overview, Corrosion Sci. 99 (2015) 1–30.

[8] G. Gece, The use of quantum chemical methods in corrosion inhibitor studies, Corrosion Sci. 50 (2008) 2981–2992.

[9] J. Haque, K. Ansari, V. Srivastava, M. Quraishi, I. Obot, Pyrimidine derivatives as novel acidizing corrosion inhibitors for N80 steel useful for petroleum industry: a combined experimental and theoretical approach, J. Ind. Eng. Chem. 49 (2017) 176–188.

[10] C. Verma, L.O. Olasunkanmi, E.E. Ebenso, M.A. Quraishi, I.B. Obot, Adsorption behavior of glucosamine-based, pyrimidine-fused heterocycles as green corrosion inhibitors for mild steel: experimental and theoretical studies, J. Phys. Chem. C 120 (2016) 11598–11611.

[11] M. Yadav, L. Gope, N. Kumari, P. Yadav, Corrosion inhibition performance of pyranopyrazole derivatives for mild steel in HCl solution: Gravimetric, electrochemical and DFT studies, J. Mol. Liq. 216 (2016) 78–86.

[12] C. Verma, M. Quraishi, A. Singh, 5-Substituted 1H-tetrazoles as effective corrosion inhibitors for mild steel in 1 M hydrochloric acid, J. Taibah Univ. Sci. 10 (2016) 718–733.

[13] P. Singh, V. Srivastava, M. Quraishi, Novel quinoline derivatives as green corrosion inhibitors for mild steel in acidic medium: electrochemical, SEM, AFM, and XPS studies, J. Mol. Liq. 216 (2016) 164–173.

[14] M. Jeeva, M. susai Boobalan, G.V. Prabhu, Adsorption and anticorrosion behavior of 1-((pyridin-2-ylamino) (pyridin-4-yl) methyl) pyrrolidine-2, 5-dione on mild steel surface in hydrochloric acid solution, Res. Chem. Intermed. 44 (2018) 425–454.

[15] P. Singh, E.E. Ebenso, L.O. Olasunkanmi, I. Obot, M. Quraishi, Electrochemical, theoretical, and surface morphological studies of corrosion inhibition effect of green naphthyridine derivatives on mild steel in hydrochloric acid, J. Phys. Chem. C 120 (2016) 3408–3419.

[16] C. Verma, M.A. Quraishi, K. Kluza, M. Makowska-Janusik, L.O. Olasunkanmi, E.E. Ebenso, Corrosion inhibition of mild steel in 1M HCl by D-glucose derivatives of dihydropyrido [2, 3-d: 6, 5-d′] dipyrimidine-2, 4, 6, 8 (1H, 3H, 5H, 7H)-tetraone, Sci. Rep. 7 (2017) 44432.

[17] C. Verma, M. Quraishi, L. Olasunkanmi, E.E. Ebenso, L-Proline-promoted synthesis of 2-amino-4-arylquinoline-3-carbonitriles as sustainable corrosion inhibitors for mild steel in 1 M HCl: experimental and computational studies, RSC Adv. 5 (2015) 85417–85430.
[18] C. Verma, L. Olasunkanmi, I. Obot, E.E. Ebenso, M. Quraishi, 5-Arylpyrimido-[4, 5-b] quinoline-diones as new and sustainable corrosion inhibitors for mild steel in 1 M HCl: a combined experimental and theoretical approach, RSC Adv. 6 (2016) 15639–15654.
[19] C. Verma, L.O. Olasunkanmi, I. Obot, E.E. Ebenso, M. Quraishi, 2, 4-Diamino-5-(phenylthio)-5 H-chromeno [2, 3-b] pyridine-3-carbonitriles as green and effective corrosion inhibitors: gravimetric, electrochemical, surface morphology and theoretical studies, RSC Adv. 6 (2016) 53933–53948.
[20] N.K. Gupta, C. Verma, R. Salghi, H. Lgaz, A. Mukherjee, M. Quraishi, New phosphonate based corrosion inhibitors for mild steel in hydrochloric acid useful for industrial pickling processes: experimental and theoretical approach, New J. Chem. 41 (2017) 13114–13129.
[21] D.K. Yadav, M.A. Quraishi, Application of some condensed uracils as corrosion inhibitors for mild steel: gravimetric, electrochemical, surface morphological, UV–visible, and theoretical investigations, Ind. Eng. Chem. Res. 51 (2012) 14966–14979.
[22] M.A. Quraishi, 2-Amino-3, 5-dicarbonitrile-6-thio-pyridines: new and effective corrosion inhibitors for mild steel in 1 M HCl, Ind. Eng. Chem. Res. 53 (2014) 2851–2859.
[23] R. González-Olvera, V. Román-Rodríguez, G.E. Negrón-Silva, A. Espinoza-Vázquez, F.J. Rodríguez-Gómez, R. Santillan, Multicomponent synthesis and evaluation of new 1, 2, 3-triazole derivatives of dihydropyrimidinones as acidic corrosion inhibitors for steel, Molecules 21 (2016) 250.
[24] M. Bahrami, S. Hosseini, P. Pilvar, Experimental and theoretical investigation of organic compounds as inhibitors for mild steel corrosion in sulfuric acid medium, Corrosion Sci. 52 (2010) 2793–2803.
[25] C. Verma, M. Quraishi, A. Singh, 2-Amino-5-nitro-4, 6-diarylcyclohex-1-ene-1, 3, 3-tricarbonitriles as new and effective corrosion inhibitors for mild steel in 1 M HCl: Experimental and theoretical studies, J. Mol. Liq. 212 (2015) 804–812.
[26] A.K. Singh, M. Quraishi, Inhibiting effects of 5-substituted Isatin-based Mannich bases on the corrosion of mild steel in hydrochloric acid solution, J. Appl. Electrochem. 40 (2010) 1293–1306.
[27] M. Jeeva, G.V. Prabhu, M.S. Boobalan, C.M. Rajesh, Interactions and inhibition effect of urea-derived Mannich bases on a mild steel surface in HCl, J. Phys. Chem. C 119 (2015) 22025–22043.
[28] K. Ansari, M. Quraishi, Experimental and computational studies of naphthyridine derivatives as corrosion inhibitor for N80 steel in 15% hydrochloric acid, Physica E: Low-Dimensional Syst. Nanostruct. 69 (2015) 322–331.
[29] P. Singh, M. Makowska-Janusik, P. Slovensky, M. Quraishi, Nicotinonitriles as green corrosion inhibitors for mild steel in hydrochloric acid: electrochemical, computational and surface morphological studies, J. Mol. Liq. 220 (2016) 71–81.
[30] K. Ansari, M. Quraishi, A. Singh, Pyridine derivatives as corrosion inhibitors for N80 steel in 15% HCl: Electrochemical, surface and quantum chemical studies, Measurement 76 (2015) 136–147.
[31] K. Ansari, M. Quraishi, A. Singh, Corrosion inhibition of mild steel in hydrochloric acid by some pyridine derivatives: an experimental and quantum chemical study, J. Ind. Eng. Chem. 25 (2015) 89–98.
[32] V. Srivastava, J. Haque, C. Verma, P. Singh, H. Lgaz, R. Salghi, M. Quraishi, Amino acid based imidazolium zwitterions as novel and green corrosion inhibitors for mild steel: Experimental, DFT and MD studies, J. Mol. Liq. 244 (2017) 340–352.
[33] K. Ansari, A.Singh Sudheer, M. Quraishi, Some pyrimidine derivatives as corrosion inhibitor for mild steel in hydrochloric acid, J. Disp. Sci. Technol. 36 (2015) 908–917.
[34] K. Ansari, M. Quraishi, A. Singh, Chromenopyridin derivatives as environmentally benign corrosion inhibitors for N80 steel in 15% HCl, J. Assoc. Arab Univ. Basic Appl. Sci. 22 (2017) 45–54.
[35] S. Kumar, D. Sharma, P. Yadav, M. Yadav, Experimental and quantum chemical studies on corrosion inhibition effect of synthesized organic compounds on N80 steel in hydrochloric acid, Ind. Eng. Chem. Res. 52 (2013) 14019–14029.
[36] R. Korde, C.B. Verma, E. Ebenso, M. Quraishi, Electrochemical and thermo dynamical investigation of 5-ethyl 4-(4-methoxyphenyl)-6-methyl-2-thioxo-1, 2, 3, 4 tetrahydropyrimidine-5-carboxylate on corrosion inhibition behavior of aluminium in 1M Hydrochloric acid medium, Int. J. Electrochem. Sci. 10 (2015) 1081–1093.

15

Green corrosion inhibitors from microwave and ultrasound irradiations

Nowadays, chemical synthesis using conventional hot-plate heatings is highly restricted because of their various disadvantages including tedious and time-consuming, nonuniform heating, low yield, tedious work-up, purification, etc. In view of this, recently various environmental friendly approaches are developed. Some common examples of environmental friendly synthetic approaches are solvent-free synthesis, MCRs, mechanochemical mixing and chemical synthesis using nonconventional microwave (MW) and ultrasound (US) heatings. In the last 30 years, these heatings have been extensively used for a variety of chemical transformations. Literature study showed that MW and US irradiations have also been extensively used for the synthesis of organic corrosion inhibitors. These compounds show excellent corrosion inhibition tendency for different metals and alloys in various electrolytes. Obviously, these compounds become effective by forming a surface protective through their adsorption at the interface of metal and electrolyte. It is important to mention that polar functional groups and electron-rich multiple bonds act as adsorption centers during the adsorption of organic compounds over the metallic surface.

Adsorption and inhibition efficiency of the organic corrosion inhibitors depends upon numerous factors [1–3]:

- (i) Molecular/chemical structure of organic compounds.
- (ii) Nature of substituents.
- (iii) Solubility.
- (iv) Environmental temperature.
- (v) Nature of hydrophilicity and hydrophobicity.
- (vi) Nature and number of polar functional groups and multiple bonds.
- (vii) Nature of electrolyte (acidic, neutral, basic, i.e., pH)
- (viii) Thermal and chemical stability of the inhibitor molecule.
- (ix) Nature of the metal–inhibitor complex.
- (x) Intermolecular force of attraction between inhibitor molecules.

15.1 Microwave heating

Recently, nonconventional MW heating is extensively used for the synthesis of biologically and industrially useful organic compounds. Various classes of organic compounds including nanomaterials, polymers, inorganic, and heterocyclic compounds are being synthesized using MW irradiation. Traditional hot-plate method of chemical transformation is limited because of its slow activation of the reactant molecules. Therefore, conventional hot-plate heating is an incompetent method of chemical transformation. Obviously, in conventional hot-plate heating, energy (heat) is utilized in the heating of a reaction vessel before reaching to the reactant and solvent molecules [4–6].

However, MW irradiation causes random and sudden increase of throughout the system. Generally, MW radiation shows direct and simultaneous interactions with the reactant and solvent molecules especially that have a permanent dipole moment. One of the greatest advantages of MW-mediated reaction is that selectivity and rate can be controlled using suitable MW indices. In the last 30 years, various chemical transformations have been carried out using MW irradiation. Fig. 15.1 represents the collection of some major chemical reactions that are carried out using MW heating. Some major advantages of MW assisted reactions are as follows [4–6]:

(i) Rapid and instantaneous activation of reactant molecules.
(ii) Uniform heating of throughout the system.
(iii) Magnified reaction rate, i.e., MW irradiation sometimes increases reaction rate up to 1000-fold.

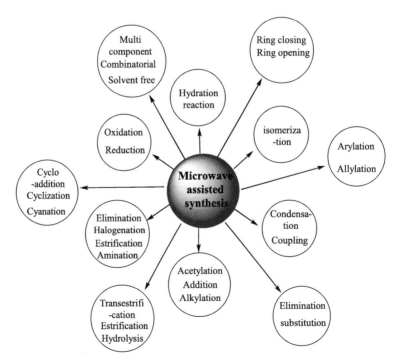

FIG. 15.1 Some common examples of microwave assisted chemical transformations.

- (iv) High synthetic efficiency and yield (high atom economy).
- (v) Reduction in waste production (waste minimization).
- (vi) High selectivity and facile automation.
- (vii) High purity of the synthesized compounds.
- (viii) Reproducibility and facile automation.
- (ix) Avoid undesired heating of the reaction vessel.
- (x) Avoid environmental heating.
- (xi) Low operating cost.
- (xii) Time-saving.
- (xiii) Simplified work-up and purification.

It is important to mention that MW heating and activation of react molecules are entirely different from that of traditional hot-plate heating. The heating and activation mechanism of reactant molecules using hot-plate and MW irradiation are presented in Fig. 15.2. It can be seen that hot-plate heating causes slow activation of reactant molecules as it results into the heating of the outer wall of the reaction vessel. However, MW irradiation penetrates inside the reaction vessel and causes instantaneous heating of the entire reaction system. Because of this only, reactions requiring several hours or days using conventional hot-plate heating can be successfully completed using MW irradiation in a few seconds or minutes. Generally, MW heating causes activation of the molecules that possess permanent dipole moment or/and have ionic conductance.

Using MW irradiation, heating of reactant molecules takes place in nanoseconds (10^{-9} s) which is insufficient to return the excited reactant molecules in their ground state (relaxation usually needs 10^{-5} s). Therefore, excited reactant molecules are unable to gain the state of equilibrium and create a state of nonequilibrium that usually results in an instantaneous high temperature (T_I). Obviously, T_I is a function of MW power input [7]. MW irradiation satisfies various principles of green chemistry and serves as an ideal environmental friendly protocol for the synthesis of biologically and industrially useful organic compounds including as corrosion inhibitors.

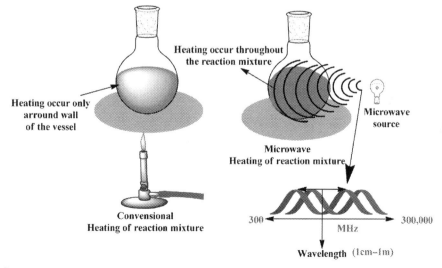

FIG. 15.2 Activation mechanism of reactant molecules using conventional and microwave heatings.

15.2 Ultrasound heating

Similar to MW irradiation, US irradiation has various advantages over the conventional hot-plate heating. Recently, various useful chemical transformations including addition, oxidation, condensation, reduction, alkylation, acylation, nitration, sulphonation, saponification, esterification, solvolysis, hydrolysis, etc., are effectively carried out using US heating method. Fig. 15.3 represents some common examples of chemical transformation carried out using US irradiation. Generally, US irradiation increases the rate of chemical reactions up to many folds.

The complete US range can be divided into three distinct regions namely, power US (20–100 kHz), high frequency US (100 kHz–1 mHz), and diagnostic US (1–500 mHz). Whenever US radiation allows it to pass through a liquid, the cyclic compression and expansion starts because of the mechanism vibration. It is important to mention that during compression cyclic a positive pressure is applied which brings liquid molecules close together and just inverse phenomenon occurs during expansion cycle [8]. These cycles result in the formation of gas-filled microbubbles, when the magnitude of pressure of the expansion cycle exceeds the tensile strength of the liquid. This is called cavitation. It is important to notice that generally pure liquids are associated with high tensile strength therefore US radiations are incapable of causing cavitation [8]. Nevertheless, most of the liquids contain some dissolved impurities including suspended and dissolved solid materials. It is noteworthy that purified water requires about 1000 atm. for cavitation whereas tap-water needed only a few atm. pressures for the same [9]. Generally, cavitation proceeds in three steps, i.e., formation of microbubbles, their rapid growth, and violent collapse. This situation is presented in Fig. 15.4. Each microbubble acts as a hot spot and their collapse produces enormous temperature (up to 5000 K) and pressure (up to 500 atm.) that are capable of driving various

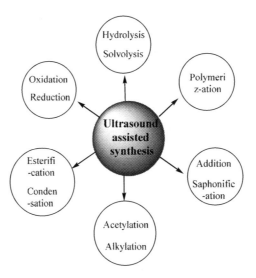

FIG. 15.3 Some common examples of ultrasound assisted chemical transformations.

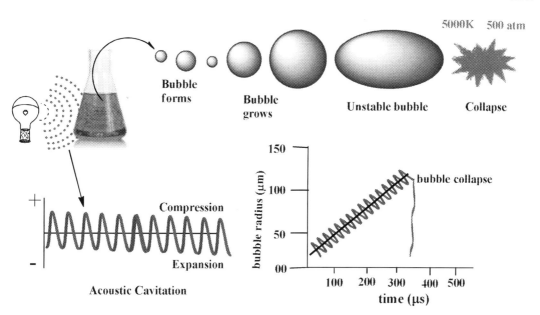

FIG. 15.4 Mechanism of microbubbles formation, their growth and collapse (cavitation) through US irradiation. *US*, ultrasound.

chemical transformations [10]. Cooling rate of US-mediated reactions is very high (10^9 K s^{-1}), which is one of the greatest advantages of US heating over MW heating [11].

15.3 Case studies

15.3.1 Green corrosion inhibitors from microwave irradiation

Nowadays, MW irradiation serves as one of the most convenient, economic, and reproducible methods for the synthesis of various industrially and biologically useful organic compounds. Organic compounds derived through MW irradiation are also extensively used as corrosion inhibitors for different metals and alloys in various electrolytes. Amar et al. [12] reported the synthesis, characterization, and corrosion inhibition effect of piperidin-1-yl-phosphonic acid (PPA) for Armco iron in 3% NaCl solution using chemical and electrochemical methods. PPA was synthesized using MW irradiation and its characterization was carried out using ^{1}H, ^{13}C and ^{31}P NMR and FT-IR spectroscopic methods. Analyses showed that PPA acts as an effective corrosion inhibitor and its inhibition effect was concentration-dependent. It was further observed that inhibition efficiency of the PPA was synergistically increased in the presence of Zn^{2+}. The PPA showed highest protection effectiveness at 20% of Zn^{2+}. The PPA alone showed the highest protection of 75.3% at 5×10^{-3}M concentration however in the presence of 20% Zn^{2+} its inhibition efficiency at the same PPA concentration increased to 89.9%. FT-IR study showed that PPA becomes effective by forming surface protective Fe^{2+}-PPA complexes and $Zn(OH)_2$. Adsorption of PPA on metallic surface and formation of the inhibitive film was further reinforced by EDX analysis.

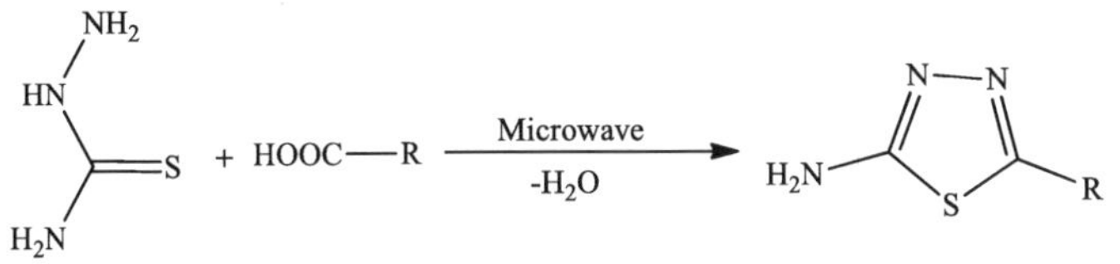

R = $-C_2H_5$, $-C_5H_{11}$, $-C_7H_{15}$, $-C_{11}H_{23}$, $-C_{13}H_{27}$

SCHEME 15.1 **Microwave assisted synthesis of 5-alkyl-1,3,4-thiadiazole derivatives.**

In another study [13], effect of alkyl chain length on corrosion inhibition effect of 2 amino 5-alkyl 1,3,4 thiadiazole derivatives for steel alloy corrosion in 1M H_2SO_4 was investigated using EIS and SEM methods. Synthesis of the investigated 5-alkyl 1,3,4 thiadiazole derivatives is presented in Scheme 15.1. The synthesized compounds were named and designated as 2 Amino 5 ethyl 1,3,4 thiadiazole (IC-1), 2 Amino 5 n propyl 1,3,4 thiadiazole (IC-3), 2 Amino 5 n penthyl 1,3,4 thiadiazole (IC-5), 2 Amino 5 hepthyl 1,3,4 thiadiazole (IC-7), 2 Amino 5 undecyl 1,3,4 thiadiazole (IC-11), and 2 Amino 5 tridecyl 1,3,4 thiadiazole (IC-13). All the tested compounds were synthesized using conventional as well as MW heating methods. Electrochemical impedance spectroscopy analysis showed that inhibition efficiency of the tested compounds increases on increasing the number of carbon up to un-decyl C-chain spacer and then decreases rapidly. Results showed that all investigated compounds become effective by adsorbing on the metallic surface and their mode of adsorption obeyed the Langmuir adsorption isotherm model.

Similarly, in another study, three triazolyl bis-amino acid derivatives were synthesized using MW irradiation and evaluated as corrosion inhibitors for mild steel in acidic solution of hydrochloric acid [14]. Scheme for the synthesis of triazolyl bis-amino acid derivatives is presented in Scheme 15.2. Corrosion inhibition effect of the compounds was determined using weight loss electrochemical and DFT-based quantum chemical calculations. Weight loss study showed that inhibition efficiency of the tested compound increases on increasing its concentration and maximum value of inhibition efficiency of 96.6% was obtained at 1×10^{-3} M concentration. Adsorption of the tested compound on mild steel surface in 1M HCl followed the Langmuir adsorption isotherm model. Potentiodynamic polarization study revealed that presence of inhibitors adversely affects the anodic and cathodic reactions and

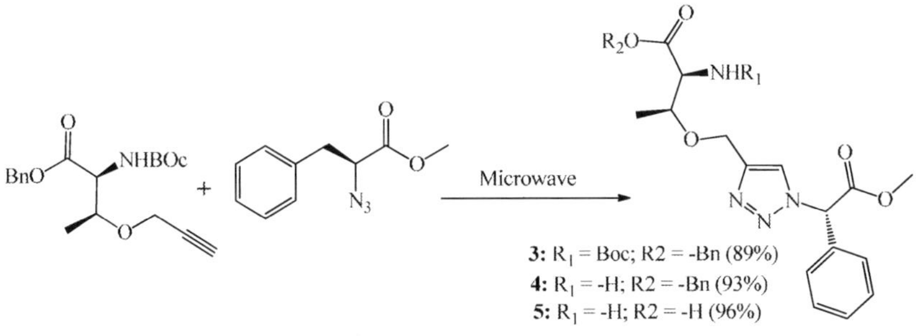

3: R_1 = Boc; R2 = -Bn (89%)
4: R_1 = -H; R2 = -Bn (93%)
5: R_1 = -H; R2 = -H (96%)

SCHEME 15.2 **Microwave assisted synthesis of triazolyl bis-amino acid derivatives.**

acted as mixed-type corrosion inhibitors. DFT study showed that the investigated compound interacted with metallic surface using charge sharing (donor-acceptor) interaction.

Ansari et al. [15] reported MW assisted synthesis of two isatin-β-thiosemicarbzone derivatives, namely, 1-Benzylidene5-(2-oxoindoline-3-ylidene) thiocarbohydrazone (TZ-1) and 1-(4-methylbenzylidene)-5-(2-oxoindolin-3-ylidene) thiocarbohydrazone (TZ-2) and evaluated them as corrosion inhibitors for mild steel in 20% H2SO4 using experimental and computational analyses. Both TZ-1 and TZ-2 showed highest inhibition efficiencies of 91.5% and 97.2%, respectively, at 300 mg L^{-1} concentration. Both TZ-1 and TZ-2 become effective by adsorbing on the metallic surface that followed the Langmuir adsorption isotherm model. EIS study showed that presence of TZ-1 and TZ-2 in the corrosive electrolyte increases the value of charge transfer resistance by adsorbing at the interface of metal and electrolyte. Thereby, TZ-1 and TZ-2 acted as interface-type corrosion inhibitors. Potentiodynamic polarization study showed that TZ-1 and TZ-2 adversely affect both anodic as well as cathodic reactions and acted as mixed-type corrosion inhibitors with slight cathodic-predominance. Adsorption of the TZ-1 and TZ-2 on metallic surface was supported by SEM-EDX analyses. DFT-based quantum chemical calculation study showed that TZ-1 and TZ-2 possess strong ability to interact with the metallic surface through its electron-rich active centers. Synthesis of the TZ-1 and TZ-2 is presented in Scheme 15.3.

In another study [16], four bis-benzimidazoles (BBI) synthesized designated as 1a-1d are synthesized using MW irradiation method. All synthesized compounds are evaluated as corrosion inhibitors for mild steel in 1M HCl. Corrosion inhibition effect of the tested compounds was measured using electrochemical, weight loss, surface, and computational analyses. Results showed that all tested compounds acted as reasonably good corrosion inhibitors and their inhibition effect increases with their concentrations. Inhibition effectiveness of the tested compounds followed the order: BBP > BBMS > BBMO > BBE. The best inhibitor (BBP) showed highest protection efficiency of 94%–96% at 1.0 mM concentration. Potentiodynamic

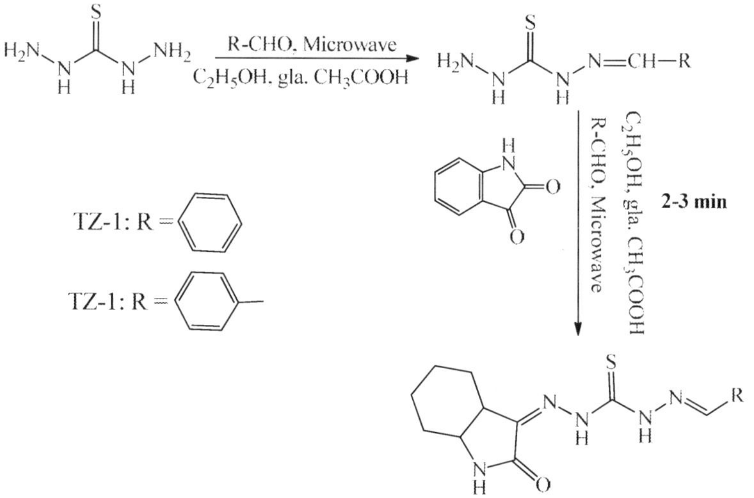

SCHEME 15.3 Microwave assisted synthesis of isatin-β-thiosemicarbzone derivatives.

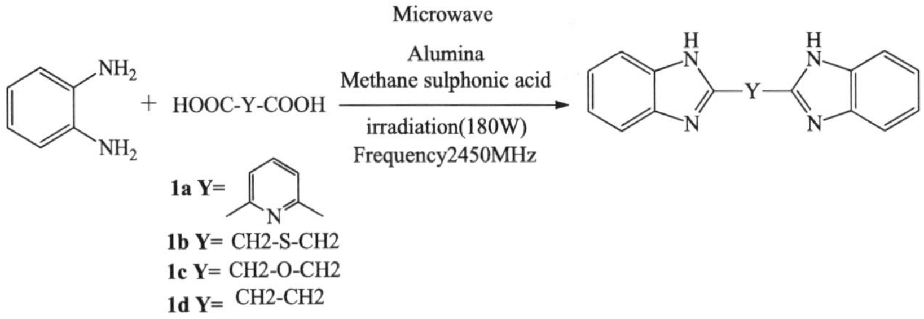

SCHEME 15.4 Microwave assisted synthesis of bis-benzimidazoles (*BBI*).

polarization study showed that all studied compounds act as mixed-type corrosion inhibitors as they adversely affect the rate of both anodic and cathodic reaction. By means of EIS study, it was observed that all evaluated compounds acted as interface-type corrosion inhibitors. Adsorption behavior of the tested compounds followed the Langmuir adsorption isotherm model. DFT study showed the compounds tested as corrosion inhibitors effectively interact with metallic surfaces using donor–acceptor interactions, i.e., charge sharing mechanism. MD simulations study revealed that all tested compounds spontaneously adsorbed on the metallic surface using their flat or horizontal orientations. Synthesis of the tested compounds is given in Scheme 15.4.

Our research group [1], reported MW catalyzed synthesis of three 2-aminobenzene-1,3-dicarbonitriles derivatives (ABDNs) and their anticorrosive effect on aluminum corrosion in 0.5 M NaOH. Corrosion inhibition effect of ABDNs was evaluated using weight loss, electrochemical, surface-, and DFT-based quantum chemical calculations. The study showed that inhibition efficiency of the ABDNs was substituent and concentration-dependent.

Generally, increase in ABDNs concentration causes subsequent increase in their inhibition efficiency. Inhibition efficiency of the evaluated compounds increases on increasing the electron-donating ability of the substituent. The inhibition efficiencies of the evaluated compounds followed the order: ABDN-3 (–OH, –OCH$_3$) > ABDN-2 (–OH) > ABDN-1 (–CH$_3$). All the tested compounds behaved as mixed-type corrosion inhibitors. Adsorption of these compounds on aluminum surface was supported using SEM and EDX analyses. DFT study showed that ABDNs effectively interact with metal surfaces using a charge-sharing mechanism. These compounds were also evaluated as corrosion for mild steel in 1M HCl in another report [17]. Tested ABDNs were synthesized using Scheme 15.5.

Various halogen substituted 1,2,3-triazole derivatives were synthesized using MW irradiation and conventional methods and tested as corrosion inhibitors for API 5L X52 steel in 1M HCl [18]. Investigated compounds were effectively synthesized in a shorter time using MW irradiation method. Study showed that inhibition efficiencies of the tested compounds followed the sequence: BPTI (–I) >BPTBr (–Br) >BPTCl(–Cl) > BPTF (–F) > BPT (–H). Replacement of the hydrogen from halogens increases their inhibition efficiency. Among the halogen substituted compounds, order of corrosion inhibition efficiency was consistent with the atomic size of halogen. Synthesis of the halogen substituted 1,2,3-triazole derivatives is presented in Scheme 15.6.

Singh et al. [19] reported the MW catalyzed synthesis of a bis-phenol polymer containing piperazine and evaluated it as corrosion inhibitor for mild steel in 1M HCl. Potentiodynamic

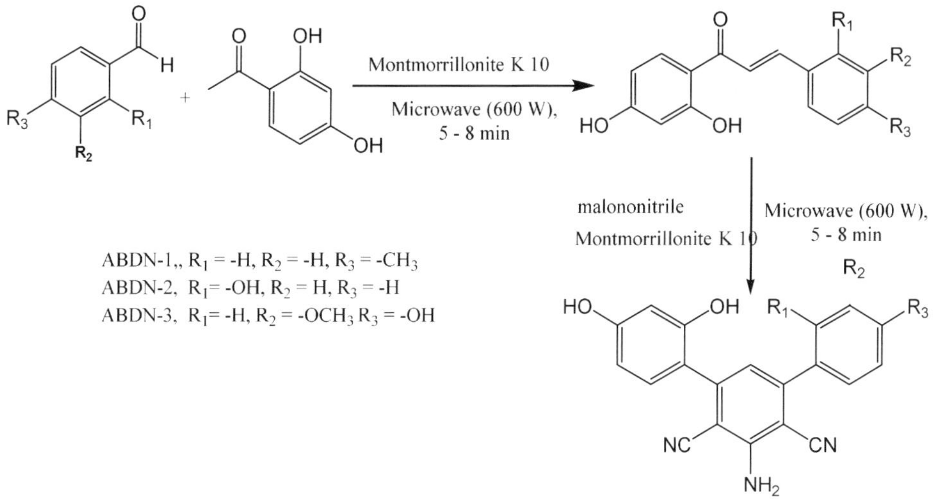

SCHEME 15.5 Microwave assisted synthesis of 2-aminobenzene-1,3-dicarbonitriles derivatives (*ABDNs*).

ABDN-1, R_1 = -H, R_2 = -H, R_3 = -CH_3
ABDN-2, R_1 = -OH, R_2 = H, R_3 = -H
ABDN-3, R_1 = -H, R_2 = -OCH_3 R_3 = -OH

polarization study showed that bis-phenol polymer acted as mixed-type corrosion inhibitors and its presence adversely affects the anodic as well as cathodic reactions. EIS study showed that bis-phenol polymer acted as interface-type corrosion inhibitor as its presence increased the value of charge transfer resistance. The bis-phenol polymer showed highest inhibition efficiency of 96.8% at 75 ppm concentration. Synthesis of bis-phenol polymer is presented in Scheme 15.7. Compounds derived from MW irradiation are also reported as effective corrosion inhibitors in other reports.

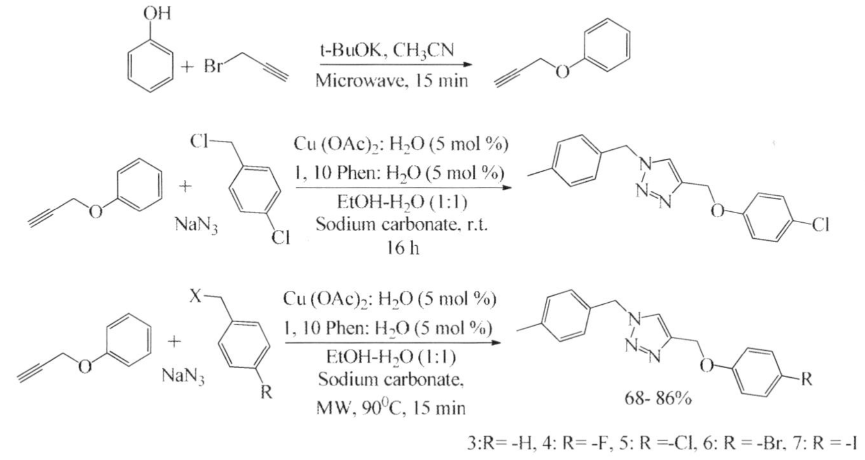

3: R = -H, 4: R = -F, 5: R = -Cl, 6: R = -Br, 7: R = -I

SCHEME 15.6 Microwave assisted synthesis of halogen substituted 1,2,3-triazole derivatives.

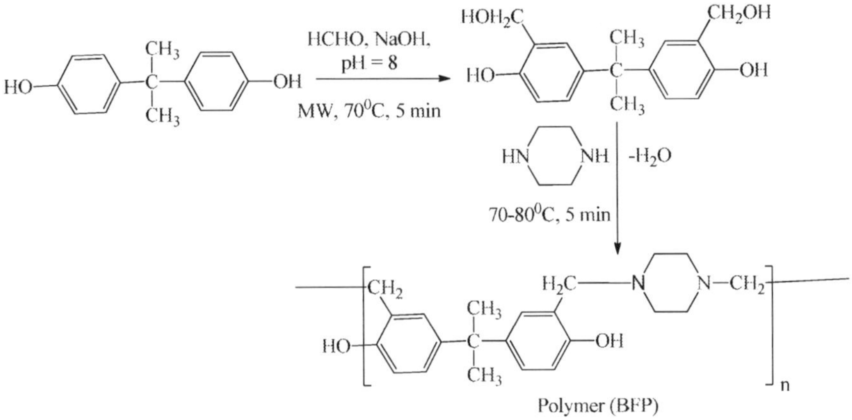

SCHEME 15.7 Microwave assisted synthesis of bis-phenol polymer.

15.3.2 Green corrosion inhibitors from ultrasound irradiation

Compounds derived from US irradiations have also been widely investigated as corrosion inhibitors for different metals and alloys in various electrolytes. Dandia et al. [20] reported the synthesis of five pyrazolo[3,4-b]pyridines, designated as APs, using US irradiation in aqueous medium using p-TSA as catalyst. All synthesized compounds were tested as corrosion inhibitors for mild steel in 1M HCl. Study showed that APS acted as excellent corrosion inhibitors and their inhibition efficiency increases with their concentration. Among the tested compounds, AP-5 showed the highest inhibition efficiency of 95.2% at 100 ppm concentration. Potentiodynamic polarization study showed that investigated compounds acted as mixed-type corrosion inhibitors with slight cathodic predominance. All tested compounds become effective by adsorbing on the metallic surface. EIS study revealed that tested compounds acted as interface-type corrosion inhibitors. Adsorption of the investigated compounds on the metal surface followed the Langmuir adsorption isotherm model. Scheme 15.8 shows the scheme for the synthesis of pyrazolo[3,4-b]pyridines.

Our research team [21] described the synthesis of a chalcone, namely, 3-4-(dimethylamino)phenyl)-1-phenylprop-2-en-1-one (designated as INH-1) and its three heterocyclic derivatives of hydrazine (INH-2), urea (INH-3). and thiourea (INH-4) using US irradiation.

All synthesized compounds were tested as corrosion inhibitors for mild steel in 1MHCl. Among the tested compounds, INH-4 showed highest inhibition efficiency of 98.26% at

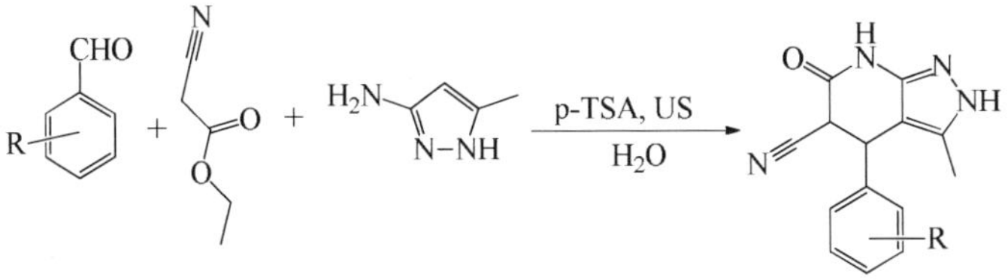

SCHEME 15.8 Microwave assisted synthesis of bis-phenol polymer.

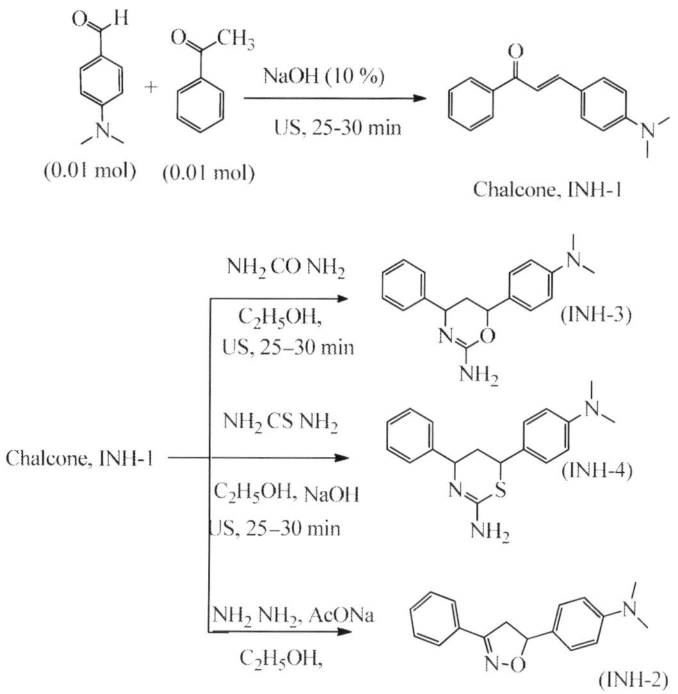

SCHEME 15.9 Microwave assisted synthesis of INHs.

25 ppm concentration. Potentiodynamic polarization study showed that INHs acted as mixed-type corrosion inhibitors and their presence in corrosive electrolyte adversely affected anodic as well as cathodic Tafel reactions. Adsorption of the INHs over the metallic surface followed the Langmuir adsorption isotherm model. Scheme 15.9 represents the scheme for the synthesis of INHs. Corrosion inhibition effect of the other organic compounds derived from US irradiation is also reported in other studies.

15.4 Summary

MW and US irradiations have emerged as a powerful tool for the synthesis of environmental friendly industrially and biologically useful organic compounds. These nonconventional heating sources have several advantages over the conventional hot-plate heating. These nonconventional heatings satisfy various principles of the green chemistry therefore compounds derived from MW and US irradiations are treated as environmental friendly. The MW and US irradiations cause sudden and instantaneous activation of the reactant molecules therefore the reactions that require several hours or days to complete can be successfully completed in a few seconds or minutes using these nonconventional heatings. Compounds derived MW and US irradiations are also used as corrosion inhibitors for different metals and alloys in various electrolytes. These organic compounds mostly act as mixed-type corrosion inhibitors as their presence suppressed the anodic as well as cathodic Tafel reactions. EIS study revealed that compounds derived from MW and US acted as interface-type corrosion inhibitors and their presence increased the value of charge transfer resistance. Adsorption of these compounds on metallic surfaces mostly follows the Langmuir adsorption isotherm model. SEM, EDX, and FT-IR studies are widely used to demonstrate the adsorption behavior of these organic compounds on the metallic surface. DFT-based quantum chemical calculations and MD-/MC-based simulations are widely used to describe the interaction of these compounds with the metallic surface.

15.5 Useful websites

https://en.wikipedia.org/wiki/Microwave#:~:text=Microwave%20is%20a%20form%20of,EHF%20(millimeter%20wave)%20bands.
https://pubs.acs.org/doi/10.1021/acssuschemeng.8b03286
https://en.wikipedia.org/wiki/Microwave_chemistry
https://www.hindawi.com/journals/jchem/2018/3132747/
https://pubmed.ncbi.nlm.nih.gov/15474961/

References

[1] C. Verma, P. Singh, I. Bahadur, E. Ebenso, M. Quraishi, Electrochemical, thermodynamic, surface and theoretical investigation of 2-aminobenzene-1, 3-dicarbonitriles as green corrosion inhibitor for aluminum in 0.5 M NaOH, J. Mol. Liq. 209 (2015) 767–778.
[2] S. Kaya, P. Banerjee, S.K. Saha, B. Tüzün, C. Kaya, Theoretical evaluation of some benzotriazole and phospono derivatives as aluminum corrosion inhibitors: DFT and molecular dynamics simulation approaches, RSC Adv. 6 (2016) 74550–74559.
[3] X. Ren, S. Xu, S. Chen, N. Chen, S. Zhang, Experimental and theoretical studies of triisopropanolamine as an inhibitor for aluminum alloy in 3% NaCl solution, RSC Adv. 5 (2015) 101693–101700.
[4] A.M. Sarotti, R.A. Spanevello, A.G. Suárez, An efficient microwave-assisted green transformation of cellulose into levoglucosenone. Advantages of the use of an experimental design approach, Green Chem. 9 (2007) 1137–1140.
[5] N. Kaval, W. Dehaen, P. Mátyus, E. Van der Eycken, Convenient and rapid microwave-assisted synthesis of pyrido-fused ring systems applying the tert-amino effect, Green Chem. 6 (2004) 125–127.
[6] M.N. Nadagouda, T.F. Speth, R.S. Varma, Microwave-assisted green synthesis of silver nanostructures, Accounts Chem. Res. 44 (2011) 469–478.
[7] B.L. Hayes, Recent advances in microwave-assisted synthesis, Aldrichim. Acta 37 (2004) 66–76.
[8] C. Gong, D.P. Hart, Ultrasound induced cavitation and sonochemical yields, J. Acoust. Soc. Am. 104 (1998) 2675–2682.
[9] A. Brotchie, R. Mettin, F. Grieser, M. Ashokkumar, Cavitation activation by dual-frequency ultrasound and shock waves, Phys. Chem. Chem. Phys. 11 (2009) 10029–10034.
[10] E.B. Flint, K.S. Suslick, The temperature of cavitation, Science 253 (1991) 1397–1399.
[11] Y. Kegelaers, O. Eulaerts, J. Reisse, N. Segebarth, On the quantitative measure of a sonochemical effect in heterogeneous sonochemistry, Eur. J. Organ. Chem. 2001 (2001) 3683–3688.
[12] H. Amar, J. Benzakour, A. Derja, D. Villemin, B. Moreau, T. Braisaz, A. Tounsi, Synergistic corrosion inhibition study of Armco iron in sodium chloride by piperidin-1-yl-phosphonic acid–Zn2+ system, Corrosion Sci. 50 (2008) 124–130.
[13] M. Palomar-Pardavé, M. Romero-Romo, H. Herrera-Hernández, M. Abreu-Quijano, N.V. Likhanova, J. Uruchurtu, J. Juárez-García, Influence of the alkyl chain length of 2 amino 5 alkyl 1, 3, 4 thiadiazole compounds on the corrosion inhibition of steel immersed in sulfuric acid solutions, Corrosion Sci. 54 (2012) 231–243.

[14] Q. Deng, H.-W. Shi, N.-N. Ding, B.-Q. Chen, X.-P. He, G. Liu, Y. Tang, Y.-T. Long, G.-R. Chen, Novel triazolyl bis-amino acid derivatives readily synthesized via click chemistry as potential corrosion inhibitors for mild steel in HCl, Corrosion Sci. 57 (2012) 220–227.

[15] K. Ansari, M. Quraishi, A. Singh, Isatin derivatives as a non-toxic corrosion inhibitor for mild steel in 20% H2SO4, Corrosion Sci. 95 (2015) 62–70.

[16] A. Dutta, S.K. Saha, P. Banerjee, D. Sukul, Correlating electronic structure with corrosion inhibition potentiality of some bis-benzimidazole derivatives for mild steel in hydrochloric acid: combined experimental and theoretical studies, Corrosion Sci. 98 (2015) 541–550.

[17] C.B. Verma, M. Quraishi, A. Singh, 2-Aminobenzene-1, 3-dicarbonitriles as green corrosion inhibitor for mild steel in 1 M HCl: Electrochemical, thermodynamic, surface and quantum chemical investigation, J. Taiwan Inst. Chem. Eng. 49 (2015) 229–239.

[18] A. Espinoza-Vázquez, F. Rodríguez-Gómez, R. González-Olvera, D. Angeles-Beltrán, D. Mendoza-Espinosa, G. Negrón-Silva, Electrochemical assessment of phenol and triazoles derived from phenol (BPT) on API 5L X52 steel immersed in 1 M HCl, RSC Adv. 6 (2016) 72885–72896.

[19] P. Singh, M. Quraishi, E.E. Ebenso, Microwave assisted green synthesis of bis-phenol polymer containing piperazine as a corrosion inhibitor for mild steel in 1M HCl, Int. J. Electrochem. Sci 8 (2013) 890–902.

[20] A. Dandia, S. Gupta, P. Singh, M. Quraishi, Ultrasound-assisted synthesis of pyrazolo [3, 4-b] pyridines as potential corrosion inhibitors for mild steel in 1.0 M HCl, ACS Sustain. Chem. Eng. 1 (2013) 1303–1310.

[21] C.B. Verma, M.J. Reddy, M.A. Quraishi, Ultrasound assisted green synthesis of 3-(4-(dimethylamino) phenyl)-1-phenylprop-2-en-1-one and its heterocyclics derived from hydrazine, urea and thiourea as corrosion inhibitor for mild steel in 1M HCl, Analyt. Bioanalyt. Electrochem. 6 (2014) 515–534.

CHAPTER 16

Green corrosion inhibitors using environmental friendly solvents

Solvents are essential candidates of chemical industry. Solvents are used in huge quantity especially in pharmaceutical and fine-chemical production. Green or environmental friendly solvents are those that have no or negligible adverse impact on the living beings and/or environment. Because of the increasing ecological awareness and strict environmental regulations, use of traditional petrochemical solvents is highly restricted. Recently, five guidelines toward the environmental friendly solvents are developed: (1) application of bio-solvents, i.e., solvents produced from the processing of agricultural crops and renewable resources as ethyl lactate derived from the processing of corn and ethanol is derived from the fermentation of starchy materials including lingo-cellulosic and sugar-containing feeds; (2) replacement of hazardous solvents with ones that are associated with superior environmental, health and safety properties such as enhanced biodegradability and minimized potential for ozone depletion; (3) replacement of traditional organic solvents with relatively harmless supercritical fluids, i.e., application of supercritical CO_2 in polymer processing (this avoids the use of highly toxic and ozone-depleting chlorofluorocarbons); (4) replacement of organic solvents with ionic liquids (ILs) that are associated with low vapor pressure, therefore less emission and contamination to air; and (5) replacement of organic solvents with water, one of the greenest solvents.

Broadly, traditional solvent can be divided into three classes: (1) hydrocarbon or paraffin solvents, (2) oxygenated solvents, such as ethers, alcohols, esters, carbonyls, glycol, and glycol-based ethers esters; (3) halogenated solvents, these solvents contain halogens especially chlorine, bromine, or iodine in their molecular structures, such as perchloroethylene. Organic solvents are widely used in various chemical transformations. Nevertheless, they are highly toxic therefore their recent application is highly restricted owing to increasing demands of green solvents. Obviously, water, ILs, supercritical CO_2, and organic solvents derived from biomass are treated as environmental friendly solvents for numerous chemical transformations [1–5]. Literature study shows that these solvents are also used for the synthesis of heterocyclic compounds to be used as corrosion inhibitors. The physical properties of some common solvents are given in Table 16.1.

TABLE 16.1 Physical properties of some common solvents.

SN.	Solvents	Chemical formula	bp°C	mp°C	Water solubility	Dipole moment	Dielectric constant	Viscosity (1/1000 Pa s.)	Density	Relative polarity
1	Acetonitrile	C_2H_3N	81.6	−46	M	3.5	37.5	0.34	0.786	0.460
2	Chloroform	$CHCl_3$	61.2	−63.5	0.8	1.0	4.8	0.54	1.498	0.259
3	Carbon disulfide	CS_2	46.3	−111.6	0.2	0	2.6	0.36	1.263	0.065
4	Diethylamine	$C_4H_{11}N$	56.3	−48	M	1.2	3.8	0.32	0.706	0.145
5	1-butanol	$C_4H_{10}O$	117.6	−89.5	7.7	1.7	17.5	2.59	0.81	0.586
6	Acetic acid	$C_2H_4O_2$	118	16.6	M	1.68	6.2	1.12	1.049	0.648
7	2-aminoethanol	C_2H_7NO	170.9	10.5	M	2.4	37.7	20.8	1.018	0.651
8	Acetone	C_3H_6O	56.2	−94.3	M	2.85	21	0.30	0.786	0.355
9	Anisole	C_7H_8O	153.7	−37.5	0.10	1.4	4.3	0.98	0.996	0.198
10	2-butanol	$C_4H_{10}O$	99.5	−114.7	18.1	1.7	17.3	3.1	0.808	0.506
11	Acetyl acetone	$C_5H_8O_2$	140.4	−23	16	3.0	23	0.6	0.975	0.571
12	Aniline	C_6H_7N	184.4	−6.0	3.4	1.6	6.8	3.8	1.022	0.420
13	Benzene	C_6H_6	80.1	5.1	0.18	0	2.3	0.60	0.879	0.111
14	Benzonitrile	C_7H_5N	205	−13	0.2	4.1	26	1.27	0.996	0.333
15	2-butanone	C_4H_8O	79.6	−86.3	25.6	2.8	18.6	0.41	0.805	0.327
16	Cyclohexanol	$C_6H_{12}O$	161.1	25.2	4.2	1.9	15		0.962	0.509
17	Chlorobenzene	C_6H_5Cl	132	−45.6	0.05	1.54	5.7	0.75	1.106	0.188
18	Cyclohexanone	$C_6H_{10}O$	155.6	−16.4	2.3	2.9	15	2.00	0.948	0.281
19	di-n-butylphthalate	$C_{16}H_{22}O_4$	340	−35	0.0011			16.6	1.049	0.272
20	t-butyl alcohol	$C_4H_{10}O$	82.2	25.5	M	1.7	12.4	3.35	0.786	0.389
21	Cyclohexane	C_6H_{12}	80.7	−16.4	0.005	0	2	0.89	0.779	0.006
22	Iso-butanol	$C_4H_{10}O$	107.9	−108.2	8.5	1.8	17.9	6.68	0.803	0.552
23	Dimethoxyethane	$C_4H_{10}O_2$	85	−58	M	1.7	7.3	1.1	0.868	0.231
24	1,2-dichloroethane	$C_2H_4Cl_2$	83.5	−35.4	0.87	1.8	10.4	0.78	1.235	0.327
25	Diglyme	$C_6H_{14}O_3$	162	−64	M	1.9	7.23	1.88	0.945	0.244

(continued)

SN.	Solvents	Chemical formula	bp°C	mp°C	Water solubility	Dipole moment	Dielectric constant	Viscosity (1/1000 Pa s.)	Density	Relative polarity
26	Dimethyl phthalate	$C_{10}H_{10}O_4$	283.8	1	0.43				1.190	0.309
27	Benzyl alcohol	C_7H_8O	205.4	−15.3	3.5	1.7	13	5.47	1.042	0.608
28	1,1-dichloroethane	$C_2H_4Cl_2$	57.3	−97	0.5	1.8	10	0.84	1.176	0.269
29	Carbon tetrachloride	CCl_4	76.7	−22.4	0.08	0	2.3	0.90	1.594	0.052
30	N,N-dimethylaniline	$C_8H_{11}N$	194.2	2.4	0.14	4.40	1.61	1.40	0.956	0.179
31	Diethylene glycol	$C_4H_{10}O_3$	245	−10	M	2.3	31.8	30.2	1.118	0.713
32	1-heptanol	$C_7H_{16}O$	176.4	−35	0.17	1.7	12	6.0	0.819	0.549
33	Pentane	C_5H_{12}	36.1	−129.7	0.004	0	1.84	0.23	0.626	0.009
34	Hexane	C_6H_{14}	69	−95	0.0014	0	1.9	0.29	0.655	0.009
35	Ethyl acetoacetate	$C_6H_{10}O_3$	180.4	−80	2.9				1.028	0.577
36	2-pentanone	$C_5H_{10}O$	102.3	−76.9	4.3	2.7	15.4	0.50	0.809	0.321
37	Pyridine	C_5H_5N	115.5	−42	M	2.2	13	0.88	0.982	0.302
38	Ehylene glycol	$C_2H_6O_2$	197	−13	M	2.3	37.7	16.1	1.115	0.790
39	2-pentanol	$C_5H_{12}O$	119.0	−50	4.5	1.7	13.7	3.5	0.810	0.488
40	Ethanol	C_2H_6O	78.5	−114.1	M	1.7	24	1.08	0.789	0.654
41	3-pentanol	$C_5H_{12}O$	115.3	−8	5.1	1.7	13.3		0.821	0.463
42	Water	H_2O	100.00	0.00	M	1.85	80.1	0.89	0.998	1.000
43	Toluene	C_7H_8	110.6	−93	0.05	0.36	2.4	0.55	0.867	0.099
44	Methylene chloride	CH_2Cl_2	39.8	−96.7	1.32	1.6	9.0	0.42	1.326	0.309
45	1-hexanol	$C_6H_{14}O$	158	−46.7	0.59	1.60	12.5	0.59	0.814	0.559
46	Ethyl benzoate	$C_9H_{10}O_2$	213	−34.6	0.07	2.0	6.0		1.047	0.228
47	Methyl acetate	$C_3H_6O_2$	56.9	−98.1	24.4	1.7	6.68	0.36	0.933	0.253
48	Dioxane	$C_4H_8O_2$	101.1	11.8	M	0.4	2	1.18	1.033	0.164
49	2-propanol	C_3H_8O	82.4	−88.5	M	1.66	19	2.07	0.785	0.546
50	3-pentanone	$C_5H_{10}O$	101.7	−39.8	3.4	2.8	17.0	0.44	0.814	0.265
51	Dimethylformamide	C_3H_7NO	153	−61	M	3.8	37	0.80	0.944	0.386
52	Methanol	CH_4O	64.6	−98	M	1.6	33	0.54	0.791	0.762

(continued)

TABLE 16.1 (Cont'd)

SN.	Solvents	Chemical formula	bp°C	mp°C	Water solubility	Dipole moment	Dielectric constant	Viscosity (1/1000 Pa s.)	Density	Relative polarity
53	1-octanol	$C_8H_{18}O$	194.4	−15	0.096	1.7	10.3	7.4	0.827	0.537
54	Methyl t-butyl ether	$C_5H_{12}O$	55.2	−109	4.8	1.4	2.6	0.36	0.741	0.124
55	Glycerin	$C_3H_8O_3$	290	17.8	M	2.7	42.5	934	1.261	0.812
56	Water, heavy	D_2O	101.3	4	M	1.84	787.3	1.10	1.107	0.991
57	1-propanol	C_3H_8O	97	−126	M	1.68	22	1.95	0.803	0.617
58	p-xylene	C_8H_{10}	138.3	13.3	0.02	0	2.27	0.65	0.861	0.074
59	Dimethyl sulfoxide	C_2H_6OS	189	18.4	M	3.9	46.7	2.00	1.092	0.444
60	Ether	$C_4H_{10}O$	34.6	−116.3	7.5	1.25	4.3	0.22	0.713	0.117
61	Ethyl acetate	$C_4H_8O_2$	77	−83.6	8.7	1.78	6.0	0.43	0.894	0.228
62	Heptane	C_7H_{16}	98	−90.6	0.0003	0	1.9	0.39	0.684	0.012
63	1-pentanol	$C_5H_{12}O$	138.0	−78.2	2.2	1.7	14	3.5	0.814	0.568
64	Tetrahydrofuran	C_4H_8O	66	−108.4	30 or M (7)	1.63	7.5	0.46	0.886	0.207

16.1 Literature survey: corrosion inhibitors using green solvents

16.1.1 Green corrosion inhibitors derived using water

Water is regarded as "universal solvent" as it is capable of dissolving more chemicals/substances than any other solvent. Polar molecules exhibit great solubility in the water however nonpolar molecules show relatively less soluble. Water acquires the molecular formula of H_2O and molecular mass of 18.01528(33) g mol^{-1}. It acquires the highest density of 1000 kg m^{-3} at 4°C. It has melting and boiling points of 0°C (273.15K or 32°F) and 99.98°C (373.13K or 211.73°F), respectively. Water contains pK_a (acidity) and pK_b (basicity) value of 15.74. Because of its high dipole moment (1.85 D), water is associated with very high dielectric constant (ε), higher than most of the common solvents. Water is treated as one of the most preferred solvents because of its following properties:

(i) Excellent solvent for a variety of chemicals.
(ii) Highly polar.
(iii) High heat capacity.
(iv) High dipole moment and dielectric constant.
(v) Wide liquidus range (0–100°C).
(vi) High heat of vaporization.
(vii) Strong cohesive and adhesive forces.
(viii) Its liquid form is denser as compared to its solid form.
(ix) Nontoxic (eco-friendly).
(x) Cost-effective and commercially available.
(xi) Biorenewable.
(xii) Nonflammable.
(xiii) Capability to dissolve CO_2 (help in reducing greenhouse gases).
(xiv) Unique redox potential.

Apart from the above advantages, using water as solvent for chemical transformations is connected with few disadvantages. Some of the common disadvantages of using water as solvent are:

(i) Energy intensive (water removal needs distillation).
(ii) Many reagents are water-sensitive and degrade in the contact with water.
(iii) Waste streams are difficult to treat.
(iv) Limited only for polar compounds.
(v) Obviously, water is a poor solvent for organics.
(vi) Acidity and basicity of the reagents and catalysts may change in water.

Regardless of the disadvantages, use of water as solvent for organic synthesis is a very popular area of research. Chemical species derived using water as solvent is extensively used as corrosion inhibitors. Our research team [6], reported the synthesis, characterization, and corrosion inhibition effect of three 2-amino-4-arylquinoline-3-carbonitriles (AACs) for mild steel in 1M HCl. AACs were synthesized by the condensation reaction of aromatic aldehydes (differing in the nature of substituents), aniline, and malononitrile using water as solvent and L-proline (20%) as catalyst. Environmental friendly behavior of AACs synthesis was

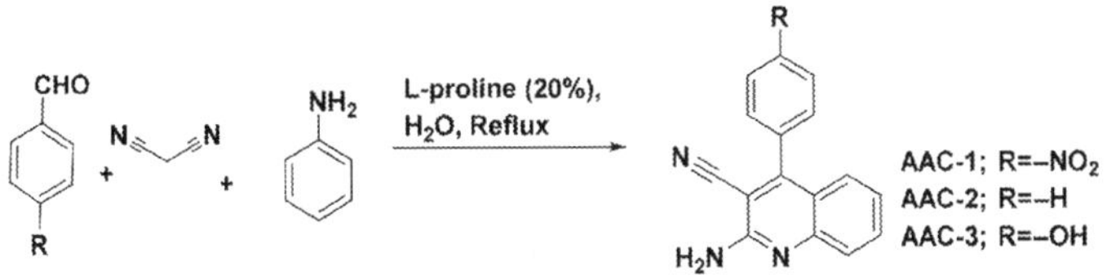

SCHEME 16.1 Scheme for the synthesis of 2-amino-4-arylquinoline-3-carbonitriles (*AACs*).

based on fulfillment of various aspects of green chemistry such as being its one step multi-component reactions (MCRs), using L-proline (a natural species) as a catalyst and more importantly using water as a green solvent. AACs show excellent corrosion inhibition efficiency for mild steel corrosion in acidic medium. Experimental and computational analyses showed that the presence of electron-withdrawing nitro-substituent decreases the corrosion inhibition efficiency while converse it true for electron-donating methyl-substituent. Scheme for synthesis of AACs is given in Scheme 16.1.

Weight loss study showed that AACs act as effective corrosion inhibitors for mild steel and their inhibition effectiveness increases with their concentration. AAC-1, AAC-2, and AAC-3 exhibited highest inhibition effectiveness of 95.21%, 96.08%, and 96.95%, respectively. Potentiodynamic polarization study reveals that AACs become effective by adsorbing at active sites of metal surface using their electron-rich adsorption centers. They inhibit both anodic oxidation and cathodic hydrogen evolution Tafel reactions and behaved as mixed-type corrosion inhibitors. However, mainly they behave as cathodic-type corrosion inhibitors. EIS analysis reveals that AACs become effective by adsorbing at the interface of metal and electrolyte and act as interface-type corrosion inhibitors. Adsorption mechanism of corrosion inhibition using AACs was supported by SEM, EDX, and AFM analyses. DFT-based quantum chemical calculations were carried out to describe the nature interactions between AACs and metallic surface. Results showed that AACs interact with metal surface using donor-acceptor interactions. DFT-based analysis further showed that contribution of frontier molecular orbitals (FMOs: HOMO & LUMO) is greatly reduced in the presence of electron-withdrawing nitro ($-NO_2$) substituent and converse is true in the presence of electron-donating methyl-substituent. A good agreement in the results of experimental and computational analyses was observed.

In another study [7], four 5-Arylpyrimido-[4,5-b]quinoline-diones (APQDs) differing in the nature of substituents were synthesized using one step MCRs of aromatic aldehydes (containing different substituents), barbituric acid, and aniline using L-proline (20%) as catalyst and water as green solvent (Scheme 16.2). APQDs were synthesized in good yield. Environmental friendly behavior of APQDs synthesis was based on fulfillment of various

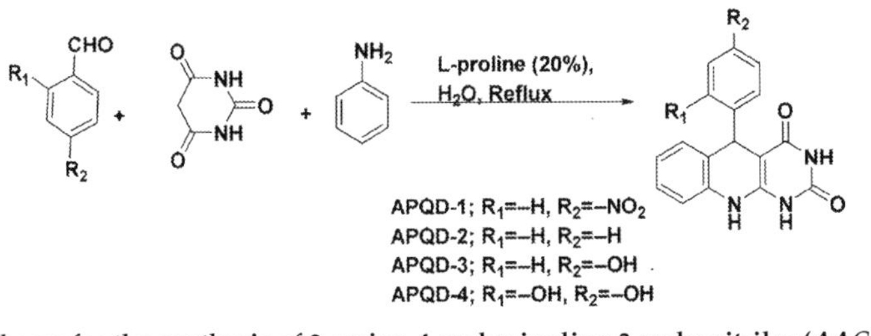

SCHEME 16.2 Scheme for the synthesis of 2-amino-4-arylquinoline-3-carbonitriles (*AACs*).

aspects of green chemistry such as being its one step MCRs, using L-proline (a natural species) as a catalyst and more importantly using water as a green solvent. Inhibition efficiencies of tested APQDs followed the order: APQD-4 (96.52%) > APQD-3 (95.65%) > APQD-2 (93.91%) > APQD-1 (92.17%). Difference in the inhibition effectiveness of the APQDs is attributed due to the presence of different substituents. Both experimental and computational analyses showed that electron withdrawing nitrosubstituent decreases inhibition efficiency and electron-donating hydroxyl-substituent(s) increases the inhibition efficiency. Weight loss analysis showed that inhibition efficiency of APQDs was concentration dependent as increase in their concentration causes subsequent increase in the inhibition efficiency. APQDs showed the highest protection efficiency at 20 mg L^{-1} concentration.

Potentiodynamic polarization study reveals that APQDs act as mixed-type but predominantly cathodic-type corrosion inhibitors for mild steel corrosion in 1M HCl. Weight loss study further showed that increase in the solution temperature causes subsequent decrease in the corrosion inhibition performance of APQDs. Obviously, increase in temperature can cause various physical and chemical changes such as increase in kinetic energy of inhibitor molecules, acid-catalyzed rearrangement, fragmentation or decomposition at high temperature.

EIS study suggests that APQDs inhibit corrosion by adsorbing at the interface of metal and electrolyte and therefore behave as interface-type corrosion inhibitors. DFT study shows that APQDs interactions with the metallic surface involve the donor–acceptor (charge transfer) reactions with the metallic surface. Molecular dynamics simulations study shows that APQDs spontaneously adsorb on metallic surface using their electron rich centers. Orientation of the APQDs on metallic surface was greatly affected by the substituents present in their molecular structures. A good agreement in the results of experimental and computational analyses was observed. Literature study suggests that organic compounds synthesized using water as solvents have also been investigated as green corrosion inhibitors in various studies. Such compounds mostly act as mixed-type and interface-type corrosion inhibitors. They become effective by adsorbing on the metallic surface following through Langmuir adsorption isotherm model.

16.1.2 Green corrosion inhibitors derived using supercritical CO_2

Supercritical carbon dioxide (CO_2) represents the fluid state of CO_2 where it is held at or above its critical temperature (31°C, 87.8°F, 304.13 K) and critical pressure (73.8 bar, 1070 psi, 72.8 atm., 7.3773 MPa). Supercritical CO_2 is widely used as solvent for the extraction of caffeine from coffee at the place of tradition toxic CH_2Cl_2. It is also used for the extraction of fatty acid triglycerides from crisps. Supercritical CO_2 is also used for dry cleaning and spray-printing. Application of Supercritical CO_2 as solvent at the place of traditional petrochemical solvents possesses various advantages including the following:

(i) Supercritical CO_2 is relatively cheap.
(ii) Nontoxic.
(iii) Easily recover and recycle.
(iv) Excellent solvent for gases (e.g., H_2)
(v) High liquidus range.
(vi) Nonflammable.
(vii) Low viscosity and fast diffusion.

However, using supercritical CO_2 as solvent for chemical transformation is associated with few disadvantages such as:

(i) Expensive instrumentation.
(ii) Supercritical CO_2 exists in liquid rage only at high partial pressure which is potentially dangerous for equipments.
(iii) Supercritical CO_2 is a relatively poor solvent.
(iv) It is susceptible to react with strong nucleophiles such as amines.

Because of the low dissolving power of supercritical CO_2 small amount (5%–10%) of ethanol (as cosolvent) is generally used to increase its polarity. Compounds derived from supercritical CO_2 are also investigated as effective inhibitors for metallic corrosion. Through, the literature on corrosion inhibition effect of compounds derived using supercritical CO_2 is a solvent is relatively lesser. Therefore, researches on the synthesis of organic compounds in the supercritical CO_2 and their subsequent implementation as corrosion inhibitors should be increased. It is important to mention that supercritical CO_2 itself is a highly corrosive electrolyte for different metals and alloys. Therefore, various attempts, especially use of heterocyclic compounds have been made to overcome the corrosion problem of supercritical CO_2 [8–10]. In supercritical CO_2 heterocyclic compounds adsorb at the interface of metal and electrolyte and form corrosion protective surface film which isolates metal from aggressive environment and protect from corrosive damage. Adsorption of heterocyclic compounds on metallic surface mostly followed the Langmuir adsorption isotherm model. Most of the molecules used as corrosion inhibitors in supercritical CO_2 act as mixed- and interface-type corrosion inhibitors.

16.1.3 Green corrosion inhibitors derived using ionic liquids

ILs are organic salts that are mainly composed of organic cations and inorganic anions. They are liquid below 100°C. Some of the ILs such as 1-ethyl-3-methylimidazolium and 1-butyl-3,5-dimethylpyridinium bromide are liquid even below room temperature and they acquire the melting points of −21°C (−6°F) and −24°C (−11°F). ILs are another very important eco-friendly class of compounds that have been extensively used as solvents for different chemical transformations. ILs are associated with various advantages that make them eco-friendly solvents for various chemical transformations. Some of the common advantages of ILs are:

(i) Low vapor pressure.
(ii) Ease synthesis.
(iii) Act as catalysts along with solvents and reagents.
(iv) Tunable viscosity and properties (designer chemicals).
(v) High liquidus range.
(vi) Excellent dissolving capability.
(vii) Recyclability.
(viii) High thermal stability.
(ix) High chemical stability.
(x) High solubility in both organic and aqueous solvents.
(xi) Strong ability to interact with the metallic surface.

Application of ILs as solvents is connected with the following advantages:

(i) Synthesis and purification of ILs are highly expensive and toxic.
(ii) Nonbiodegradable.
(iii) Use of toxic organic solvents require for purification, usually through distillation and extraction process.
(iv) Synthesis of ILs mainly utilizes haloalkanes (toxic).

Compounds derived using ILs as corrosion inhibitors are used as corrosion inhibitors. They are also used as catalysts for the synthesis of various organic compounds to be used for various industrial and biological applications. ILs are themselves extensively used as corrosion inhibitors for different metals and alloys in various electrolytes. Various research and review articles describing the anticorrosive effect of ILs have been published [11,12]. Literature study shows that imidazolium-based ILs are most frequently utilized as corrosion inhibitors. It can also be seen that literature on corrosion inhibitors derived using ILs as solvents is relatively less. Therefore, researches on the synthesis of corrosion inhibitors in ILs and their subsequent implementation should be enhanced. A detail description on the corrosion inhibition effect of the ILs is given in Chapter 13. It can be seen that most of the ILs act as mixed-type corrosion inhibitors as they become effective by retarding both anodic as well as cathodic Tafel reactions. They adsorb on the active sites of the metal surface and decrease the value of corrosion current density. Their adsorption on metallic surface mostly follows Langmuir adsorption isotherm model. Through, EIS study it is observed that most of the ILs act as interface-type corrosion inhibitors as their adsorption at the interface of metal and electrolyte increases the value of charge transfer resistance.

16.1.4 Green corrosion inhibitors derived polyethylene glycol

Polyethylene glycols (PEGs) are associated with various advantageous features that make them suitable solvents for various chemical transformations. PEGs are regarded as environmental friendly solvents because of their association with various benefits including less environmental toxicity, low vapor pressure, inexpensive behavior, nonflammability, and minimized environmental [13]. PEGs are available in various molecular weights ranging from 200 to tens of thousands [13,14]. Generally, low molecular weight PEGs, having molecular weight of 200–600 D are liquid at or below room temperature and PEGs having molecular weight of 600–800 D are water-soluble viscous materials. PEGs with molecular weight of more than 800D exist in the solid state. It is important to mention that solubility of the PEGs get decreased on increasing their molecular weight. For example, PEGs-200-600D are soluble in water in all proportion whereas PEG-2000 is only 60% soluble (at 20°C). Owing to their environmental-friendly behavior, they are widely used as solvents for various chemical transformations. The chemical species derived using PEGs as solvents can be used as effective corrosion inhibitors for different metals and alloys.

PEGs are ideal materials to be used as corrosion inhibitors because of their various advantages such as high surface area and reasonably high solubility in the polar electrolytes. Ashassi-Sorkhabi and Ghalebsaz-Jeddi [15] described the corrosion inhibition power of PEGs 200–1000 gmol-1 for carbon steel in 3N H_2SO_4 using electrochemical and chemical methods. PEGs show more than 90% efficiency at 10^{-1}M concentration. Literature survey showed that

PEGs are widely used as inhibitors for metal in H_2SO_4 media [16–18]. They are also investigated as corrosion inhibitors in NaCl [19–21] and MOH (metal oxide) [22–24] solutions. In most of the study it is derived that PEGs act as mixed-type corrosion inhibitors as they inhibit both anodic as well as cathodic Tafel reactions. EIS studies show that PEGs mostly act as interface-type corrosion inhibitors and increase the value of charge transfer resistance through their adsorption at the interface of metal and electrolyte. Adsorption of PEGs mostly followed the Langmuir adsorption isotherm model.

16.2 Summary

Solvents are used in huge quantity in chemical industry, especially in pharmaceutical and fine-chemical production. Most of the traditional solvents are toxic to human being and the surrounding environment. Recently, because of the increasing ecological awareness and strict environmental regulations use of these solvents is highly restricted. In view of this, several environmental friendly alternative solvents, especially water, ILs, supercritical CO_2, and PEG are being used as solvents. These solvents are associated with several advantages such as cost-effectivity, commercial availability, nonflammability, nontoxicity, nonbioaccumulation, etc. Chemical compounds derived using these solvents are/can be used as effective corrosion inhibitors for different metals and alloys in various electrolytes. Literature study suggests that ILs and PEGs, along with being used as solvents they are used as corrosion inhibitors themselves. These compounds become effective by retarding both anodic and cathodic Tafel reactions. They mostly act as mixed-type corrosion inhibitors. They inhibit metallic corrosion by adsorbing at the interface of metal and electrolyte and behave as interface-type corrosion inhibitors. Their adsorption increases the value of charge transfer resistance for corrosion process. Although, water, ILs, supercritical CO_2, and PEG are well-established environmental friendly solvents however literatures on the corrosion inhibition effect of compounds derived using them as solvents are relatively lesser. Therefore, use of these chemical species as solvents for the synthesis of corrosion inhibitors should be explored.

16.3 Useful websites

chm.bris.ac.uk/webprojects2004/vickery/green_solvents.htm#:~:text=Green%20solvents%20were%20developed%20as,the%20ester%20of%20lactic%20acid

https://www.sciencedirect.com/topics/chemistry/green-solvent

https://pubs.rsc.org/en/content/articlelanding/2007/gc/b617536h#!divAbstract

References

[1] T. Welton, Ionic liquids in green chemistry, Green Chem. 13 (2011) 225.
[2] S. Mallakpour, M. Dinari, Ionic Liquids as Green Solvents: Progress and Prospects, Springer, Dordrecht, 2012, pp. 1–32 Green Solvents II.
[3] D. Lozowski, Supercritical CO2: a green solvent, Chem. Eng. 117 (2010) 15.
[4] S.P. Nalawade, F. Picchioni, L. Janssen, Supercritical carbon dioxide as a green solvent for processing polymer melts: Processing aspects and applications, Progr. Poly. Sci. 31 (2006) 19–43.
[5] U. Ejaz, M. Sohail, Ionic liquids: green solvent for biomass pretreatment. In: Nanotechnology-Based Industrial Applications of Ionic Liquids, Springer, Cham, 2020, pp. 27–36.

[6] C. Verma, M. Quraishi, L. Olasunkanmi, E.E. Ebenso, L-Proline-promoted synthesis of 2-amino-4-arylquinoline-3-carbonitriles as sustainable corrosion inhibitors for mild steel in 1 M HCl: experimental and computational studies, RSC Adv. 5 (2015) 85417–85430.

[7] C. Verma, L. Olasunkanmi, I. Obot, E.E. Ebenso, M. Quraishi, 5-Arylpyrimido-[4, 5-b] quinoline-diones as new and sustainable corrosion inhibitors for mild steel in 1 M HCl: a combined experimental and theoretical approach, RSC Adv. 6 (2016) 15639–15654.

[8] H. Cen, J. Cao, Z. Chen, X. Guo, 2-Mercaptobenzothiazole as a corrosion inhibitor for carbon steel in supercritical CO2-H2O condition, Appl. Surface Sci. 476 (2019) 422–434.

[9] Y. Xiang, Z. Long, C. Li, H. Huang, X. He, Inhibition of N80 steel corrosion in impure supercritical CO2 and CO2-saturated aqueous phases by using Imino inhibitors, Int. J. Greenhouse Gas Control 63 (2017) 141–149.

[10] B. Hou, Q. Zhang, Y. Li, G. Zhu, H. Liu, G. Zhang, A pyrimidine derivative as a high efficiency inhibitor for the corrosion of carbon steel in oilfield produced water under supercritical CO2 conditions, Corrosion Sci. 164 (2020) 108334.

[11] C. Verma, E.E. Ebenso, M. Quraishi, Ionic liquids as green and sustainable corrosion inhibitors for metals and alloys: an overview, J. Mol. Liq. 233 (2017) 403–414.

[12] G. Vastag, A. Shaban, M. Vraneš, A. Tot, S. Belić, S. Gadžurić, Influence of the N-3 alkyl chain length on improving inhibition properties of imidazolium-based ionic liquids on copper corrosion, J. Mol. Liq. 264 (2018) 526–533.

[13] J. Chen, S.K. Spear, J.G. Huddleston, R.D. Rogers, Polyethylene glycol and solutions of polyethylene glycol as green reaction media, Green Chem. 7 (2005) 64–82.

[14] F. Bailey Jr, J. Koleske, Poly (ethylene oxide), Academic Press, New York, NY, 1976.

[15] H. Ashassi-Sorkhabi, N. Ghalebsaz-Jeddi, Inhibition effect of polyethylene glycol on the corrosion of carbon steel in sulphuric acid, Mater. Chem. Phys. 92 (2005) 480–486.

[16] M. Mobin, M.A. Khan, Adsorption and corrosion inhibition behavior of polyethylene glycol and surfactants additives on mild steel in H 2 SO 4, J. Mater. Eng. Perform. 23 (2014) 222–229.

[17] A. Algaber, E.M. El-Nemma, M.M. Saleh, Effect of octylphenol polyethylene oxide on the corrosion inhibition of steel in 0.5 M H2SO4, Mater. Chem. Phys. 86 (2004) 26–32.

[18] S. Deng, X. Li, J. Sun, Adsorption and inhibition effect of polyethylene glycol 20000 on steel in H_2SO_4 solution, Clean. World 4 (2011).

[19] H. Boudellioua, Y. Hamlaoui, L. Tifouti, F. Pedraza, Comparison between the inhibition efficiencies of two modification processes with PEG–ceria based layers against corrosion of mild steel in chloride and sulfate media, J. Mater. Eng. Perform. 26 (2017) 4402–4414.

[20] H. Boudellioua, Y. Hamlaoui, L. Tifouti, F. Pedraza, Effects of polyethylene glycol (PEG) on the corrosion inhibition of mild steel by cerium nitrate in chloride solution, Appl. Surface Sci. 473 (2019) 449–460.

[21] S. Božović, S. Martinez, V. Grudić, A novel environmentally friendly synergistic mixture for steel corrosion inhibition in 0.51 M NaCl, Acta Chimica Slovenica 66 (2019) 112.

[22] S.A. El Wanees, A.A. El Aal, E.A. El Aal, Effect of polyethylene glycol on pitting corrosion of cadmium in alkaline solution, British Corrosion J. 28 (1993) 222–226.

[23] Y. Ein-Eli, M. Auinat, D. Starosvetsky, Electrochemical and surface studies of zinc in alkaline solutions containing organic corrosion inhibitors, J. Power Sources 114 (2003) 330–337.

[24] D. Gelman, I. Lasman, S. Elfimchev, D. Starosvetsky, Y. Ein-Eli, Aluminum corrosion mitigation in alkaline electrolytes containing hybrid inorganic/organic inhibitor system for power sources applications, J. Power Sources 285 (2015) 100–108.

CHAPTER 17

Plant extracts as green corrosion inhibitors

Due to increasing ecological awareness and environmental regulations, the use of organic molecules of natural and biological origin for different industrial and biological applications is gaining particular attention. Recently, traditional toxic corrosion inhibitors have been replaced by chemicals derived from plants (called phytochemicals). Plants contain different complex phytochemicals in their different parts. The use of phytochemicals or plant extracts as corrosion inhibitors can be treated as one of the greenest corrosion inhibition practices. Generally, each plant extract contain various phytochemicals that can act as corrosion inhibitors. These phytochemicals contain various electron-rich polar functional groups such as hydroxyl (–OH), amine (–NH_2), amide (–$CONH_2$), ester (–$COOC_2H_5$), acid chloride (–COCl), etc., and aromatic ring(s) through which they easily get adsorbed and behave as strong anticorrosive materials [1–5]. Literature study reveals that extracts of different parts such as fruit, leave, root, peel, bark, flower, seed, and sometimes entire plant are used as effective corrosion inhibitors for various metals and alloys in different electrolytic media.

Obviously, leaves extracts exhibit relatively higher inhibition efficiency than that of the other extracts. This is because leaves are the primary source of phytochemicals in the plant where their synthesis takes place [6,7]. Plant extracts can be prepared either in organic or aqueous or their mixed solvents therefore the extracts can be classified as organic extracts and aqueous extracts type. Literature study suggests that both types of plant extracts are expensively employed as corrosion inhibitors [8–11]. Reasonably, the use of aqueous plant extracts would be preferred as they exhibit relatively more solubility and high protection effectiveness as compared to the organic extracts. Generally, the preparation of both types of extracts involves similar steps. The first step involves the collection, cleaning, washing, and drying of the plant materials to be used for the preparation of extract. The second step involves the crushing and powdering of the plant materials. The third step involves stirring the power of plant materials in an appropriate (organic or aqueous) solvent. The last step involves the concentrating of the extracts and their storage at low temperatures. Common steps involve in the preparation of an extract are illustrated in Fig. 17.1.

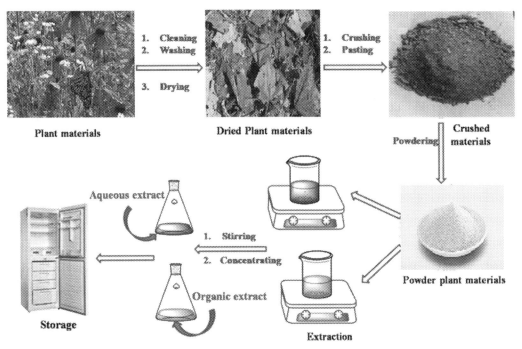

FIG. 17.1 Diagrammatic illustration of common steps involve in the preparation of a plant extract.

Plant extracts are ideal environmental-friendly materials to replace the tradition toxic corrosion inhibitors because of their following benefits:

- (i) Ease of fabrication and application.
- (ii) Environmental friendly (nontoxic, nonbioaccumulative, and biodegradable).
- (iii) High corrosion inhibition efficiency, as each plant extract contains various complex phytochemicals that exhibit strong binding with the metallic surface.
- (iv) Useful for different metal/electrolyte systems.

However, despite being environmental-friendly and associated with various advantages, use of plant extracts as corrosion inhibitors is associated with few limitations. Some of the common limitations are:

- (i) Sometimes fabrication of some plant extracts involves quite tedious processes (several steps).
- (ii) Organic extraction cannot be treated as an environmental-friendly approach, as it is associated with consumption and discharge of huge amount of toxic (organic) solvents.
- (iii) Most of the phytochemicals are temperature dependent and undergo degradation at high temperature.
- (iv) Phytochemicals can undergo decomposition, degradation, or rearrangement in high acid or basic conditions (of electrolyte).

(v) Organic extracts show limited solubility in the polar aqueous electrolytes that limits their application.
(vi) Some of the plant extracts exhibit small corrosion inhibition effectiveness and need some external additives to increase their inhibition effect through synergism [12–18].

17.1 Mechanism of corrosion inhibition using plant extracts

Plant extracts become effective by the adsorption and formation of protective film by the phytochemicals. Obviously, phytochemicals are complex molecules and they contain several electrons rich centers (active sites) through which they can adsorb on the metallic surface [19–21]. These phytochemicals may get adsorbed using physisorption or chemisorption or physiochemisorption mechanism [22–24]. It is important to mention that physisorption occurs through the electrostatic interactions between charged inhibitor (phytochemical) molecules and charged metallic surfaces. On the other hand, chemisorption occurs through the charge transfer reactions between the metal surface and phytochemical molecules [25,26].

In the acidic solutions, heteroatoms (mainly, N, O, and S) easily get protonated, because of the presence of their unshared electron pair(s), and exist in their cationic (protonated) forms. On the other side, the metallic surface becomes negatively charged because of the adsorption of negatively charged counter ions. These oppositely charged species attracted each other through electrostatic force of attraction (physisorption) [27–29]. However, once the protonated heteroatoms return their neutral form by the intake of electrons, they transfer their nonbonding electrons to the d-orbitals of surface metallic atoms through coordination bonding. This process of electron transfer from phytochemicals to metallic d-orbitals is known as a donation. Nevertheless, metals are already electron-rich species, such type of electron donation causes interelectronic repulsion that renders metal to transfer its electron to the empty orbitals of the phytochemical molecules. This process of electron transfer is known as a retro-(back) donation. It is important to mention that the greater the donation greater will be retro-donation [27–29]. Mechanism of metallic corrosion (anodic and cathodic Tafel reactions) with and without phytochemicals (PHYTs) can be presented as follows [27–29]:

(a) Mechanism of corrosion in the absence of PHYTs
    (i) Anodic reactions (oxidative dissolution)

$$Fe + H_2O \leftrightarrow FeOH_{ads} + H^+ + e^- \tag{17.1}$$

$$FeOH_{ads} \leftrightarrow FeOH^+ + e^- \ (\text{Slowest/rate determining step}) \tag{17.2}$$

$$FeOH^+ + H^+ \leftrightarrow Fe^2 + H_2O \tag{17.3}$$

    (ii) Cathodic reactions (reductive hydrogen evolution)

$$Fe + H^+ \leftrightarrow (FeH)^+_{ads} \tag{17.4}$$

$$(FeH)^+_{ads} + e^- \leftrightarrow (FeH)_{ads} \tag{17.5}$$

$$(FeH)_{ads} + H^+ + e^- \leftrightarrow Fe + H_2 \tag{17.6}$$

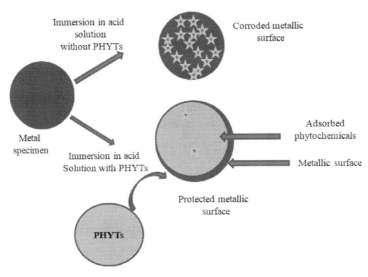

FIG. 17.2 Protection of metallic surface from corrosive damage through phytochemicals adsorption.

(b) Mechanism of corrosion in the presence of PHYTs

$$Fe + H_2O \leftrightarrow Fe(H_2O)_{ads} \tag{17.7}$$

$$Fe(H_2O)_{ads} + PHYTs \leftrightarrow (FeOH)^-_{ads} + H^+PHYTs \tag{17.8}$$

$$Fe(H_2O)_{ads} + PHYTs \leftrightarrow Fe-PHYTs_{ads} + H_2O \tag{17.9}$$

$$(FeOH)^-_{ads} \leftrightarrow (FeOH)_{ads} + e^- \text{ (Slowest/rate determining step)} \tag{17.10}$$

$$Fe-PHYTs_{ads} \leftrightarrow Fe-PHYTs^+_{ads} + e^- \tag{17.11}$$

$$Fe-PHYTs_{ads} + FeOH_{ads} \leftrightarrow Fe-PHYTs_{ads} + FeOH^+_{ads} \tag{17.12}$$

After their adsorbance, phytochemicals build an inhibitive film that isolates the metal surfaces from further aggressive attack [12,30]. Protection of metallic surface from corrosive damage is presented in Fig. 17.2.

17.2 Plant extracts as corrosion inhibitors: literature survey

17.2.1 Plant extracts as corrosion inhibitors in HCl based electrolytes

Hydrochloric acid solutions of different concentrations are extensively used as electrolytes for industrial as well as academic purposes [31–34]. Literature investigation shows that lower concentrations of HCl solutions are useful for academic purposes while highly concentrated HCl solutions are used for industrial purposes [35,36]. Some of the common industrial processes that involve the consumption of HCl solutions are acid pickling, acid descaling, acid cleaning (of rusts and scales), and industrial oil-well acidification [37–39]. HCl solutions are highly corrosive at all concentrations therefore various measures have been developed to overcome the problem of corrosion.

Plant extracts have also been used widely as corrosion inhibitors for different metals in hydrochloric acid solutions. Muthukumarasamy and coworkers [40] demonstrated the corrosion inhibition effect of *T. fragrans* for mild steel corrosion in an acidic solution of hydrochloric acid (1M HCl). Through potentiodynamic polarization study, it was observed that invested plant extract acted as a mixed-type corrosion inhibitor and it became effective by retarding both anodic as well as cathodic Tafel reactions. It was further observed that the extract become effective by adsorbing on the metallic surface following through the Langmuir adsorption isotherm model. *T. fragrans* extract exhibited the highest protection effectiveness of 81% at 500 ppm concentration. Adsorption of the *T. fragrans* extracts on mild steel surface mainly followed the physisorption mechanism.

In another study, the corrosion inhibition effect of *Lavandula mairei* (LM) was investigated for mild steel in 1M hydrochloric acid solution [41]. The corrosion inhibition effect of the LM extract was determined using various experimental and computational methods. The highest inhibition efficiency of 92% was observed at 0.4g L^{-1} concentration. Potentiodynamic polarization study shows that LM extract becomes effective by retarding both anodic and cathodic Tafel reactions and following through a mixed-type corrosion inhibition mechanism. Electrochemical impedance spectroscope (EIS) analysis showed that LM extract becomes effective by adsorbing at the interface of metal and electrolyte and behaves as an interface-type corrosion inhibitor. The presence of the LM in the corrosive medium increases the charge transfer resistance value for the corrosion process.

Adsorption mechanism of corrosion inhibition was supported by XPS, SEM, and UV-vis spectroscope analyses. SEM analysis reveals a significant improvement in the surface morphology of the protected metallic specimen. Quercetin-3-glucuronide was identified as one of the major phytochemicals and undertaken for DFT-based computational study. DFT study was carried for a mono-protonated and neutral form of quercetin-3-glucuronide and it was observed that quercetin-3-glucuronide interacts with the metallic surface using donor–acceptor interactions. Frontier molecular orbitals (optimized, HOMO, and LUMO) of quercetin-3-glucuronide are given in Fig. 17.3. It can be seen that HOMO and LUMO are distributed over the entire part of quercetin-3-glucuronide molecule indicating that entire segments of the molecules are participating in the charge (electron) sharing. Molecular dynamics (MD) simulations study suggests that quercetin-3-glucuronide adsorbs spontaneously over the metallic surface using its electron-rich centers. For simulation Fe (110) surface was chosen. Negative values of interaction energy (E_{int}) indicate that quercetin-3-glucuronide spontaneously adsorbs on the metallic surface.

Table 17.1 summarizes the collection of some major reports in which different plant extracts have been used as corrosion inhibitors in hydrochloric acid solution. Investigation of the table showed that most of the extracts act as mixed-type corrosion inhibitors. Their adsorption mostly followed the Langmuir adsorption isotherm model. Adsorption of the plant extracts on the metallic surface was studied using various surface monitoring techniques, especially using EDX, AFM, SEM, XPS, XRD, FT-IR, and UV-vis methods.

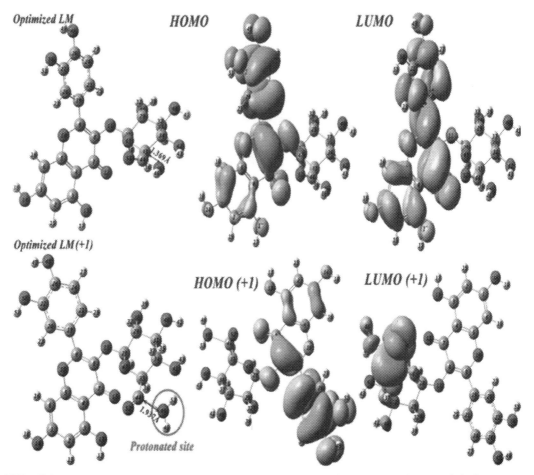

FIG. 17.3 Molecular orbitals of neutral as well as protonated form of Quercetin-3-glucuronide [41].

TABLE 17.1 Collection on some major reports on plant extracts as corrosion inhibitors in hydrochloric acid solutions.

S. No.	Name/scientific names of plants	Electrolyte/metal	Conc./Efficiency (%EI)	Isotherm model/ electrochemical	References
1	*Tunbergia fragrans*	1M HCl/Mild steel	500 ppm/81%	Langmuir/Mixed type	[40]
2	*Lavandula mairei* extract	1M HCl/Mild steel	0.4 g/L/92%	Langmuir/Mixed type	[41]
3	Magnolia grandiflora	1M HCl/Q235 steel	500 ppm/85.0%	Langmuir/Mixed type	[42]
4	Tea factory solid waste extract	1 M HCl/Boiler quality steel	500 mg/L /84.53%	Langmuir/Mixed type	[43]
5	Rosmarinus officinalis	1M HCl/XC48 steel	0.5g/L/87.5%	Langmuir/Mixed type	[44]
6	Ficus racemosa leaf extract	0.1M HCl/Carbon steel	1500 ppm/93.11 ± 1.99%	Langmuir/Mixed type	[45]
7	Ziziphora leaves extract	1M HCl/Mild steel	800 ppm/93%	Langmuir/Mixed type	[46]

(*continued*)

17.2 Plant extracts as corrosion inhibitors: literature survey

S. No.	Name/scientific names of plants	Electrolyte/metal	Conc./Efficiency (%EI)	Isotherm model/ electrochemical	References
8	Hyalomma tick extract	1 M HCl/Carbon steel in	3 g/L/92%	Langmuir/Mixed type	[47]
9	Aloysia citrodora leaves extract	1M HCl/Mild steel	800 ppm/94%	Langmuir/Mixed type	[48]
10	Citrus reticulata peels extract	1M HCl/Carbon steel	1g/L/93.3%	Langmuir/Mixed type	[49]
11	Lemon balm extract	1M HCl/Carbon steel	200 ppm (LB)+600 ppm Zn^{2+}/93%	Mixed type	[50]
12	Euphorbia heterophylla L. extract	1.5 M HCl/Mild steel	–	Flory–Huggins isotherm	[51]
13	Portulaca grandiflora	0.5M HCl/N80 carbon steel	2 ml/L/95%	Langmuir/Mixed type	[52]
14	Rosa canina fruit extract	1M HCl/Mild steel	800 ppm/86%	Mixed type	[53]
15	Musa acuminata	5M HCl/Mild steel	4g/L/90.0%	Mixed type	[54]
16	Chinese gooseberry fruit shell extract	1M HCl/Mild steel	1000 ppm/94%	Langmuir/Mixed type	[55]
17	Poppy extract	1M HCl/Mild steel	600 ppm/97.64%	Mixed type	[56]
18	Pterocarpus santalinoides	1M HCl/Carbon steel	0.7g/L/85.25%	Mixed type	[57]
19	Garcinia indica (Binda) fruit rind extract	0.5 & 1 M HCl/Mild steel	4 (V/V%)/93.94%	Langmuir/Mixed type	[58]
20	Borage flower extract	1M HCl/Mild steel	800 ppm/91%	Langmuir/Mixed type	[59]
21	Bee pollen extract	1M HCl/Copper	7 g/L/94.5%	Langmuir/Mixed type	[60]
22	Dardagan Fruit extract	1 M HCl/Carbon steel in	3000 ppm/92.1%	Langmuir/Mixed type	[61]
23	Houttuynia cordata leaf extract	0.1M HCl/Carbon steel	1500 ppm/98.3%	Mixed type	[62]
24	Luffa cylindrica leaf extract	0.5M HCl/Mild steel	1 g/L/87.89 ±0.325%	Langmuir/Mixed type	[63]
25	Punica granatum L (pomegranate bark)	1M HCl/Mild steel	1 g/L/88%	Langmuir/Mixed type	[64]
26	Tithonia Diversifolia flower Extract	1M HCl/Mild steel	0.7 V/V%/92.15%	Langmuir & Temkin/Mixed type	[65]
27	Prunus dulcis peels extract	0.1 M HCl/Mild steel	11.8 mg/L/94%	Langmuir/Mixed type	[66]
28	Garlic extract	0.5M HCl/ISI 304 S steel	8 ccL-1 /88.6%	Langmuir/Mixed type	[67]
29	Zea mays hairs waste extract	1M HCl/Mild steel	0.15 mM/87.92%	Langmuir/Mixed type	[68]
30	Petroselium Sativum extract	1.2M HCl/Mild steel	5 g/L/92.39%	Langmuir/Mixed type	[69]

(continued)

TABLE 17.1 (Cont'd)

S. No.	Name/scientific names of plants	Electrolyte/metal	Conc./Efficiency (%EI)	Isotherm model/ electrochemical	References
31	Eriobotrya japonica Lindl leaves	1M HCl/Mild steel	800 ppm	Mixed type	[70]
32	Taxus baccata extract	1M HCl/Carbon steel	600 ppm/83.28%	Langmuir/Mixed type	[71]
33	Cissus quadrangularis plant extract	1M HCl/Mild steel	1100 ppm/89.8%	Langmuir/Mixed type	[72]
34	Pigeon pea leaf extract	1.2M HCl/Mild steel	0.9 g/L/87.13%	Langmuir/Mixed type	[73]
35	Pineapple stem extract (Bromelain)	1M HCl/Carbon steel	1000 ppm/97.6%	Langmuir/Mixed type	[74]
36	Laurus nobilis extract	1M HCl/Carbon steel	400 ppm/92%	Langmuir/Mixed type	[75]
37	Ficus tikoua leaves extract	1M HCl/Carbon steel	1000 ppm/95.8%	Langmuir/Mixed type	[76]
38	Borassus flabellifer	1M HCl/Aluminum	0.4 gL-1/66.8%	Langmuir/Mixed type	[77]
39	Clove essential oil extract	0.5 M HCl & H_2SO_4/Carbon steel	2.5g/L/95.48% (H_2SO_4)	Cathodic-type	[78]
40	Primula vulgaris flower aqueous extract	1M HCl/Carbon steel	1000 ppm/95.5%	Langmuir/Cathodic-type	[79]
41	Sargassum muticum extract	Carbon steel 1M HCl	1 gL-1/97%	Mixed type	[80]
42	Peganum harmala seed extract	1M HCl/Mild steel	800 ppm/95%	Langmuir/Mixed type	[81]
43	Lemon Balm extract	1M HCl/Mild steel	800 ppm/95%	Mixed type	[82]
44	Mangifera indica (mango) leaves	1M HCl/Mil steel	1000 ppm/92%	Langmuir/Mixed type	[83]
45	Gentiana olivieri extracts	0.5M HCl/Mild steel	800 mg/L/93.7%	Langmuir/Mixed type	[84]
46	Ipomea staphylina Leaf extract	1M HCl/Mild steel	800 ppm/92.77%	Langmuir/Mixed type	[85]
47	Haematostaphis barteri Leaves Extract	1M HCl/Mild steel	40 g/L/73.74%	Langmuir & Temkin	[86]
48	Urtica dioica leaves extract	1M HCl/Mild steel	800 ppm/92%	Langmuir/Mixed type	[87]
49	Mustard seed extract	1M HCl/Mild steel	200 mg/L/97%	Langmuir/Mixed type	[88]
50	Rollinia occidentalis	1M HCl/Carbon steel	1 g/L/85.7%	Langmuir/Mixed type	[89]

(continued)

S. No.	Name/scientific names of plants	Electrolyte/metal	Conc./Efficiency (%EI)	Isotherm model/ electrochemical	References
51	Papaya peel extract	1M HCl/Aluminum alloy	2 g/L/98.1%	Langmuir/Mixed type	[90]
52	Sweet melon peel extract	1M HCl/Mild steel	0.5 g/L/81.26%	Langmuir/Mixed type	[91]
53	Pisum sativum	1M HCl/Mild steel	400 mg/L/91%	Langmuir/Mixed type	[92]
54	Griffonia simplicifolia	1M HCl/J55 steel	1000 ppm/91.73%	Temkin/Mixed type	[93]
55	Lychee fruit	0.5M HCl/Mild steel	600 ppm/97.95%	Langmuir/Mixed type	[94]
56	Zizyphus Lotuse	1M HCl/Copper	1 g/L/93%	Cathodic-type	[95]
57	Lecaniodiscus cupaniodes	0.5M HCl/Mild steel	4 ml/L/93.46%		[96]
58	Xanthium strumarium	1M HCl/Carbon steel	10 ml/L/94.82%	Langmuir	[97]
59	Sunflower seed hull extract	1M HCl/Mild steel	400 ppm/98%	Langmuir/Mixed type	[98]
60	Glycine max, Cuscuta reflexa & Spirogyra	1M HCl/Mild steel	2g/L/94.05% (SGAE)	Langmuir/Mixed type	[99]
61	Longan seed and peel	0.5 M HCl/Mild steel	600 ppm/92.35%	Langmuir/Mixed type	[100]
62	Calendula officinalis flower heads extract	1M HCl/Mild steel	500 ppm/94.88%	Langmuir/Mixed type	[101]
63	Glycyrrhiza glabra leaves extract	1M HCl/Mild steel	800 ppm	Langmuir/Mixed type	[102]
64	Thymus vulgaris	1M HCl/304L SS	2% Inh conc./49%	Mixed type	[103]
65	Citrullus lanatus fruit (CLF) extract	1M HCl/Mild steel	800 ppm/91%	Langmuir/Mixed type	[104]
66	Ficus hispida leaves	1M HCl/Mild steel	250 ppm/90%	Langmuir/Cathodic type	[105]
67	Ginkgo leaf extract	1M HCl/X70 steel	200 mg/L/90%	Langmuir/Cathodic type	[106]
68	Tamarindus indiaca aqueous extract	1M HCl/Mild steel	800 ppm/93%	Langmuir/Mixed type	[107]
69	Aquilaria subintergra	1M HCl/Mild steel	10-4M/62%	Langmuir/Mixed type	[108]

17.2.2 Plant extracts as corrosion inhibitors in H_2SO_4-based electrolytes

Sulphuric or sulfuric acid (H_2SO_4) solutions of different concentrations are extensively used as electrolytes for different metals and alloys for academic and industrial purposes [31]. Similar to the HCl-based solution, highly concentrated sulfuric acid solutions are used for industrial purposes while whereas lower concentrations of sulfuric acid solutions are used for academic purposes [109–111]. Plant extracts are widely used as corrosion inhibitors for different metals and alloys in sulfuric acid solutions. In a study, `Singh and coworkers [112] investigated the inhibition effect of the *Litchi Chinensis* (litchi) for mild steel corrosion in 0.5 H_2SO_4. Various chemical, surface, and electrochemical methods were used to demonstrate the corrosion inhibition effect of the litchi extract. The litchi extract showed the highest protection effectiveness of 95.7% at 3 g L^{-1} concentration. Potentiodynamic polarization study shows that litchi extract acts as a mixed-type corrosion inhibitor. SEM analysis showed that significant improvement in the surface morphology of the metal was observed in the presence of litchi extract.

Table 17.2 represents the corrosion inhibition effect of the other plant extracts for different metals and alloys in sulfuric acid solution. Generally, plant extracts act as mixed-type corrosion inhibitors and they become effective by retarding both anodic and cathodic half-cell reactions. Using EIS analysis, it is observed that most of the plant extracts become effective by increasing the value of charge transfer resistance for the corrosion process through their adsorption at the interface of metal and electrolyte. Adsorption of plant extracts on the metallic surface in sulfuric acid solution mainly follows the Langmuir adsorption isotherm model. Adsorption of plant extracts on the metallic surface in sulfuric acid solutions is supported

TABLE 17.2 Collection on some major reports on plant extracts as corrosion inhibitors in hydrochloric acid solutions.

S. No.	Name/scientific names of plants	Electrolyte/metal	Conc./Efficiency (%EI)	Isotherm model/ Electrochemical	References
1	Idesia polycarpa fruit extract	0.5M H_2SO_4/Copper	300 mg/L/93.8%	Langmuir/Mixed type	[113]
2	Salvia officinalis L.	0.5M H_2SO_4/Carbon steel	4 g/L/91.62%	Mixed type	[114]
3	Papaya leaves extract	0.5M H_2SO_4/Copper	150 ppm/92.9%	Langmuir/Mixed type	[115]
4	*Litchi Chinensis* (litchi)	0.5M H_2SO_4/Mild steel	3 g/L/95.7%	Langmuir/Mixed type	[112]
5	Essential oil extracts	1M H_2SO_4/Mild steel	6 g/L/86.58%	Langmuir/Mixed type	[116]
6	Sida cordifolia	0.5M H_2SO_4/Mild steel	500 mg/L/98.83%	Langmuir/Mixed type	[117]
7	Saraca ashoka	0.5M H_2SO_4/Mild steel	100 ppm/95.48%	Langmuir/Mixed type	[118]
8	Spathodea Campanulata	1N H_2SO_4/Mild steel	1.5% inh conc./82.74%	Langmuir/Mixed type	[119]
9	Ficus religiosa	0.5M H_2SO_4/Mild steel	500 mg/L/92.26%	Langmuir/Mixed type	[120]
10	Dryopteris cochleata	1M H_2SO_4/Aluminum	2400 ppm/83.24%	Freundlich/Mixed type	[121]
11	Lannea coromandelica	1M H_2SO_4/Mild steel	250 ppm/88.5%	Langmuir/Mixed type	[122]
12	Magnolia kobus DC. (M. kobus),	1M H_2SO_4/Mild steel	500 ppm/95.01%	Mixed type	[123]
13	Cuscuta reflexa	0.5M H_2SO_4/Mild steel	500 mg/L/95.47%	Langmuir	[124]
14	Myristica fragrans	0.5M H_2SO_4/Mild steel	500 mg/L/87.81%	Langmuir/Mixed type	[125]

by surface investigation methods, especially through SEM, EDX, AFM, XRD, XPD, UV-vis, and FT-IR methods. Through computational analyses, it was observed that plant extracts mostly interact with the metallic surface using donor–acceptor interactions. Most of the plant extracts spontaneously interact with the metal surface using donor–acceptor mechanism.

17.2.3 Plant extracts as corrosion inhibitors in other electrolytes

Plant extracts are also evaluated as corrosion inhibitors for the corrosive media other than hydrochloric acid and sulfuric acid solutions. A summary of major recent reports on corrosion inhibition effect of plant extracts in media other than HCl and sulfuric acid is given in Table 17.3. It can be seen that plant extracts are widely used as corrosion inhibitors for metals

TABLE 17.3 Collection on some major reports on plant extracts as corrosion inhibitors in the media other than HCl and H_2SO_4.

S. No.	Name/scientific names of plants	Electrolyte/metal	Conc./Efficiency (%EI)	Isotherm model/ electrochemical	References
1	Ricinus communis	3.5% NaCl Carbon steel	100 ppm/87.5%	Temkin/Mixed type	[138]
2	Leaves extracts	1M NaOH/Aluminum	2 g/L/97% (RS)	Mixed type	[133]
3	Bark extracts	1M NaOH/Aluminum	0.6 g/L/85.3% (MO)	Langmuir/Mixed type	[132]
4	Centaurea cyanus	Saline/Carbon steel	10 ppm/63.95%	Langmuir/Mixed type	[129]
5	Apium graveolens L.	0.25M NaOH/Aluminum	1.5 g/L/93.33%	Langmuir	[134]
6	Allium sativum (garlic extract)	MIC/Carbon steel	100 ppm/81%	–	[136]
7	Cichorium intybus L	3.5% NaCl/Mild steel	200 ppm/93%	Mixed type	[128]
8	Urtica Dioica leaves extract	3.5% NaCl/Carbon steel	–	Mixed type	[126]
9	Nettle leaves extract	3.5% NaCl/Mild steel	400 ppm/95%	Mixed type	[127]
10	Cistus ladanifer leaves extract	Biotic/304L SS	99.36%	Mixed type	[135]
11	Neem extract	MIC/API 5LX CS	150 ppm/81%	–	[137]
12	Cascabela Thevetia	(3.5% NaCl+Na2S)/Carbon steel	300 ppm/95%	Freundlich/Mixed type	[139]
13	Bagassa guianensis	3% NaCl/Zinc	100 ppm/97%	Mixed type	[101]
14	Chromolaena odorata stems extract	1M NaCl/Mild steel	1000 ppm/99.83%	Langmuir/Mixed type	[140]
15	Platanus acerifolia leaves	Chloride media/Reinforced concrete		Mixed type	[141]
16	Olive leaf extract	Seawater/Reinforced concrete	91.9%	Mixed type	[142]
17	Myrmecodia Pendans Extract	3.5% NaCl/Carbon steel	400 ppm/79.70%	Langmuir/Mixed type	[143]

and alloys in NaCl-based [126–131] and sodium hydroxide-based solutions [132–134]. Plant extracts are also evaluated as corrosion inhibitors for microbially influenced corrosion [135–137]. It can be seen that in such electrolytes, plant extracts behaved as mixed-type corrosion inhibitors and become effective by retarding both anodic as well as cathodic Tafel reactions. They inhibit corrosion by adsorbing on the metallic surface following mostly through the Langmuir adsorption isotherm model. EIS analyses confirm the interfacial behavior of the plant extracts.

17.3 Summary

From the ongoing discussion, it is clear that plant extracts are ideal environmental-friendly materials to replace the traditional toxic corrosion inhibitors. Because of their natural and biological origin, they are regarded as eco-friendly alternatives. Literature study shows that various reports have been published on extracts as green corrosion inhibitors. Plant extracts are successfully used as corrosion inhibitors for different metals and alloys in various electrolytes including hydrochloric acid, sulfuric acid, and sodium chloride based electrolytes. Mostly, plant extracts become effective by retarding both anodic and cathodic reactions and behaving as mixed-type corrosion inhibitors. Their adsorption on the metallic surface mostly followed the Langmuir adsorption isotherm model, though few other isotherm models such as Temkin and Freundlich isotherm models have also been identified. EIS study shows that plant extracts mostly act as interface-type corrosion inhibitors as they become effective by increasing the value of charge transfer resistance. Through computational analyses, it was observed that plant extracts mostly interact with the metallic surface using donor–acceptor interactions.

17.4 Important websites

https://www.sciencedirect.com/topics/medicine-and-dentistry/plant-extract
https://www.intechopen.com/books/plant-extracts/introductory-chapter-plant-extracts
https://www.gea.com/en/pharma-healthcare/nutraceuticals/plant-extracts.jsp
http://www.berkem.com/en/expertise-en/plant-extraction

References

[1] G. Ji, S.K. Shukla, P. Dwivedi, S. Sundaram, R. Prakash, Inhibitive effect of Argemone mexicana plant extract on acid corrosion of mild steel, Ind. Eng. Chem. Res. 50 (2011) 11954–11959.

[2] P.M. Krishnegowda, V.T. Venkatesha, P.K.M. Krishnegowda, S.B. Shivayogiraju, Acalypha torta leaf extract as green corrosion inhibitor for mild steel in hydrochloric acid solution, Ind. Eng. Chem. Res. 52 (2013) 722–728.

[3] P.B. Raja, M. Fadaeinasab, A.K. Qureshi, A.A. Rahim, H. Osman, M. Litaudon, K. Awang, Evaluation of green corrosion inhibition by alkaloid extracts of Ochrosia oppositifolia and isoreserpiline against mild steel in 1 M HCl medium, Ind. Eng. Chem. Res. 52 (2013) 10582–10593.

[4] S.A. Umoren, Z.M. Gasem, I.B. Obot, Natural products for material protection: inhibition of mild steel corrosion by date palm seed extracts in acidic media, Ind. Eng. Chem. Res. 52 (2013) 14855–14865.

[5] E.E. Oguzie, K.L. Oguzie, C.O. Akalezi, I.O. Udeze, J.N. Ogbulie, V.O. Njoku, Natural products for materials protection: corrosion and microbial growth inhibition using Capsicum frutescens biomass extracts, ACS Sustain. Chem. Eng. 1 (2013) 214–225.

[6] E.C.C. Baly, I.M. Heilbron, W.F. Barker, CX.—Photocatalysis. Part I. The synthesis of formaldehyde and carbohydrates from carbon dioxide and water, J. Chem. Soc. Trans. 119 (1921) 1025–1035.

[7] M. Schreiner, S. Huyskens Keil, Phytochemicals in fruit and vegetables: health promotion and postharvest elicitors, Crit. Rev. Plant Sci. 25 (2006) 267–278.

[8] K. Xhanari, M. Finšgar, M.K. Hrnčič, U. Maver, Ž. Knez, B. Seiti, Green corrosion inhibitors for aluminium and its alloys: a review, RSC Adv. 7 (2017) 27299–27330.

[9] M.A. Dar, A review: plant extracts and oils as corrosion inhibitors in aggressive media, Ind. Lubr. Tribol. 64 (2011) 227–233.

[10] C.O. Akalezi, C.E. Ogukwe, C.K. Enenebaku, E.E. Oguzie, Corrosion inhibition of aluminium pigments in aqueous alkaline medium using plant extracts, Environ. Pollut. 1 (2012) 45.

[11] S. Mo, H.-Q. Luo, N.-B. Li, Plant extracts as "green" corrosion inhibitors for steel in sulphuric acid, Chem. Pap. 70 (2016) 1131–1143.

[12] P. Okafor, V. Osabor, E. Ebenso, Eco-friendly corrosion inhibitors: inhibitive action of ethanol extracts of Garcinia kola for the corrosion of mild steel in H2SO4 solutions, Pigm. Resin Technol. 36 (2007) 299–305.

[13] I. Obot, S. Umoren, N. Obi-Egbedi, Corrosion inhibition and adsorption behaviour for aluminuim by extract of Aningeria robusta in HCl solution: Synergistic effect of iodide ions, J. Mater. Environ. Sci. 2 (2011) 60–71.

[14] E.E. Oguzie, Evaluation of the inhibitive effect of some plant extracts on the acid corrosion of mild steel, Corros. Sci. 50 (2008) 2993–2998.

[15] U. Eduok, S. Umoren, A. Udoh, Synergistic inhibition effects between leaves and stem extracts of Sida acuta and iodide ion for mild steel corrosion in 1 M H2SO4 solutions, Arabian J. Chem. 5 (2012) 325–337.

[16] O. Benali, H. Benmehdi, O. Hasnaoui, C. Selles, R. Salghi, Green corrosion inhibitor: inhibitive action of tannin extract of Chamaerops humilis plant for the corrosion of mild steel in 0.5 M H2SO4, J. Mater. Environ. Sci 4 (2013) 127–138.

[17] A. Fouda, K. Shalabi, M. Shaaban, Synergistic effect of potassium iodide on corrosion inhibition of carbon steel by achillea santolina extract in hydrochloric acid solution, J. Bio- Tribo-Corros. 5 (2019) 71.

[18] K. Orubite-Okorosaye, I. Jack, M. Ochei, O. Akaranta, Synergistic effect of potassium iodide on corrosion inhibition of mild steel in HCl medium by extracts of Nypa fruticans' Wurmb, JASEM 11 (2007) 27–31.

[19] J. Buchweishaija, Phytochemicals as green corrosion inhibitors in various corrosive media: a review, Tanz. J. Sci. 35 (2009) 77–92.

[20] N. Palaniappan, I. Cole, F. Caballero-Briones, S. Manickam, K.J. Thomas, D. Santos, Experimental and DFT studies on the ultrasonic energy-assisted extraction of the phytochemicals of Catharanthus roseus as green corrosion inhibitors for mild steel in NaCl medium, RSC Adv. 10 (2020) 5399–5411.

[21] M. Al-Otaibi, A. Al-Mayouf, M. Khan, A. Mousa, S. Al-Mazroa, H. Alkhathlan, Corrosion inhibitory action of some plant extracts on the corrosion of mild steel in acidic media, Arabian J. Chem. 7 (2014) 340–346.

[22] C. Verma, E. Ebenso, M. Quraishi, Alkaloids as green and environmental benign corrosion inhibitors: An overview, Int. J. Corros. Scale Inhib. 8 (2019) 512–528.

[23] C. Verma, D.K. Verma, E.E. Ebenso, M.A. Quraishi, Sulfur and phosphorus heteroatom-containing compounds as corrosion inhibitors: an overview, Heteroat. Chem. 29 (2018) e21437.

[24] N. Gupta, M. Quraishi, P. Singh, V. Srivastava, K. Srivastava, C. Verma, A. Mukherjee, Curcumine longa: green and sustainable corrosion inhibitor for aluminum in HCl medium, Anal. Bioanal. Electrochem. 9 (2017) 245–265.

[25] E.A. Noor, A.H. Al-Moubaraki, Thermodynamic study of metal corrosion and inhibitor adsorption processes in mild steel/1-methyl-4 [4′(-X)-styryl pyridinium iodides/hydrochloric acid systems, Mater. Chem. Phys. 110 (2008) 145–154.

[26] N. Kovačević, A. Kokalj, Chemistry of the interaction between azole type corrosion inhibitor molecules and metal surfaces, Mater. Chem. Phys. 137 (2012) 331–339.

[27] D.K. Yadav, M.A. Quraishi, Application of some condensed uracils as corrosion inhibitors for mild steel: gravimetric, electrochemical, surface morphological, UV–visible, and theoretical investigations, Ind. Eng. Chem. Res. 51 (2012) 14966–14979.

[28] J. Fu, H. Zang, Y. Wang, S. Li, T. Chen, X. Liu, Experimental and theoretical study on the inhibition performances of quinoxaline and its derivatives for the corrosion of mild steel in hydrochloric acid, Ind. Eng. Chem. Res. 51 (2012) 6377–6386.

[29] D.K. Yadav, M. Quraishi, Electrochemical investigation of substituted pyranopyrazoles adsorption on mild steel in acid solution, Ind. Eng. Chem. Res. 51 (2012) 8194–8210.
[30] F.A. Ansari, M. Quraishi, Inhibitive performance of gemini surfactants as corrosion inhibitors for mild steel in formic acid, Portugaliae Electrochimica Acta 28 (2010) 321–335.
[31] M.A. Quraishi, D.S. Chauhan, V.S. Saji, Heterocyclic Organic Corrosion Inhibitors: Principles and Applications, Elsevier, UK, 2020.
[32] L.-F. Li, P. Caenen, J.-P. Celis, Effect of hydrochloric acid on pickling of hot-rolled 304 stainless steel in iron chloride-based electrolytes, Corros. Sci. 50 (2008) 804–810.
[33] J. Hudson, J.C. Bell, N.I. Jaeger, Potentiostatic current oscillations of cobalt electrodes in hydrochloric acid/chromic acid electrolytes, Ber. Bunsenges. Phys. Chem. 92 (1988) 1383–1387.
[34] K. Lund, H.S. Fogler, C. McCune, J. Ault, Acidization—II. The dissolution of calcite in hydrochloric acid, Chem. Eng. Sci. 30 (1975) 825–835.
[35] R. Myer, C.F. Pickett, Method of and composition for removing rust and scale, Google Patents, 1953, US2631950A, United States.
[36] H. Hans-Gunther, Z. Wilfried, Method of removing rust from ironcontaining materials, particularly for the cleaning of boiler plants, Google Patents, 1963, US3085915A, United States.
[37] B. Tang, W. Su, J. Wang, F. Fu, G. Yu, J. Zhang, Minimizing the creation of spent pickling liquors in a pickling process with high-concentration hydrochloric acid solutions: Mechanism and evaluation method, J. Environ. Manage. 98 (2012) 147–154.
[38] M. Tomaszewska, M. Gryta, A. Morawski, Recovery of hydrochloric acid from metal pickling solutions by membrane distillation, Sep. Purif. Technol. 22 (2001) 591–600.
[39] R. Gueccia, A.R. Aguirre, S. Randazzo, A. Cipollina, G. Micale, Diffusion dialysis for separation of hydrochloric acid, iron and zinc ions from highly concentrated pickling solutions, Membranes 10 (2020) 129.
[40] K. Muthukumarasamy, S. Pitchai, K. Devarayan, L. Nallathambi, Adsorption and corrosion inhibition performance of *Tunbergia fragrans* extract on mild steel in acid medium, Mater. Today: Proc. 33 (2020) 4054–4058.
[41] A. Berrissoul, A. Ouarhach, F. Benhiba, A. Romane, A. Zarrouk, A. Guenbour, B. Dikici, A. Dafali, Evaluation of *Lavandula mairei* extract as green inhibitor for mild steel corrosion in 1 M HCl solution. Experimental and theoretical approach, J. Mol. Liq. 313 (2020) 113493.
[42] S. Chen, S. Chen, B. Zhu, C. Huang, W. Li, Magnolia grandiflora leaves extract as a novel environmentally friendly inhibitor for Q235 steel corrosion in 1 M HCl: Combining experimental and theoretical researches, J. Mol. Liq. 311 (2020) 113312.
[43] A. Pal, C. Das, A novel use of solid waste extract from tea factory as corrosion inhibitor in acidic media on boiler quality steel, Ind. Crops Prod. 151 (2020) 112468.
[44] A. Belakhdar, H. Ferkous, S. Djellali, R. Sahraoui, H. Lahbib, Y.B. Amor, A. Erto, M. Balsamo, Y. Benguerba, Computational and experimental studies on the efficiency of Rosmarinus officinalis polyphenols as green corrosion inhibitors for XC48 steel in acidic medium, Colloids Surf. A 606 (2020) 125458.
[45] H. Anh, N. Vu, L. Huyen, N. Tran, H. Thu, L. Bach, Q. Trinh, S.P. Vattikuti, N. Nam, Ficus racemosa leaf extract for inhibiting steel corrosion in a hydrochloric acid medium, Alex. Eng. J. 59 (2020) 4449–4462.
[46] A. Dehghani, G. Bahlakeh, B. Ramezanzadeh, M. Ramezanzadeh, Potential role of a novel green eco-friendly inhibitor in corrosion inhibition of mild steel in HCl solution: detailed macro/micro-scale experimental and computational explorations, Constr. Build. Mater. 245 (2020) 118464.
[47] M.A. Bidi, M. Azadi, M. Rassouli, A new green inhibitor for lowering the corrosion rate of carbon steel in 1 M HCl solution: Hyalomma tick extract, Mater. Today Commun. 24 (2020) 100996.
[48] A. Dehghani, G. Bahlakeh, B. Ramezanzadeh, M. Ramezanzadeh, Aloysia citrodora leaves extract corrosion retardation effect on mild-steel in acidic solution: molecular/atomic scales and electrochemical explorations, J. Mol. Liq. 310 (2020) 113221.
[49] E. Ituen, E. Ekemini, L. Yuanhua, R. Li, A. Singh, Mitigation of microbial biodeterioration and acid corrosion of pipework steel using Citrus reticulata peels extract mediated copper nanoparticles composite, Int. Biodeterior. Biodegrad. 149 (2020) 104935.
[50] N. Asadi, M. Ramezanzadeh, G. Bahlakeh, B. Ramezanzadeh, Theoretical MD/DFT computer explorations and surface-electrochemical investigations of the zinc/iron metal cations interactions with highly active molecules from Lemon balm extract toward the steel corrosion retardation in saline solution, J. Mol. Liq. 310 (2020) 113220.
[51] O.A. Akinbulumo, O.J. Odejobi, E.L. Odekanle, Thermodynamics and adsorption study of the corrosion inhibition of mild steel by Euphorbia heterophylla L. Extract in 1.5 M HCl, Results in Materials 5 (2020) 100074.

[52] A.A. Fadhil, A.A. Khadom, S.K. Ahmed, H. Liu, C. Fu, H.B. Mahood, Portulaca grandiflora as new green corrosion inhibitor for mild steel protection in hydrochloric acid: quantitative, electrochemical, surface and spectroscopic investigations, Surf. Interfac. 20 (2020) 100595.

[53] Z. Sanaei, M. Ramezanzadeh, G. Bahlakeh, B. Ramezanzadeh, Use of Rosa canina fruit extract as a green corrosion inhibitor for mild steel in 1 M HCl solution: a complementary experimental, molecular dynamics and quantum mechanics investigation, J. Ind. Eng. Chem. 69 (2019) 18–31.

[54] H. Kumar, V. Yadav, Musa acuminata, (Green corrosion inhibitor) as anti-pit and anti-cracking agent for mild steel in 5M hydrochloric acid solution, Chem. Data Collect. 29 (2020) 100500.

[55] A. Dehghani, G. Bahlakeh, B. Ramezanzadeh, A detailed electrochemical/theoretical exploration of the aqueous Chinese gooseberry fruit shell extract as a green and cheap corrosion inhibitor for mild steel in acidic solution, J. Mol. Liq. 282 (2019) 366–384.

[56] M. Ramezanzadeh, G. Bahlakeh, B. Ramezanzadeh, Probing molecular adsorption/interactions and anti-corrosion performance of poppy extract in acidic environments, J. Mol. Liq. 304 (2020) 112750.

[57] C.C. Ahanotu, I.B. Onyeachu, M.M. Solomon, I.S. Chikwe, O.B. Chikwe, C.A. Eziukwu, Pterocarpus santalinoides leaves extract as a sustainable and potent inhibitor for low carbon steel in a simulated pickling medium, Sustain. Chem. Pharm. 15 (2020) 100196.

[58] A. Thomas, M. Prajila, K. Shainy, A. Joseph, A green approach to corrosion inhibition of mild steel in hydrochloric acid using fruit rind extract of Garcinia indica (Binda), J. Mol. Liq. 312 (2020) 113369.

[59] A. Dehghani, G. Bahlakeh, B. Ramezanzadeh, M. Ramezanzadeh, Potential of Borage flower aqueous extract as an environmentally sustainable corrosion inhibitor for acid corrosion of mild steel: electrochemical and theoretical studies, J. Mol. Liq. 277 (2019) 895–911.

[60] R.K. Ahmed, S. Zhang, Bee pollen extract as an eco-friendly corrosion inhibitor for pure copper in hydrochloric acid, J. Mol. Liq. 316 (2020) 113849.

[61] A. Sedik, D. Lerari, A. Salci, S. Athmani, K. Bachari, İ. Gecibesler, R. Solmaz, Dardagan Fruit extract as eco-friendly corrosion inhibitor for mild steel in 1 M HCl: Electrochemical and surface morphological studies, J. Taiwan Inst. Chem. Eng. 107 (2020) 189–200.

[62] N.S.H. Vu, P.M.Q. Binh, V.A. Dao, V.T.H. Thu, P. Van Hien, C. Panaitescu, N.D. Nam, Combined experimental and computational studies on corrosion inhibition of Houttuynia cordata leaf extract for steel in HCl medium, J. Mol. Liq. 315 (2020) 113787.

[63] O. Ogunleye, A. Arinkoola, O. Eletta, O. Agbede, Y. Osho, A. Morakinyo, J. Hamed, Green corrosion inhibition and adsorption characteristics of Luffa cylindrica leaf extract on mild steel in hydrochloric acid environment, Heliyon 6 (2020) e03205.

[64] A. Marsoul, M. Ijjaali, F. Elhajjaji, M. Taleb, R. Salim, A. Boukir, Phytochemical screening, total phenolic and flavonoid methanolic extract of pomegranate bark (Punica granatum L): Evaluation of the inhibitory effect in acidic medium 1 M HCl, Mater. Today: Proc. 27 (2020) 3193–3198.

[65] P. Divya, S. Subhashini, A. Prithiba, R. Rajalakshmi, Tithonia diversifolia flower extract as green corrosion inhibitor for mild steel in acid medium, Mater. Today: Proc. 18 (2019) 1581–1591.

[66] S. Pal, H. Lgaz, P. Tiwari, I.-M. Chung, G. Ji, R. Prakash, Experimental and theoretical investigation of aqueous and methanolic extracts of Prunus dulcis peels as green corrosion inhibitors of mild steel in aggressive chloride media, J. Mol. Liq. 276 (2019) 347–361.

[67] M.P. Asfia, M. Rezaei, G. Bahlakeh, Corrosion prevention of AISI 304 stainless steel in hydrochloric acid medium using garlic extract as a green corrosion inhibitor: electrochemical and theoretical studies, J. Mol. Liq. 315 (2020) 113679.

[68] S. Aourabi, M. Driouch, M. Kadiri, F. Mahjoubi, M. Sfaira, B. Hammouti, K. Emran, Valorization of Zea mays hairs waste extracts for antioxidant and anticorrosive activity of mild steel in 1 M HCl environment, Arabian J. Chem. 13 (2020) 7183–7198.

[69] M. Benarioua, A. Mihi, N. Bouzeghaia, M. Naoun, Mild steel corrosion inhibition by Parsley (Petroselium Sativum) extract in acidic media, Egypt. J. Pet. 28 (2019) 155–159.

[70] S. Nikpour, M. Ramezanzadeh, G. Bahlakeh, B. Ramezanzadeh, M. Mahdavian, Eriobotrya japonica Lindl leaves extract application for effective corrosion mitigation of mild steel in HCl solution: experimental and computational studies, Constr. Build. Mater. 220 (2019) 161–176.

[71] K. Hanini, B. Merzoug, S. Boudiba, I. Selatnia, H. Laouer, S. Akkal, Influence of different polyphenol extracts of Taxus baccata on the corrosion process and their effect as additives in electrodeposition, Sustain. Chem. Pharm. 14 (2019) 100189.

[72] M. Mobin, M. Basik, M. Shoeb, A novel organic-inorganic hybrid complex based on Cissus quadrangularis plant extract and zirconium acetate as a green inhibitor for mild steel in 1 M HCl solution, Appl. Surf. Sci. 469 (2019) 387–403.

[73] V. Anadebe, O. Onukwuli, M. Omotioma, N. Okafor, Experimental, theoretical modeling and optimization of inhibition efficiency of pigeon pea leaf extract as anti-corrosion agent of mild steel in acid environment, Mater. Chem. Phys. 233 (2019) 120–132.

[74] M. Mobin, M. Basik, J. Aslam, Pineapple stem extract (Bromelain) as an environmental friendly novel corrosion inhibitor for low carbon steel in 1 M HCl, Measurement 134 (2019) 595–605.

[75] A. Dehghani, G. Bahlakeh, B. Ramezanzadeh, M. Ramezanzadeh, Experimental complemented with microscopic (electronic/atomic)-level modeling explorations of Laurus nobilis extract as green inhibitor for carbon steel in acidic solution, J. Ind. Eng. Chem. 84 (2020) 52–71.

[76] Q. Wang, B. Tan, H. Bao, Y. Xie, Y. Mou, P. Li, D. Chen, Y. Shi, X. Li, W. Yang, Evaluation of Ficus tikoua leaves extract as an eco-friendly corrosion inhibitor for carbon steel in HCl media, Bioelectrochemistry 128 (2019) 49–55.

[77] R. Nathiya, S. Perumal, V. Murugesan, V. Raj, Evaluation of extracts of Borassus flabellifer dust as green inhibitors for aluminium corrosion in acidic media, Mater. Sci. Semicond. Process. 104 (2019) 104674.

[78] R.T. Loto, T. Olukeye, E. Okorie, Synergistic combination effect of clove essential oil extract with basil and atlas cedar oil on the corrosion inhibition of low carbon steel, S. Afr. J. Chem. Eng. 30 (2019) 28–41.

[79] M.T. Majd, S. Asaldoust, G. Bahlakeh, B. Ramezanzadeh, M. Ramezanzadeh, Green method of carbon steel effective corrosion mitigation in 1 M HCl medium protected by Primula vulgaris flower aqueous extract via experimental, atomic-level MC/MD simulation and electronic-level DFT theoretical elucidation, J. Mol. Liq. 284 (2019) 658–674.

[80] I. Nadi, Z. Belattmania, B. Sabour, A. Reani, A. Sahibed-Dine, C. Jama, F. Bentiss, Sargassum muticum extract based on alginate biopolymer as a new efficient biological corrosion inhibitor for carbon steel in hydrochloric acid pickling environment: Gravimetric, electrochemical and surface studies, Int. J. Biol. Macromol. 141 (2019) 137–149.

[81] G. Bahlakeh, B. Ramezanzadeh, A. Dehghani, M. Ramezanzadeh, Novel cost-effective and high-performance green inhibitor based on aqueous Peganum harmala seed extract for mild steel corrosion in HCl solution: detailed experimental and electronic/atomic level computational explorations, J. Mol. Liq. 283 (2019) 174–195.

[82] N. Asadi, M. Ramezanzadeh, G. Bahlakeh, B. Ramezanzadeh, Utilizing Lemon Balm extract as an effective green corrosion inhibitor for mild steel in 1M HCl solution: a detailed experimental, molecular dynamics, Monte Carlo and quantum mechanics study, J. Taiwan Inst. Chem. Eng. 95 (2019) 252–272.

[83] M. Ramezanzadeh, G. Bahlakeh, Z. Sanaei, B. Ramezanzadeh, Corrosion inhibition of mild steel in 1 M HCl solution by ethanolic extract of eco-friendly Mangifera indica (mango) leaves: electrochemical, molecular dynamics, Monte Carlo and ab initio study, Appl. Surf. Sci. 463 (2019) 1058–1077.

[84] E. Baran, A. Cakir, B. Yazici, Inhibitory effect of Gentiana olivieri extracts on the corrosion of mild steel in 0.5 M HCl: electrochemical and phytochemical evaluation, Arabian J. Chem. 12 (2019) 4303–4319.

[85] R. Thilgavathi, P. Sandhiya, A. Prithiba, R. Rajalakshmi, Application of ipomea staphylina leaf as an eco-friendly biomass for the corrosion inhibition of mild steel in 1M HCl, Mater. Today: Proc. 18 (2019) 1633–1647.

[86] A. Ishak, F.V. Adams, J.O. Madu, I.V. Joseph, P.A. Olubambi, Corrosion inhibition of mild steel in 1M hydrochloric acid using Haematostaphis barteri leaves extract, Procedia Manuf. 35 (2019) 1279–1285.

[87] M. Ramezanzadeh, G. Bahlakeh, Z. Sanaei, B. Ramezanzadeh, Studying the Urtica dioica leaves extract inhibition effect on the mild steel corrosion in 1 M HCl solution: Complementary experimental, ab initio quantum mechanics, Monte Carlo and molecular dynamics studies, J. Mol. Liq. 272 (2018) 120–136.

[88] G. Bahlakeh, A. Dehghani, B. Ramezanzadeh, M. Ramezanzadeh, Highly effective mild steel corrosion inhibition in 1 M HCl solution by novel green aqueous Mustard seed extract: Experimental, electronic-scale DFT and atomic-scale MC/MD explorations, J. Mol. Liq. 293 (2019) 111559.

[89] P.E. Alvarez, M.V. Fiori-Bimbi, A. Neske, S.A. Brandán, C.A. Gervasi, Rollinia occidentalis extract as green corrosion inhibitor for carbon steel in HCl solution, J. Ind. Eng. Chem. 58 (2018) 92–99.

[90] N. Chaubey, V.K. Singh, M. Quraishi, Papaya peel extract as potential corrosion inhibitor for Aluminium alloy in 1 M HCl: Electrochemical and quantum chemical study, Ain Shams Eng. J. 9 (2018) 1131–1140.

[91] M.T. Saeed, M. Saleem, S. Usmani, I.A. Malik, F.A. Al-Shammari, K.M. Deen, Corrosion inhibition of mild steel in 1 M HCl by sweet melon peel extract, J. King Saud Univ. Sci. 31 (2019) 1344–1351.

[92] M. Srivastava, P. Tiwari, S. Srivastava, A. Kumar, G. Ji, R. Prakash, Low cost aqueous extract of Pisum sativum peels for inhibition of mild steel corrosion, J. Mol. Liq. 254 (2018) 357–368.

[93] E. Ituen, O. Akaranta, A. James, S. Sun, Green and sustainable local biomaterials for oilfield chemicals: Griffonia simplicifolia extract as steel corrosion inhibitor in hydrochloric acid, Sustain. Mater. Techno. 11 (2017) 12–18.

[94] L.L. Liao, S. Mo, H.Q. Luo, N.B. Li, Corrosion protection for mild steel by extract from the waste of lychee fruit in HCl solution: experimental and theoretical studies, J. Colloid Interface Sci. 520 (2018) 41–49.

[95] A. Jmiai, B. El Ibrahimi, A. Tara, M. Chadili, S. El Issami, O. Jbara, A. Khallaayoun, L. Bazzi, Application of Zizyphus Lotuse-pulp of Jujube extract as green and promising corrosion inhibitor for copper in acidic medium, J. Mol. Liq. 268 (2018) 102–113.

[96] O. Joseph, O. Fayomi, O. Adenigba, Effect of Lecaniodiscus cupaniodes extract in corrosion inhibition of normalized and annealed mild steels in 0.5 M HCl, Energy Procedia 119 (2017) 845–851.

[97] A.A. Khadom, A.N. Abd, N.A. Ahmed, Xanthium strumarium leaves extracts as a friendly corrosion inhibitor of low carbon steel in hydrochloric acid: kinetics and mathematical studies, South Afr. J. Chem. Eng. 25 (2018) 13–21.

[98] H. Hassannejad, A. Nouri, Sunflower seed hull extract as a novel green corrosion inhibitor for mild steel in HCl solution, J. Mol. Liq. 254 (2018) 377–382.

[99] D. Verma, F. Khan, I. Bahadur, M. Salman, M. Quraishi, C. Verma, E.E. Ebenso, Inhibition performance of Glycine max, Cuscuta reflexa and Spirogyra extracts for mild steel dissolution in acidic medium: density functional theory and experimental studies, Results Phys. 10 (2018) 665–674.

[100] L.L. Liao, S. Mo, H.Q. Luo, N.B. Li, Longan seed and peel as environmentally friendly corrosion inhibitor for mild steel in acid solution: experimental and theoretical studies, J. Colloid Interface Sci. 499 (2017) 110–119.

[101] M. Lebrini, F. Suedile, P. Salvin, C. Roos, A. Zarrouk, C. Jama, F. Bentiss, Bagassa guianensis ethanol extract used as sustainable eco-friendly inhibitor for zinc corrosion in 3% NaCl: Electrochemical and XPS studies, Surf. Interfac. 20 (2020) 100588.

[102] E. Alibakhshi, M. Ramezanzadeh, G. Bahlakeh, B. Ramezanzadeh, M. Mahdavian, M. Motamedi, Glycyrrhiza glabra leaves extract as a green corrosion inhibitor for mild steel in 1 M hydrochloric acid solution: experimental, molecular dynamics, Monte Carlo and quantum mechanics study, J. Mol. Liq. 255 (2018) 185–198.

[103] A. Ehsani, M. Mahjani, M. Hosseini, R. Safari, R. Moshrefi, H.M. Shiri, Evaluation of Thymus vulgaris plant extract as an eco-friendly corrosion inhibitor for stainless steel 304 in acidic solution by means of electrochemical impedance spectroscopy, electrochemical noise analysis and density functional theory, J. Colloid Interface Sci. 490 (2017) 444–451.

[104] A. Dehghani, G. Bahlakeh, B. Ramezanzadeh, M. Ramezanzadeh, A combined experimental and theoretical study of green corrosion inhibition of mild steel in HCl solution by aqueous Citrullus lanatus fruit (CLF) extract, J. Mol. Liq. 279 (2019) 603–624.

[105] P. Muthukrishnan, P. Prakash, B. Jeyaprabha, K. Shankar, Stigmasterol extracted from Ficus hispida leaves as a green inhibitor for the mild steel corrosion in 1 M HCl solution, Arabian J. Chem. 12 (2019) 3345–3356.

[106] Y. Qiang, S. Zhang, B. Tan, S. Chen, Evaluation of Ginkgo leaf extract as an eco-friendly corrosion inhibitor of X70 steel in HCl solution, Corros. Sci. 133 (2018) 6–16.

[107] A. Dehghani, G. Bahlakeh, B. Ramezanzadeh, M. Ramezanzadeh, Electronic/atomic level fundamental theoretical evaluations combined with electrochemical/surface examinations of Tamarindus indiaca aqueous extract as a new green inhibitor for mild steel in acidic solution (HCl 1 M), J. Taiwan Inst. Chem. Eng. 102 (2019) 349–377.

[108] H.L.Y. Sin, A.A. Rahim, C.Y. Gan, B. Saad, M.I. Salleh, M. Umeda, Aquilaria subintergra leaves extracts as sustainable mild steel corrosion inhibitors in HCl, Measurement 109 (2017) 334–345.

[109] A. Chekioua, R. Delimi, Purification of H2SO4 of pickling bath contaminated by Fe (II) ions using electrodialysis process, Energy Procedia 74 (2015) 1418–1433.

[110] J. Zhai, C.H. Jiang, J. Wu, Waste Steel-pickling Sulphuric Acid Removal Treated by Freezing Crystallization-Acid Retardation Coupling Technology, Advanced Materials Research, Trans. Tech. Publ. (2014) 2915–2918.

[111] A.A. Baba, F. Adekola, O.S. Awobode, B.R. Adekunle, S. Pradhan, A. Biswal, Analysis of heavy metals from spent pickling liquor of sulfuric acid, Int. J. Chem. 21 (2011) 231–240.

[112] M.R. Singh, P. Gupta, K. Gupta, The litchi (*Litchi Chinensis*) peels extract as a potential green inhibitor in prevention of corrosion of mild steel in 0.5 M H2SO4 solution, Arabian J. Chem. 12 (2019) 1035–1041.

[113] X. Zhang, W. Li, G. Yu, X. Zuo, W. Luo, J. Zhang, B. Tan, A. Fu, S. Zhang, Evaluation of Idesia polycarpa Maxim fruits extract as a natural green corrosion inhibitor for copper in 0.5 M sulfuric acid solution, J. Mol. Liq. 318 (2020) 114080.

[114] Z. Khiya, M. Hayani, A. Gamar, S. Kharchouf, S. Amine, F. Berrekhis, A. Bouzoubae, T. Zair, F. El Hilali, Valorization of the Salvia officinalis L. of the Morocco bioactive extracts: phytochemistry, antioxidant activity and corrosion inhibition, J. King Saud Univ. Sci. 31 (2019) 322–335.

[115] B. Tan, B. Xiang, S. Zhang, Y. Qiang, L. Xu, S. Chen, J. He, Papaya leaves extract as a novel eco-friendly corrosion inhibitor for Cu in H2SO4 medium, J. Colloid Interface Sci. 582 (2020) 918–931.

[116] R.T. Loto, O. Olowoyo, Corrosion inhibition properties of the combined admixture of essential oil extracts on mild steel in the presence of SO42– anions, S. Afr. J. Chem. Eng. 26 (2018) 35–41.

[117] A. Saxena, D. Prasad, R. Haldhar, G. Singh, A. Kumar, Use of Sida cordifolia extract as green corrosion inhibitor for mild steel in 0.5 M H2SO4, J. Environ. Chem. Eng. 6 (2018) 694–700.

[118] A. Saxena, D. Prasad, R. Haldhar, G. Singh, A. Kumar, Use of Saraca ashoka extract as green corrosion inhibitor for mild steel in 0.5 M H2SO4, J. Mol. Liq. 258 (2018) 89–97.

[119] B. Prathibha, V. Vasudha, H. Nagaswarupa, Spathodea Campanulata as a Corrosion Inhibitor for Mild Steel in 1N H2SO4 Media, Mater. Today: Proc. 5 (2018) 22595–22604.

[120] R. Haldhar, D. Prasad, A. Saxena, R. Kumar, Experimental and theoretical studies of Ficus religiosa as green corrosion inhibitor for mild steel in 0.5 M H2SO4 solution, Sustain. Chem. Pharm. 9 (2018) 95–105.

[121] R. Nathiya, V. Raj, Evaluation of Dryopteris cochleata leaf extracts as green inhibitor for corrosion of aluminium in 1 M H2SO4, Egypt. J. Pet. 26 (2017) 313–323.

[122] P. Muthukrishnan, B. Jeyaprabha, P. Prakash, Adsorption and corrosion inhibiting behavior of Lannea coromandelica leaf extract on mild steel corrosion, Arabian J. Chem. 10 (2017) S2343–S2354.

[123] I.-M. Chung, R. Malathy, R. Priyadharshini, V. Hemapriya, S.-H. Kim, M. Prabakaran, Inhibition of mild steel corrosion using Magnolia kobus extract in sulphuric acid medium, Mater. Today Commun. (2020) 101687.

[124] A. Saxena, D. Prasad, R. Haldhar, Investigation of corrosion inhibition effect and adsorption activities of Cuscuta reflexa extract for mild steel in 0.5 M H2SO4, Bioelectrochemistry 124 (2018) 156–164.

[125] R. Haldhar, D. Prasad, A. Saxena, Myristica fragrans extract as an eco-friendly corrosion inhibitor for mild steel in 0.5 M H2SO4 solution, J. Environ. Chem. Eng. 6 (2018) 2290–2301.

[126] S. Abrishami, R. Naderi, B. Ramezanzadeh, Fabrication and characterization of zinc acetylacetonate/Urtica Dioica leaves extract complex as an effective organic/inorganic hybrid corrosion inhibitive pigment for mild steel protection in chloride solution, Appl. Surf. Sci. 457 (2018) 487–496.

[127] M. Ramezanzadeh, Z. Sanaei, G. Bahlakeh, B. Ramezanzadeh, Highly effective inhibition of mild steel corrosion in 3.5% NaCl solution by green Nettle leaves extract and synergistic effect of eco-friendly cerium nitrate additive: experimental, MD simulation and QM investigations, J. Mol. Liq. 256 (2018) 67–83.

[128] Z. Sanaei, T. Shahrabi, B. Ramezanzadeh, Synthesis and characterization of an effective green corrosion inhibitive hybrid pigment based on zinc acetate-Cichorium intybus L leaves extract (ZnA-CIL. L): Electrochemical investigations on the synergistic corrosion inhibition of mild steel in aqueous chloride solutions, Dyes Pigm. 139 (2017) 218–232.

[129] F.E.-T. Heakal, M. Deyab, M. Osman, A. Elkholy, Performance of Centaurea cyanus aqueous extract towards corrosion mitigation of carbon steel in saline formation water, Desalination 425 (2018) 111–122.

[130] H. Gadow, M. Motawea, H. Elabbasy, Investigation of myrrh extract as a new corrosion inhibitor for α-brass in 3.5% NaCl solution polluted by 16 ppm sulfide, RSC Adv. 7 (2017) 29883–29898.

[131] J. Halambek, I. Cindrić, A.N. Grassino, Evaluation of pectin isolated from tomato peel waste as natural tin corrosion inhibitor in sodium chloride/acetic acid solution, Carbohydr. Polym. 234 (2020) 115940.

[132] N. Chaubey, V.K.Singh Savita, M. Quraishi, Corrosion inhibition performance of different bark extracts on aluminium in alkaline solution, J. Assoc. Arab Univ. Basic Appl. Sci. 22 (2017) 38–44.

[133] N. Chaubey, D.K. Yadav, V.K. Singh, M. Quraishi, A comparative study of leaves extracts for corrosion inhibition effect on aluminium alloy in alkaline medium, Ain Shams Eng. J. 8 (2017) 673–682.

[134] A.H. Al-Moubaraki, A.A. Al-Howiti, M.M. Al-Dailami, E.A. Al-Ghamdi, Role of aqueous extract of celery (Apium graveolens L.) seeds against the corrosion of aluminium/sodium hydroxide systems, J. Environ. Chem. Eng. 5 (2017) 4194–4205.

[135] Y. Lekbach, D. Xu, S. El Abed, Y. Dong, D. Liu, M.S. Khan, S.I. Koraichi, K. Yang, Mitigation of microbiologically influenced corrosion of 304L stainless steel in the presence of Pseudomonas aeruginosa by Cistus ladanifer leaves extract, Int. Biodeterior. Biodegrad. 133 (2018) 159–169.

[136] P. Parthipan, P. Elumalai, J. Narenkumar, L.L. Machuca, K. Murugan, O.P. Karthikeyan, A. Rajasekar, Allium sativum (garlic extract) as a green corrosion inhibitor with biocidal properties for the control of MIC in carbon steel and stainless steel in oilfield environments, Int. Biodeterior. Biodegrad. 132 (2018) 66–73.

[137] P. Parthipan, J. Narenkumar, P. Elumalai, P.S. Preethi, A.U.R. Nanthini, A. Agrawal, A. Rajasekar, Neem extract as a green inhibitor for microbiologically influenced corrosion of carbon steel API 5LX in a hypersaline environments, J. Mol. Liq. 240 (2017) 121–127.

[138] S. Palanisamy, G. Maheswaran, A.G. Selvarani, C. Kamal, G. Venkatesh, Ricinus communis–a green extract for the improvement of anti-corrosion and mechanical properties of reinforcing steel in concrete in chloride media, J. Build. Eng. 19 (2018) 376–383.

[139] A. Fouda, A. Emam, R. Refat, M. Nageeb, Cascabela thevetia plant extract as corrosion inhibitor of carbon steel in polluted sodium chloride solution, J. Anal Pharm. Res. 6 (2017) 00168.

[140] M. Muzakir, F. Nwosu, S. Amusat, Mild Steel Corrosion Inhibition in a NaCl Solution by Lignin Extract of Chromolaena odorata, Portugaliae Electrochimica Acta 37 (2019) 359–372.

[141] Q. Liu, Z. Song, H. Han, S. Donkor, L. Jiang, W. Wang, H. Chu, A novel green reinforcement corrosion inhibitor extracted from waste Platanus acerifolia leaves, Constr. Build. Mater. 260 (2020) 119695.

[142] M.B. Harb, S. Abubshait, N. Etteyeb, M. Kamoun, A. Dhouib, Olive leaf extract as a green corrosion inhibitor of reinforced concrete contaminated with seawater, Arabian J. Chem. 13 (2020) 4846–4856.

[143] A. Pradityana, A. Shahab, L. Noerochim, D. Susanti, Inhibition of corrosion of carbon steel in 3.5% NaCl solution by Myrmecodia Pendans extract, Int. J. Corros. (2016) 2016.

18

Chemical medicines (drugs) as green corrosion inhibitors

Chemical medicines or drugs molecules can be regarded as biotolerable, nonbioaccumulative, and biodegradable environmental friendly alternatives to be used as various industrial applications. They exert a negligible adverse impact on the living organisms and the surrounding environment. Generally, chemical medicines are complex molecules that acquire various electron-rich centers, called adsorption centers, through which they can easily adsorb on the metallic surface and act as strong corrosion inhibitors [1–4]. Although, chemical medicines are quite complex molecules however they contain various polar substituents such as –OH (hydroxyl), –COOH (carboxyl), –COOC$_2$H$_5$ (ester), –COCl (acid halide) –OMe (methoxy), –CONH$_2$ (amide), –O– (ether), –CN (nitrile), –NO$_2$ (nitro), –NH$_2$ (1°-amine), >NH (2°-amine) and >N– (3°-amine), >C=O (carbonyl), >C=S (thiocarbonyl), etc., through which they easily get dissolve in the polar aqueous electrolytes. These polar functional groups also act as adsorption centers during their interactions with the metallic substrate. Chemical medicines are associated with the following advantages:

(i) Eco-friendly (biotolerable, nonbioaccumulative, and biodegradable).
(ii) Exhibit reasonable high corrosion inhibition potential because of their complex structures at relatively low concentration.
(iii) High solubility in the polar electrolytes (due to polar substituents).
(iv) Natural and biological origin (especially ayurvedic medicines).
(v) Commercial availability.
(vi) Wide-range of anticorrosive activity (useful for different metal/electrolyte systems).
(vii) Ease of fabrication and application.
(viii) Cost-effective (because of natural availability).
(ix) Minimized adverse effect on living organisms and environment.
(x) Expired chemical medicines can also be used (waste utilization).

The use of chemical medicines as corrosion inhibitors is related to various shortcomings. Some of the shortcomings of chemical medicines are:

(i) Synthesis and purification of synthetic drugs are highly expensive processes.
(ii) The chemical structure of medicines may degrade (hydrolysis) in highly corrosive electrolytes.

(iii) Syntheses of medicines are associated with the consumption and discharge of toxic chemicals, solvents and expensive catalysts.
(iv) Some of the medicines are not soluble in aqueous electrolytes and they need the addition of a small fraction of organic solvents, called cosolvents.
(v) Some of the medicines are toxic and cause severe health problems.

18.1 Chemical medicines as corrosion inhibitors: literature survey

18.1.1 Chemical medicines as corrosion inhibitors in HCl electrolytes

Chemical medicines are widely used as effective corrosion inhibitors against metallic corrosion in hydrochloric acid solutions. It is again important to notice that high concentrations of HCl solutions are useful for industrial purposes and their low concentrations are useful for the academic purposes [5,6]. More so, lower concentrations of hydrochloric acid solutions are also widely used for the removal surface of oxide layers and cleaning of rusts and scales. Noticeably, pure metals are highly reactive and thermodynamically unstable, therefore readily undergo a chemical reaction with the constituents of the surrounding environment to form metal oxides, sulfides, carbonates, etc. Acid pickling, descaling, and oil-well acidification are the common industrial processes in which highly concentrated hydrochloric acid solutions are used. During industrially cleaning of surface impurities, a huge amount of metallic constituent is lost because of the aggressiveness of the cleaning solutions. Therefore, these chemical cleaning processes employ the addition of some external chemical species, called as corrosion inhibitors. Organic compounds containing polar functional groups and aromatic rings serve as the most effective inhibitors against metallic corrosion [7,8]. Obviously, they become effective by adsorbing on the metallic surface using their electron-rich centers, called adsorption centers. Adsorption of these compounds results in the formation of corrosion protective film of organic compounds.

Nevertheless, traditional organic compounds used as corrosion inhibitors are toxic and nonenvironmental friendly in nature. Therefore, because of the huge demands for sustainable development, these toxic chemicals should be replaced by relatively more effective and environmental friendly alternatives. In view of this, chemical medicines that have been established as a most effective and environmental sustainable class of compounds are widely used against metallic corrosion. Because of their complex structures, chemical medicines provide excellent surface coverage and corrosion inhibition effectiveness. Literature study reveals that various research and review articles are available on describing the corrosion inhibition properties of chemical medicines [1,9]. This chapter aims to describe the recent development in using chemical medicines as corrosion inhibitors. Recent researches in the field of corrosion science and engineering are devoted to the utilization of chemical species originated from nature [10–12]. These include the consumption of plant extracts, amino acids (AAs), carbohydrates, natural polymers and chemical medicines. Literature investigation shows that chemical medicines are widely used as corrosion inhibitors for various metals and alloys in HCl-based electrolytes.

Liang and coworkers [13] demonstrated the corrosion inhibition effectiveness of few drugs including oxacillin, amoxicillin, penicillin V and penicillin G for mild steel in1 M HCl. The inhibition effectiveness of these chemical medicines was evaluated using weight loss, electrochemical (PDP & EIS), and computational (DFT & MDS) methods. Inhibition efficiencies of the investigated chemical medicines followed the order: oxacillin > amoxicillin > penicillin V > penicillin G. Using weight loss or gravimetric method it was observed that inhibition effectiveness of the investigated drugs increases on increasing their concentrations. Potentiodynamic polarization (PDP) study suggested that oxacillin, amoxicillin, penicillin V and penicillin G become effective by retarding both anodic and cathodic Tafel reactions. Thereby, they act as mixed-type corrosion inhibitors. It is important to mention that presence of oxacillin, amoxicillin, penicillin V and penicillin G in the corrosive medium of 1M HCl decreases the values of corrosion current densities. This observation suggests that oxacillin, amoxicillin, penicillin V, and penicillin G become effective by blocking of the surface active sites responsible for the corrosion. EIS study suggested that oxacillin, amoxicillin, penicillin V, and penicillin G act as interface-type corrosion inhibitors as they become effective by increasing the value of charge transfer resistance through their adsorption. Computational studies carried by DFT and MDS methods suggest that oxacillin, amoxicillin, penicillin V, and penicillin G interact spontaneously with the metallic surface using donor–acceptor (charge sharing) mechanism. HOMO and LUMO electron densities were distributed over the entire segments of the oxacillin, amoxicillin, penicillin V, and penicillin G molecules which indicate the entire parts of these molecules involved in the charge sharing. Three penicillin-based chemical medicines, symbolized as penicillin I, II and III were investigated as corrosion inhibitors for aluminium in 1M HCl solution [14]. Chemical, electrochemical, and computational based studies were undertaken to study the corrosion inhibition power of investigated chemical medicines. Through, PDP analysis it was observed that penicillin I, II, and III become effective by inhibiting both anodic and cathodic reactions and behaving as mixed-type corrosion inhibitors. Inhibition effectiveness of the studies penicillin-based chemical medicines followed the sequence: III > II > I. Investigated drugs become effective by adsorbing on the metallic surface and their adsorption on the metallic surface was supported by SEM and UV-vis analyses. Computational studies carried by DFT and MDS methods suggest that studied chemical medicines interact spontaneously with the metallic surface using donor-acceptor (charge sharing) mechanism. Table 18.1 describes the collection of some major recent reports on the corrosion inhibition effect of chemical medicines in hydrochloric acid solutions.

TABLE 18.1 Summary of some major reports on chemical medicines as corrosion inhibitors in hydrochloric acid solutions.

S. No.	Name of medicines	Metal/electrolyte	Conc. and efficiency	Nature of adsorption	References
1	Oxacillin, Amoxicillin, Penicillin V & Penicillin G	Mild steel/1 M HCl	1.0 mM/75.4%	Mixed type	[13]
		Aluminum/1 M HCl	0.14 mM/74.42%	Langmuir/Mixed type	[14]
		Mild steel/1 M HCl	1 mM/96.0%	Langmuir/Mixed type	[15]
2	Penicillin G (I), methicillin (II) & nafcillin (III)	Mild steel/0.5 M HCl	10^{-3} M/80.0%	Langmuir/Mixed type	[16]
		Mild steel/1 M HCl	5×10^{-3} M/75.85%	Langmuir	[17]
3	Cloxacillin	Mild steel/1 M HCl	15×10^{-4} M/81.19%	Temkin/Mixed type	[18]
		Mild steel/1 M HCl	300 ppm/95.8%	Langmuir/Mixed type	[19]

(continued)

TABLE 18.1 (Cont'd)

S. No.	Name of medicines	Metal/electrolyte	Conc. and efficiency	Nature of adsorption	References
4	Dicloxacillin	Aluminium/2 M HCl	1000 ppm/89.28%	Langmuir	[20]
8	Cefazolin	Mild steel/1.0 M HCl	10.9×10^{-4} M /3.9%	Langmuir/Mixed type	[21]
5	Flucloxacillin	Mild steel/1 M HCl	400 ppm/76.07%	Langmuir/Mixed type	[22]
6	Amoxicillin	Mild steel/1 M HCl	1.0×10^{-2} M/80.3%	Langmuir/Mixed type	[23]
		Mild steel/1 M HCl	1800 ppm/94.47%	Langmuir/Mixed type	[24]
		Mild steel/0.1 M HCl	0.5 g/L 68.30%	Langmuir	[25]
7	Cefatrexyl	Mild steel/0.1 M HCl	1 mM /96.10%	Langmuir	[26]
9	Cefalexin	Mild steel/1 M HCl	400 ppm/92.1%	Langmuir/Mixed type	[27]
		Mild steel/0.5 M HCl	800 ppm/94%	Langmuir	[28]
10	Cefadroxil	Aluminum/1M HCl	2 mM/93.22%	Langmuir/Mixed type	[29]
		Mild steel/1M HCl	11.0×10^{-4} M/96%	Langmuir/Mixed type	[30]
11	Ceftriaxone	Mild steel/1M HCl	400 ppm/90%	Langmuir/Mixed type	[31]
		Nickel/0.5M H_2SO_4 &1M HCl	10^{-5} M (1M HCl)/73%	Mixed type	[32]
		Mild steel/1M HCl	1.00×10^{-2} M/94.1%	Langmuir/Mixed type	[23]
		Carbon steel/0.5M HCl	50 ppm/93.37%	Langmuir/Mixed type	[33]
12	Cefotaxime	Mild steel/0.5 M HCl	300 ppm/95.8%	Langmuir/Mixed type	[19]
		Carbon steel /0.5-1M HCl & H_2SO_4	3 mM/75.01%	Langmuir	[34]
		Copper/0.1M HCl	12×10^{-5} M/90.2%	Langmuir/Mixed type	[35]
13	Ceftazidime	Mild steel/1M HCl	1.83×10^{-4} M/96%	Langmuir/Mixed type	[36]
14	Cefixime	Mild steel/1M HCl	8.8×10^{-4} M/90%	Langmuir/Mixed type	[37]
		Aluminum/1M HCl	2 mM/90.41 %	Langmuir/Mixed type	[38]
		Mil steel/0.25-1.0M HCl	400 ppm	–	[39]
15	Cefepime	X80 steel/1M HCl	10 mM/93.2%	Langmuir/Mixed type	[40]
16	Cefoperazone	Carbon steel/0.5 M HCl	7×10^{-4} M/98.5%	Langmuir/Mixed type	[41]
17	Ceftobiprole	Mild steel/1 M HCl	9.31×10^{-4} M/92.2%	Langmuir/Mixed type	[42]
18	Ofloxacin	Mild steel/1 M HCl	1×10^{-4} M/91.0%	Mixed type	[43]
19	Ciprofloxacin	Mild steel/0.1 M HCl	2.57×10^{-3} M/86.0%	Langmuir	[44]
		Carbon steel/1 M HCl	300 ppm/91%	Langmuir/Mixed type	[45]
20	Norfloxacin	Mild steel/HCl, HNO_3 & H_2SO_4	1100 ppm/91.54% (1M HCl)	Langmuir	[46]
		Mild steel/3%HCl	300 ppm/93.2%	Langmuir/Mixed type	[47]

(continued)

18.1 Chemical medicines as corrosion inhibitors: literature survey

S. No.	Name of medicines	Metal/electrolyte	Conc. and efficiency	Nature of adsorption	References
21	Sparfloxacin	0.1, 2.0, and 2.0 M HCl/Mild steel	12 mM/97.47%	Langmuir	[48]
22	Primaquine	Mild steel/1 M HCl	0.4 mM/98%	Langmuir/Mixed type	[49]
23	Doxycycline	Mild steel/1 M HCl	9.02×10^{-4} M/95%	Langmuir/Mixed type	[50]
24	Azithromycin	Aluminum alloy/2024 0.1N HCl	450 ppm/91%	Langmuir/Mixed type	[51,52]
25	Sulfaguanidine & Sulfadiazine	Mild steel/1M HCl	0.75 mM/67.2%	Mixed type	[53]
26		Mild steel/1M HCl	5 mM/94%	Mixed type	
27	Sulfacetamide	Mild steel/1M HCl	200 ppm/94.22%	Langmuir	[54]
		Mild steel/1M HCl	10 mM/84.7%	–	[55]
28	Dapsone	Mild steel/1 M HCl and 0.5M H_2SO_4	400 ppm/93.33% (HCl)	Langmuir/Cathodic/ HCl & anodic/ H_2SO_4	[56]
		Mild steel/1M HCl	500 ppm/95%	Langmuir/Mixed type	[57]
		Mild steel/1M HCl	0.602 mM/99.2%	Langmuir/Mixed type	[58]
29	Streptomycin	Mild steel/1M HCl	500 ppm/88.5%	Langmuir/Mixed type	[59]
		Mild steel /1M HCl	8.6×10^{-4} M/92%	Langmuir	[60]
30	Ketoconazole	Mild steel/1M HCl	100 ppm/97%	Langmuir/Mixed type	[61]
		Mild steel /1M HCl	100 ppm/93.57%	Langmuir/Mixed type	[62]
31	Fluconazole	Mild steel /1M HCl	0.30 mM/69%	–	[63]
		Aluminum/0.1M HCl	1×10^{-4} M/82.4%	Temkin	[64]
		Aluminum/0.1M HCl	8×10^{-5} M/76.5%	Dubinin–Radushkevich	[65]
32	Clotrimazole	Zinc /0.1M HCl	500 ppm/90%	Langmuir/Mixed type	[66]
		Mild steel/1M HCl	1×10^{-4} M/90.9%	Langmuir	[67]
33	Mebendazole	Mild steel/1M HCl	2.54×10^{-4} M/96.2%	Langmuir/Mixed type	[68]
34	Diethylcarbamazine	Mild steel/1M HCl	6.27×10^{-4} M/94.86%	Langmuir/Cathodic type	[69]
35	Rhodanine	304 SS/1M HCl	1×10^{-3} M/93.22%	Langmuir/Mixed type	[70]
36	Tramadole	Mild steel /1M HCl	100 ppm/96.12%	Langmuir/Mixed type	[71]
		Aluminum/1M HCl	500 ppm/98.4%	Langmuir/Mixed type	[72]
37	Meclizine	Aluminum/1M HCl	200 ppm/95.4%	Langmuir/Mixed type	[73]
38	Famotidine	Copper/3M HCl	0.4 mM/94.78%	–	[74]
39	Pheniramine	Mild steel/1M HCl	0.0833 mM/98.1%	Langmuir/Mixed type	[75]
		Mild steel/0.5M HCl	500 ppm/75.34%	Langmuir/Mixed type	[76]
40	Fexofenadine	Mild steel/1M HCl	3.0×10^{-4} M/97%	Langmuir/Mixed type	[77]
41	Ziprasidone	Mild steel/1.0 M HCl & 0.5 M H_2SO_4	100 ppm/96.9%	Langmuir/Mixed type	[78]

(continued)

TABLE 18.1 (Cont'd)

S. No.	Name of medicines	Metal/electrolyte	Conc. and efficiency	Nature of adsorption	References
42	Risperidone	Mild steel/1M HCl	5×10^{-6} M/94%	Langmuir/Mixed type	[79]
43	Atenolol	Mild steel/1M HCl	300 ppm/93.8%	Langmuir/Mixed type	[80]
		Zinc/1M HCl	500 ppm/93%	Langmuir/Mixed type	[81]
44	Tinidazole	Mild steel/1M HCl	400 ppm/90%	Langmuir/Mixed type	[82]
		Mild steel/3% HCl	400 ppm/62%	Langmuir/Mixed type	[83]
45	Metformin	Carbon steel/15% HCl	500 ppm/78.79%	Langmuir/Mixed type	[84]
46	Pioglitazone	Mild steel/1M HCl	14×10^{-4} M/79.51%	Mixed type	[85]
47	D-penicillamine	Mild steel/1M HCl	5 mM/43%	Langmuir/Mixed type	[86]
48	Rosuvastatin	Mild steel/1M HCl & 0.5 H_2SO_4	600 ppm/92%	Langmuir/Mixed type	[87]
49	Cephapirin'	Carbon steel/2M HCl	600 ppm/83.0%	Mixed type	[88]

18.1.2 Chemical medicines as corrosion inhibitors in H_2SO_4 electrolytes

Sulfuric acid based electrolytes are extensively employed for different academic and academic purposes to remove of metal oxide layers and clean the metallic surface. Obviously, similar to the hydrochloric acid solutions, lower concentration solutions of sulfuric acid solutions are useful for academic research purposes and highly concentrated sulfuric acid solutions are useful for industrial cleaning processes [89–91]. However, during metallic surface cleaning a huge amount of metallic component is lost because of corrosion. Therefore, these cleaning processes require the use of some external additives to inhibit the corrosive damage. Organic compounds, especially heterocyclic compounds containing polar functional groups and aromatic ring(s), are used as the most effective corrosion inhibitors against metallic corrosion in H_2SO_4 media [92–94]. Most of the established corrosion inhibitors are toxic in nature therefore they should be replaced by relatively more environmental friendly alternatives. Therefore, there is various research articles have been published dealing with the anticorrosive effect of chemical medicines in sulfuric acid-based electrolytes.

Eddy and coworkers [95] reported the inhibition power of penicillin G for mild steel corrosion in the different concentrations (1-2.5M) of H_2SO_4. Inhibition power of the penicillin G was determined using gasometric and thermometric methods. The penicillin G showed the highest inhibition efficiency of 90% at 15×10^{-4} M concentration. Investigation suggests that penicillin G acts as an effective corrosion inhibitor and its protection power increases on increasing its concentration and decreases with rise in the temperature. Penicillin G becomes effective by adsorbing on the metallic surface using through Langmuir adsorption isotherm model. Elsewhere, Eddy and Odoemelam [96] reported the inhibition of mild steel corrosion in 0.1M H_2SO_4 solution using potassium salt of penicillin V. similar observations were derived. A summary of some corrosion inhibition effect of some common chemical medicines used as corrosion inhibitors is presented in Table 18.2.

TABLE 18.2 Summary of some major reports on chemical medicines as corrosion inhibitors in sulfuric acid solutions.

S. No.	Name of medicine	Metal & electrolyte	Conc. & efficiency (%EI)	Nature of adsorption	References
1	Penicillin G	Mild steel/1–2.5M H_2SO_4	15×10^{-4} M/90.0% (1M HCl)	Langmuir	[95]
2	Penicillin V	Mild steel/1–2.5M H_2SO_4	15×10^{-4} M/52.33%	Langmuir	[96]
3	Cloxacillin	Mild steel/0.1M H_2SO_4	12×10^{-4} M/88.390%	Langmuir	[97]
		Mild steel/2M H_2SO_4	15×10^{-4} M/89.12%	Langmuir	[98]
4	Amoxicillin	Mild steel/1M H_2SO_4	15×10^{-4} M/66.57%	Langmuir/Mixed type	[99]
5	Ceftazidime	Mild steel/0.5M H_2SO_4	250 ppm/90.2%	Langmuir/Mixed type	[100]
6	Ofloxacin	Copper/1M HNO_3 & 0.5M H_2SO_4	1mM/85.0% (HNO_3) & 90.7% (H_2SO_4)	Langmuir/Cathodic type	[101]
7	Norfloxacin	Copper/1.0M HNO_3 & 0.5 M H_2SO_4	1mM/94.9% (HNO_3)	Langmuir/Mixed type	[102]
8	Erythromycin	Zinc/0.01–0.04M H_2SO_4	5×10^{-5} M/82.67% (0.1M H_2SO_4)	Langmuir	[103]
9	Clarithromycin	Zinc/0.01–0.04M H_2SO_4	5×10^{-5} M/69.23% (0.1M H_2SO_4)	Langmuir	[104]
10	Azithromycin	Mild steel, copper, and zinc/0.5M H_2SO_4	1×10^{-2} M/80% (steel in 0.5 M H_2SO_4)	Langmuir/Mixed type	[52]
11	Sulfamethoxazole	Brass/H_2SO_4	10^{-3} M/84.56%	Mixed type	[105]
12	Mebendazole	Carbon steel/H_2SO_4	1×10^{-4} M/85.31%	Langmuir/Mixed type	[106]
13	Famotidine	C-steel/0.5M H_2SO_4	6×10^{-6} M/90.8%	Freundlich model	[107]

Gravimetric analysis showed inhibition efficiency of the tested medicines increases on increasing their concentration and decreases with rise in the temperature. PDP analyses showed that most of the chemical medicines become effective by retarding both anodic as well as cathodic reactions and behave as mixed-type corrosion inhibitors. Mostly, they become effective by adsorbing on the metallic surface following through the Langmuir adsorption isotherm model. Adsorption of the plant extracts on the metallic surface was studied using various surface monitoring techniques, especially using EDX, AFM, SEM, XPS, XRD, FT-IR, and UV-vis methods. Computational studies carried out using DFT and MD (or MC) simulations suggest that in sulfuric acid-based electrolytes, chemical medicines interact spontaneously with the metallic surface using donor–acceptor (charge sharing) mechanism.

18.1.3 Chemical medicines as corrosion inhibitors in other electrolytes

Besides, hydrochloric acid, and sulfuric acid based electrolytes, chemical medicines are also used as effective corrosion inhibitors for different metals and alloys in sodium chloride

TABLE 18.3 Summary of some major reports on chemical medicines as corrosion inhibitors in the electrolytes other than HCl and H_2SO_4.

S. No.	Name of medicine	Metal & electrolyte	Conc. & efficiency (%EI)	Nature of adsorption	References
1	Penicillin G	304 SS/3.5 wt% NaCl	5.0 mM/87.1%	Mixed type	[111]
2	Azithromycin	Copper/0.9% NaCl	5×10^{-3} M/95.76%.	Langmuir/Mixed type	[51]
3	Amoxicillin	AISI 1020 steel/0.05 mol l−1 NaCl	NA	Interface type	[112]
4	Sulfamethazine	Mild steel/3.5% NaCl	Coating	NA	[53]
5	Streptomycin	Carbon-steel/Sea water	0.52 mM/40.48%	Langmuir/Mixed type	[113]
6	Ketoconazole	Bronze/3.5% M NaCl + 0.1M Na2SO4	10 ppm/66%	Langmuir/Mixed type	[114,115]
7	Penicillin G	Mild steel/3.0 M H_3PO_4	1 mM/95.0%	Langmuir/Mixed type	[115]
8	Penicillin V	Carbon steel/ $Na_2S + Na_2SO_4$ (pH- 3)	500 ppm/92.2%	Langmuir/Mixed type	[116]
9	Dicloxacillin	AA6063/0.5 M HNO_3	58.9%	Langmuir/Mixed type	[117]
10	Cefadroxil	Copper/1M HNO_3	2 mM/94%	Langmuir/Mixed type	[118]
11	Ciprofloxacin	Copper/1M HNO_3	0.5 & 1 mM	Langmuir/Mixed type	[119]
12	Sulfamethazine	Carbon steel/2M H_3PO_4	5×10^{-4} M/90%	Thermodynamic model/Mixed type	[120]
13	Tinidazole	Zinc/0.1 M KOH	50 mg/L/89.39%	Anodic type	[121]

(NaCl), phosphoric acid (H_3PO_4), nitric acid (HNO_3), and hydroxide (e.g., KOH) based electrolytes [108–110]. Javidi and Omidvar recently investigated the inhibition effectiveness of penicillin G as co-inhibitor with sodium tungstate (Na_2WO_4) for 304 stainless steel corrosion in 3.5% NaCl solution [111]. Summary of some other major reports on corrosion inhibition effectiveness of chemical medicines in the medium other than HCl and H_2SO_4 are presented in Table 18.3. PDP analyses showed that most of the chemical medicines behave as mixed-type corrosion inhibitors. Mostly, they become effective by adsorbing on the metallic surface following mostly through the Langmuir adsorption isotherm. DFT-based quantum chemical calculations and MD (MC) simulations are widely used to support the experimental results.

18.2 Emerging trends in corrosion inhibition using chemical medicines

Although chemical medicines are environmental friendly alternatives to be used as corrosion inhibitors, their following attempts are being made to improve their cost-effectivity and inhibition performance.

(i) *Derivatization*: generally, drugs are highly complex molecules and show limited solubility in polar aqueous electrolytes. Therefore, attempts are being made to add polar substituents to increase their solubility in such electrolytes. These polar substituents can also act as adsorption centers during the interaction of medicines with the metallic surface.

(ii) *Synergism*: few of the chemical medicines show limited corrosion inhibition effectiveness. Therefore, recent attempts are being made to increase their corrosion inhibition performance using inorganic salts such as KCl, KBr, KI, and $ZnCl_2$ [122,123].

(iii) *Use of expired chemical medicines*: most of the chemical medicines are highly expensive because of their multistep synthesis and costly purification. Therefore, attempts are being made to use expired drugs as corrosion inhibitors [124–126]. The use of expired chemical medicines as corrosion inhibitors opens a new cost-effective door for waste (expired drugs) utilization.

18.3 Summary

Chemical medicines are established as environmental friendly alternatives to traditional toxic corrosion inhibitors. Chemical medicines are associated with several eco-friendly properties such as biotolerability, nonbioaccumulation, and biodegradability. Literature study showed that chemical medicines are effectively used as corrosion inhibitors for different metals and alloys in hydrochloric acid, sulfuric acid, sodium chloride, phosphoric acid, nitric acid, and hydroxide based electrolytes. They become effective by adsorbing on the metallic surface following through the Langmuir adsorption isotherm model. DPD study shows that most of the chemical medicines behave as mixed-type corrosion inhibitors and act by retarding both anodic and cathodic Tafel reactions. EIS study reveals that chemical medicines become effective by adsorbing at the interface of metal and electrolyte and increasing the charge transfer resistance for the corrosion process. Despite several benefits, the utilization of chemical medicines as corrosion inhibitors are associated with several shortcomings. Some major shortcomings are their susceptibility to hydrolyze, highly expensive, and limited solubility in the polar electrolytes. Recently, several attempts such as derivatization of chemical medicines, synergism, and use of expired drugs are being made to improve their cost-effectivity and inhibition performance.

References

[1] G. Gece, Drugs: a review of promising novel corrosion inhibitors, Corrosion Sci. 53 (2011) 3873–3898.
[2] M. Abdallah, Rhodanine azosulpha drugs as corrosion inhibitors for corrosion of 304 stainless steel in hydrochloric acid solution, Corrosion Sci. 44 (2002) 717–728.
[3] R. Pathak, P. Mishra, Drugs as corrosion inhibitors: a review, Int. J. Sci. Res. 5 (2016) 671–677.
[4] K. Xhanari, M. Finšgar, M.K. Hrnčič, U. Maver, Ž. Knez, B. Seiti, Green corrosion inhibitors for aluminium and its alloys: a review, RSC Adv. 7 (2017) 27299–27330.
[5] L.-F. Li, P. Caenen, J.-P. Celis, Effect of hydrochloric acid on pickling of hot-rolled 304 stainless steel in iron chloride-based electrolytes, Corrosion Sci. 50 (2008) 804–810.
[6] G. Peng, K. Chen, H. Fang, H. Chao, S. Chen, EIS study on pitting corrosion of 7150 aluminum alloy in sodium chloride and hydrochloric acid solution, Mater. Corrosion 61 (2010) 783–789.
[7] B. Sanyal, Organic compounds as corrosion inhibitors in different environments—a review, Progr. Organ. Coat. 9 (1981) 165–236.

[8] M. Özcan, İ. Dehri, M. Erbil, Organic sulphur-containing compounds as corrosion inhibitors for mild steel in acidic media: correlation between inhibition efficiency and chemical structure, Appl. Surf. Sci. 236 (2004) 155–164.

[9] C. Verma, D. Chauhan, M. Quraishi, Drugs as environmentally benign corrosion inhibitors for ferrous and nonferrous materials in acid environment: an overview, J. Mater. Environ. Sci. 8 (2017) 4040–4051.

[10] C. Verma, L. Olasunkanmi, E.E. Ebenso, M. Quraishi, Substituents effect on corrosion inhibition performance of organic compounds in aggressive ionic solutions: a review, J. Mol. Liq. 251 (2018) 100–118.

[11] C. Verma, H. Lgaz, D. Verma, E.E. Ebenso, I. Bahadur, M. Quraishi, Molecular dynamics and Monte Carlo simulations as powerful tools for study of interfacial adsorption behavior of corrosion inhibitors in aqueous phase: a review, J. Mol. Liq. 260 (2018) 99–120.

[12] C. Verma, E.E. Ebenso, M. Quraishi, Molecular structural aspects of organic corrosion inhibitors: influence of–CN and–NO2 substituents on designing of potential corrosion: inhibitors for aqueous media, J. Mol. Liq. 316 (2020) 113874.

[13] Y. Liang, C. Wang, J. Li, L. Wang, J. Fu, The Penicillin derivatives as corrosion inhibitors for mild steel in hydrochloric acid solution: experimental and theoretical studies, Int. J. Electrochem. Sci 10 (2015) 8072–8086.

[14] S.M. Habibi-Khorassani, M. Shahraki, M. Noroozifar, M. Darijani, M. Dehdab, Z. Yavari, Inhibition of aluminum corrosion in acid solution by environmentally friendly antibacterial corrosion inhibitors: experimental and theoretical investigations, Protect. Metals Phys. Chem. Surf. 53 (2017) 579–590.

[15] M. Gholamhosseinzadeh, R. Farrahi-Moghaddam, Electrochemical Investigation of the effect of penicillin G benzathine as a green corrosion inhibitor for mild steel, Progr. Color, Colorants Coat. 12 (2019) 15–23.

[16] A. Eddib, M. Hamdani, Electrochemical studies of ampicillin as corrosion inhibitor for stainless steel in hydrochloric acid solution, Moroccan J. Chem. 2 (2014) 2165–2174.

[17] I. Adejoro, F. Ojo, S. Obafemi, Corrosion inhibition potentials of ampicillin for mild steel in hydrochloric acid solution, J. Taibah Univ. Sci. 9 (2015) 196–202.

[18] H. Kumar, S. Karthikeyan, Inhibition of mild steel corrosion in hydrochloric acid solution by cloxacillin drug, J. Mater. Environ. Sci. 3 (2012) 925–934.

[19] S.K. Shukla, M. Quraishi, Cefotaxime sodium: a new and efficient corrosion inhibitor for mild steel in hydrochloric acid solution, Corrosion Sci. 51 (2009) 1007–1011.

[20] M. Abdallah, Antibacterial drugs as corrosion inhibitors for corrosion of aluminium in hydrochloric solution, Corrosion Sci. 46 (2004) 1981–1996.

[21] A.K. Singh, M. Quraishi, Effect of Cefazolin on the corrosion of mild steel in HCl solution, Corrosion Sci. 52 (2010) 152–160.

[22] M. Alfakeer, M. Abdallah, A. Fawzy, Corrosion inhibition effect of expired ampicillin and flucloxacillin drugs for mild steel in aqueous acidic medium, Int. J. Electrochem. Sci. 15 (2020) 3283–3297.

[23] X. Pang, M. Gong, Y. Zhang, Q. Wei, B. Hou, Corrosion inhibition and mechanism of mild steel in hydrochloric acid by ceftriaxone and amoxicillin, Sci. China Chem. 54 (2011) 1529–1536.

[24] B. Fadila, A. Sihem, A. Sameh, G. Kardas, A study on the inhibition effect of expired amoxicillin on mild steel corrosion in 1N HCl, Mater. Res. Express 6 (2019) 046419.

[25] A.A. Siaka, N. Eddy, S. Idris, L. Magaji, Z. Garba, I. Shabanda, Quantum Chemical Studies of corrosion inhibition and adsorption potentials of Amoxicillin on mild steel in HCl solution, Int. J. Modern Chem. 4 (2013) 1–10.

[26] M. Morad, Inhibition of iron corrosion in acid solutions by Cefatrexyl: behaviour near and at the corrosion potential, Corrosion Sci. 50 (2008) 436–448.

[27] S.K. Shukla, M. Quraishi, Cefalexin drug: a new and efficient corrosion inhibitor for mild steel in hydrochloric acid solution, Mater. Chem. Phys. 120 (2010) 142–147.

[28] S. Ouchenane, K. Abderrahim, S. Abderrahmane, M. Bououdina, In-depth investigation of cefalexin's action mechanism as Al-Cu alloy corrosion inhibitor in 0.5 M HCl medium, Mater. Res. Express 5 (2018) 106508.

[29] N. Diki, K.V. Bohoussou, M.G.-R. Kone, A. Ouedraogo, A. Trokourey, Cefadroxil drug as corrosion inhibitor for aluminum in 1 M HCl medium: experimental and theoretical studies, IOSR J. Appl. Chem. 11 (2018) 24–36.

[30] S.K. Shukla, M. Quraishi, E.E. Ebenso, Adsorption and corrosion inhibition properties of cefadroxil on mild steel in hydrochloric acid, Int. J. Electrochem. Sci. 6 (2011) 2912–2931.

[31] S.K. Shukla, M. Quraishi, Ceftriaxone: a novel corrosion inhibitor for mild steel in hydrochloric acid, J. Appl. Electrochem. 39 (2009) 1517–1523.

[32] D.A. Duca, M.L. Dan, N. Vaszilcsin, Ceftriaxone as corrosion inhibitor for nickel in acid solutions, Adv. Eng. Forum, Trans. Tech. Publ. 27 (2018) 74–82.

[33] H.Z. AL-Sawaad, Evaluation of the ceftriaxone as corrosion inhibitor for carbon steel alloy in 0.5 M of hydrochloric acid, Int. J. Electrochem. Sci 8 (2013) 3105–3120.

[34] B. Lin, S. Zheng, J. Liu, Y. Xu, Corrosion inhibition effect of cefotaxime sodium on mild steel in acidic and neutral media, Int. J. Electrochem. Sci. 15 (2020) 2335–2353.

[35] M.N. El-Haddad, Inhibitive action and adsorption behavior of cefotaxime drug at copper/hydrochloric acid interface: electrochemical, surface and quantum chemical studies, RSC Adv. 6 (2016) 57844–57853.

[36] A.K. Singh, S.K. Shukla, M. Singh, M. Quraishi, Inhibitive effect of ceftazidime on corrosion of mild steel in hydrochloric acid solution, Mater. Chem. Phys. 129 (2011) 68–76.

[37] I. Naqvi, A. Saleemi, S. Naveed, Cefixime: a drug as efficient corrosion inhibitor for mild steel in acidic media. Electrochemical and thermodynamic studies, Int. J. Electrochem. Sci. 6 (2011) 146–161.

[38] G.K. Gbassi, A. Ouedraogo, M. Berte, A. Trokourey, Aluminum corrosion inhibition by cefixime drug: experimental and DFT studies, J. Electrochem. Sci. Eng. 8 (2018) 303–320.

[39] S.H. Aljbour, Modeling of corrosion kinetics of mild steel in hydrochloric acid in the presence and absence of a drug inhibitor, Port. Electrochim. Acta 34 (2016) 407–416.

[40] N. Iroha, L. Nnanna, Electrochemical and adsorption study of the anticorrosion behavior of cefepime on pipeline steel surface in acidic solution, J. Mater. Environ. Sci. 10 (2019) 898–908.

[41] A.A. Nazeer, H. El-Abbasy, A. Fouda, Adsorption and corrosion inhibition behavior of carbon steel by cefoperazone as eco-friendly inhibitor in HCl, J. Mater. Eng. Perform. 22 (2013) 2314–2322.

[42] A.K. Singh, M. Quraishi, Adsorption properties and inhibition of mild steel corrosion in hydrochloric acid solution by ceftobiprole, J. Appl. Electrochem. 41 (2011) 7–18.

[43] P. Xuehui, R. Xiangbin, F. KUANG, X. Jiandong, H. Baorong, Inhibiting effect of ciprofloxacin, norfloxacin and ofloxacin on corrosion of mild steel in hydrochloric acid, Chinese J. Chem. Eng. 18 (2010) 337–345.

[44] I.A. Akpan, N.-A.O. Offiong, Inhibition of mild steel corrosion in hydrochloric acid solution by Ciprofloxacin drug, Int. J. Corrosion (2013) 2013.

[45] A. Fouda, M. Eissa, A. El-Hossiany, Ciprofloxacin as eco-friendly corrosion inhibitor for carbon steel in hydrochloric acid solution, Int. J. Electrochem. Sci. 13 (2018) 11096–11112.

[46] A.K. Kalra, N.K. Johar, K. Bhrara, G. Singh, Evaluation of Norfloxacin and Ofloxacin as corrosion inhibitors for mild steel in different acids: weight loss data, Int. J. Creat. Res. Thoughts (IJCRT) 6 (2018) 1148–1157.

[47] F.R. Xu, S.T. Zhang, X. Li, Corrosion Inhibition of 45# mild steel in acidic solution by some norfloxacin, Adv. Mater. Res., Trans. Tech. Publ. 194–196 (2011) 44–51.

[48] N. Eddy, S. Odoemelam, A. Mbaba, Inhibition of the corrosion of mild steel in HCl by sparfloxacin, African J. Pure Appl. Chem. 2 (2008) 132–138.

[49] I. Ahamad, S. Khan, K. Ansari, M. Quraishi, Primaquine: a pharmaceutically active compound as corrosion inhibitor for mild steel in hydrochloric acid solution, J. Chem. Pharm. Res 3 (2011) 703–717.

[50] S.K. Shukla, M. Quraishi, The effects of pharmaceutically active compound doxycycline on the corrosion of mild steel in hydrochloric acid solution, Corrosion Sci. 52 (2010) 314–321.

[51] Ž.Z. Tasić, M.B.P. Mihajlović, M.B. Radovanović, M.M. Antonijević, Electrochemical investigations of copper corrosion inhibition by azithromycin in 0.9% NaCl, J. Mol. Liq. 265 (2018) 687–692.

[52] O. Abdullatef, Chemical and electrochemical studies on the corrosion of mild-steel, copper and zinc in 0. 5 MH2so4 solution in presence of Azithromycin as effective corrosion inhibitor, J. Adv. Chem. 11 (2015) 642–3655.

[53] M. El-Naggar, Corrosion inhibition of mild steel in acidic medium by some sulfa drugs compounds, Corrosion Sci. 49 (2007) 2226–2236.

[54] A. Samide, B. Tutunaru, C. Negrila, I. Trandafir, A. Maxut, Effect of sulfacetamide on the composition of corrosion products formed onto carbon steel surface in hydrochloric acid, Digest J. Nanomater. Biostruct. 6 (2011) 663–673.

[55] D.K. Verma, F. Khan, Corrosion inhibition of mild steel by using sulpha drugs in phosphoric acid medium: a combined experimental and theoretical approach, Chem. Sci. Int. J. (2016) 1–8.

[56] A. Singh, A. Kumar Singh, M.A Quraishi, Dapsone: a novel corrosion inhibitor for mild steel in acid media, Open Electrochem. J. 2 (2010) 43–51.

[57] A. Singh, J. Avyaya, E.E. Ebenso, M. Quraishi, Schiff's base derived from the pharmaceutical drug Dapsone (DS) as a new and effective corrosion inhibitor for mild steel in hydrochloric acid, Res. Chem. Intermed. 39 (2013) 537–551.

[58] P. Singh, D. Chauhan, S. Chauhan, G. Singh, M. Quraishi, Chemically modified expired Dapsone drug as environmentally benign corrosion inhibitor for mild steel in sulphuric acid useful for industrial pickling process, J. Mol. Liq. 286 (2019) 110903.

[59] S.K. Shukla, A.K. Singh, I. Ahamad, M. Quraishi, Streptomycin: a commercially available drug as corrosion inhibitor for mild steel in hydrochloric acid solution, Mater. Letters 63 (2009) 819–822.

[60] S.K. Shukla, E.E. Ebenso, Corrosion inhibition, adsorption behavior and thermodynamic properties of streptomycin on mild steel in hydrochloric acid medium, Int. J. Electrochem. Sci 6 (2011) 3277–3291.

[61] H. Yang, M. Zhang, A. Singh, Investigation of inhibition effect of ketoconazole on mild steel corrosion in hydrochloric acid, Int. J. Electrochem. Sci 13 (2018) 9131–9144.

[62] P. Nouri, M. Attar, An imidazole-based antifungal drug as a corrosion inhibitor for steel in hydrochloric acid, Chem. Eng. Commun. 203 (2016) 505–515.

[63] T.Jebakumar Immanuel Edison, M. Sethuraman, Electrochemical investigation on adsorption of fluconazole at mild steel/HCl acid interface as corrosion inhibitor, Int. Schol. Res. Notices 2013 (2013) 1–8.

[64] I. Obot, N. Obi-Egbedi, Fluconazole as an inhibitor for aluminium corrosion in 0.1 M HCl, Colloids Surf. A: Physicochem. Eng. Aspects 330 (2008) 207–212.

[65] I. Obot, N. Obi-Egbedi, S. Umoren, E. Ebenso, Adsorption and kinetic studies on the inhibition potential of fluconazole for the corrosion of Al in HCl solution, Chem. Eng. Commun. 198 (2011) 711–725.

[66] A. Guruprasad, H. Sachin, G. Swetha, B. Prasanna, Corrosion inhibition of zinc in 0.1 M hydrochloric acid medium with clotrimazole: experimental, theoretical and quantum studies, Surfaces Interf. 19 (2020) 100478.

[67] B. Obot, N. Obi-Egbedi, S. Umoren, Experimental and theoretical investigation of clotrimazole as corrosion inhibitor for aluminium in hydrochloric acid and effect of iodide ion addition, Der Pharma Chemica 1 (2009) 151–166.

[68] I. Ahamad, M. Quraishi, Mebendazole: new and efficient corrosion inhibitor for mild steel in acid medium, Corrosion Sci. 52 (2010) 651–656.

[69] A.K. Singh, M. Quraishi, Inhibitive effect of diethylcarbamazine on the corrosion of mild steel in hydrochloric acid, Corrosion Sci. 52 (2010) 1529–1535.

[70] R. Solmaz, G. Kardas, B. Yazici, M. Erbil, Inhibition effect of Rhodanine for corrosion of mild steel in hydrochloric acid solution, Protect. Metals 41 (2005) 581–585.

[71] P. Dohare, D. Chauhan, A. Sorour, M. Quraishi, DFT and experimental studies on the inhibition potentials of expired Tramadol drug on mild steel corrosion in hydrochloric acid, Mater. Disc. 9 (2017) 30–41.

[72] M. Abdallah, E. Gad, M. Sobhi, J.H. Al-Fahemi, M. Alfakeer, Performance of tramadol drug as a safe inhibitor for aluminum corrosion in 1.0 M HCl solution and understanding mechanism of inhibition using DFT, Egypt. J. Petrol. 28 (2019) 173–181.

[73] J.I. Bhat, V.D. Alva, A study of aluminium corrosion inhibition in acid medium by an antiemitic drug, Trans. Indian Inst. Metals 64 (2011) 377–384.

[74] N. Raghavendra, L.V. Hublikar, A.S. Bhinge, P.J. Ganiger, Expired famotidine as an effective corrosion inhibitor for copper in 3 M HCl solution: weight loss and atomic absorption spectroscopy investigations, Int. J. Res. Appl. Sci. Eng. Technol. (IJRASET) 7 (2019) 2289–2295.

[75] I. Ahamad, R. Prasad, M. Quraishi, Inhibition of mild steel corrosion in acid solution by Pheniramine drug: Experimental and theoretical study, Corrosion Sci. 52 (2010) 3033–3041.

[76] A. Christy, A. Aloysius, R. Ramanathan, N. Anthony, G. Sundaram, Anticorrosion activity of pheniramine maleate on mild steel in hydrochloric acid, J. Emerg. Technol. Innov. Res. (JETIR) 5 (2018) 432–438.

[77] I. Ahamad, R. Prasad, M.A. Quraishi, Experimental and theoretical investigations of adsorption of fexofenadine at mild steel/hydrochloric acid interface as corrosion inhibitor, J. Solid State Electrochem. 14 (2010) 2095–2105.

[78] S. Nataraja, T. Venkatesha, H. Tandon, B. Shylesha, Quantum chemical and experimental characterization of the effect of ziprasidone on the corrosion inhibition of steel in acid media, Corrosion Sci. 53 (2011) 4109–4117.

[79] H. Lgaz, R. Salghi, I.H. Ali, Corrosion inhibition behavior of 9-hydroxyrisperidone as a green corrosion inhibitor for mild steel in hydrochloric acid: electrochemical, DFT and MD simulations studies, Int. J. Electrochem. Sci 13 (2018) 250–264.

[80] G. Karthik, M. Sundaravadivelu, Studies on the inhibition of mild steel corrosion in hydrochloric acid solution by atenolol drug, Egypt. J. Petrol. 25 (2016) 183–191.

[81] A. Alwash, D. Fadhil, A. Ali, F. Abdul-Hameed, E. Yousif, Inhibitive effect of atenolol on the corrosion of zinc in hydrochloric acid, Rasayan J. Chem. 10 (2017) 922–928.

[82] I. Reza, A. Saleemi, S. Naveed, Corrosion inhibition of mild steel in HCl solution by Tinidazole, Polish J. Chem. Technol. 13 (2011) 67–71.

[83] L. Zhang, S.T. Zhang, X. Li, Electrochemical and quantum chemical investigations of tinidazole as corrosion inhibitor for mild steel in 3% HCl solution, Adv. Mater. Res., Trans. Tech. Publ. 194–196 (2011) 157–164.

[84] K. Haruna, T.A. Saleh, M. Quraishi, Expired metformin drug as green corrosion inhibitor for simulated oil/gas well acidizing environment, J. Mol. Liq. 315 (2020) 113716.

[85] H. Kumar, S. Karthikeyan, P. Vivekanand, S. Rajakumari, Pioglitazone (PGZ) drug as potential inhibitor for the corrosion of mild steel in hydrochloric acid medium, Mater. Today: Proc. (2020).

[86] R. Farahati, S.M. Mousavi-Khoshdel, A. Ghaffarinejad, H. Behzadi, Experimental and computational study of penicillamine drug and cysteine as water-soluble green corrosion inhibitors of mild steel, Progr. Organ. Coat. 142 (2020) 105567.

[87] M. Gholamhosseinzadeh, H. Aghaie, M.S. Zandi, M. Giahi, Rosuvastatin drug as a green and effective inhibitor for corrosion of mild steel in HCl and H2SO4 solutions, J. Mater. Res. Technol. 8 (2019) 5314–5324.

[88] M.N. El-Haddad, A. Fouda, A. Hassan, Data from chemical, electrochemical and quantum chemical studies for interaction between Cephapirin drug as an eco-friendly corrosion inhibitor and carbon steel surface in acidic medium, Chem. Data Collect. 22 (2019) 100251.

[89] R. Fuchs-Godec, M.G. Pavlović, Synergistic effect between non-ionic surfactant and halide ions in the forms of inorganic or organic salts for the corrosion inhibition of stainless-steel X4Cr13 in sulphuric acid, Corrosion Sci. 58 (2012) 192–201.

[90] G.K. Gomma, Corrosion of low-carbon steel in sulphuric acid solution in presence of pyrazole—halides mixture, Mater. Chem. Phys. 55 (1998) 241–246.

[91] W. Chen, H.Q. Luo, N.B. Li, Inhibition effects of 2, 5-dimercapto-1, 3, 4-thiadiazole on the corrosion of mild steel in sulphuric acid solution, Corrosion Sci. 53 (2011) 3356–3365.

[92] M. Bahrami, S. Hosseini, P. Pilvar, Experimental and theoretical investigation of organic compounds as inhibitors for mild steel corrosion in sulfuric acid medium, Corrosion Sci. 52 (2010) 2793–2803.

[93] P. Okafor, V. Osabor, E. Ebenso, Eco-friendly corrosion inhibitors: inhibitive action of ethanol extracts of Garcinia kola for the corrosion of mild steel in H2SO4 solutions, Pigm. Resin Technol. 36 (2007) 299–305.

[94] L.M. Vračar, D. Dražić, Adsorption and corrosion inhibitive properties of some organic molecules on iron electrode in sulfuric acid, Corrosion Sci. 44 (2002) 1669–1680.

[95] N. Eddy, S. Odoemelam, P. Ekwumemgbo, Inhibition of the corrosion of mild steel in H2SO4 by penicillin G, Sci. Res. Essays 4 (2009) 033–038.

[96] N. Eddy, S. Odoemelam, Inhibition of the corrosion of mild steel in acidic medium by penicillin V potassium, Adv. Nat. Appl. Sci. 2 (2008) 225–232.

[97] N.O. Eddy, E.E. Ebenso, Adsorption and quantum chemical studies on cloxacillin and halides for the corrosion of mild steel in acidic medium, Int. J. Electrochem. Sci. 5 (2010) 731–750.

[98] H. Kumar, S. Karthikeyan, P. Vivekanand, P. Kamaraj, The inhibitive effect of cloxacillin on mild steel corrosion in 2 N Sulphuric acid medium, Mater. Today: Proc. (2020).

[99] S.H. Kumar, S. Karthikeyan, Amoxicillin as an efficient green corrosion inhibitor for mild steel in 1M sulphuric acid, J. Mater. Environ. Sci. 4 (2013) 675–984.

[100] A.K. Singh, S.K. Shukla, M. Quraishi, Corrosion behaviour of mild steel in sulphuric acid solution in presence of ceftazidime, Int. J. Electrochem. Sci 6 (2011) 5802–5814.

[101] P. Thanapackiama, E. Subramaniama, K. Hemalathaa, B. Gayathria, Electrochemical study of inhibition of corrosion of copper by ofloxacin in acid media, J. Environ. Nanotechnol 8 (2019) 75–88.

[102] P. Thanapackiam, K. Mallaiya, S. Rameshkumar, S. Subramanian, Inhibition of corrosion of copper in acids by norfloxacin, Anti-Corrosion Methods Mater. 64 (2017) 92–102.

[103] N. Eddy, S. Odoemelam, E. Ogoko, B. Ita, Inhibition of the Corrosion of Zinc in 0.01-0.04 M H2SO4 by Erythromycin, Port. Electrochim. Acta 28 (2010) 15–26.

[104] E. Ogoko, S. Odoemelam, B. Ita, N. Eddy, Adsorption and Inhibitive Properties of Clarithromycin for the Corrosion of Zn in 0.01 to 0.05 M H2SO4, Port. Electrochim. Acta 27 (2009) 713–724.

[105] T. Ramdé, S. Rossi, L. Bonou, Corrosion inhibition action of Sulfamethoxazole for brass in acidic media, Int. J. Electrochem. Sci. 11 (2016) 6819–6829.

[106] F. Edoziuno, A. Adediran, B. Odoni, M. Oki, P. Ikubanni, O. Omodara, Performance of Methyl-5-Benzoyl-2-Benzimidazole Carbamate (Mebendazole) as corrosion inhibitor for mild steel in dilute sulphuric acid, Sci. World J. 2020 (2020) 1–11.

[107] S. Abd El Wanees, M. Elmorsi, T. Fayed, A. Fouda, S. Elyan, Inhibition effect of famotidine towards the corrosion of C-steel in sulphuric acid solution, Chem. Process Eng. Res. 46 (2016) 42–57.

[108] S. Day, M. Whalen, K. King, G. Hust, L. Wong, J. Estill, R. Rebak, Corrosion behavior of alloy 22 in oxalic acid and sodium chloride solutions, Corrosion 60 (2004) 804–814.

[109] Z.Y. Chen, S. Zakipour, D. Persson, C. Leygraf, Combined effects of gaseous pollutants and sodium chloride particles on the atmospheric corrosion of copper, Corrosion 61 (2005) 1022–1034.

[110] C. Verma, E.E. Ebenso, M. Quraishi, Corrosion inhibitors for ferrous and non-ferrous metals and alloys in ionic sodium chloride solutions: a review, J. Mol. Liq. 248 (2017) 927–942.

[111] M. Javidi, R. Omidvar, Synergistic inhibition behavior of sodium tungstate and penicillin G as an eco-friendly inhibitor on pitting corrosion of 304 stainless steel in NaCl solution using Design of Experiment, J. Mol. Liq. 291 (2019) 111330.

[112] M. Scholant, E. Coutinho, S. Dias, D. Azambuja, S. Silva, S. Tamborim, Corrosion inhibition of AISI 1020 steel based on tungstate anion and amoxicillin as corrosion inhibitors in 0.05 mol l−1 NaCl solution or inserted into cellulose acetate films, Surface Interf. Anal. 47 (2015) 192–197.

[113] M. Dehdab, Z. Yavari, M. Darijani, A. Bargahi, The inhibition of carbon-steel corrosion in seawater by streptomycin and tetracycline antibiotics: an experimental and theoretical study, Desalination 400 (2016) 7–17.

[114] D. Millan-Ocampo, J. Hernandez-Perez, J. Porcayo-Calderon, J. Flores-De los Ríos, L. Landeros-Martínez, V. Salinas-Bravo, J. Gonzalez-Rodriguez, L. Martinez, Experimental and theoretical study of ketoconazole as corrosion inhibitor for bronze in NaCl+ Na2SO4 solution, Int. J. Electrochem. Sci 12 (2017) 11428–11445.

[115] F. Soltaninejad, M. Shahidi, Investigating the effect of penicillin G as environment-friendly corrosion inhibitor for mild steel in H3PO4 solution, Prog. Color, Colorants Coatings 11 (2018) 137–147.

[116] A.M. Farimani, H. Hassannejad, A. Nouri, A. Barati, Using oral penicillin as a novel environmentally friendly corrosion inhibitor for low carbon steel in an environment containing hydrogen sulfide corrosive gas, J. Natural Gas Sci. Eng. (2020) 103262.

[117] O.S.I. Fayomi, I.G. Akande, A.P.I. Popoola, H. Molifi, Potentiodynamic polarization studies of Cefadroxil and Dicloxacillin drugs on the corrosion susceptibility of aluminium AA6063 in 0.5 M nitric acid, J. Mater. Res. Technol. 8 (2019) 3088–3096.

[118] J.S.N.G.T., Yao, A. Trokourey, Thermodynamic and DFT studies on the behavior of cefadroxil drug as effective corrosion inhibitor of copper in one molar nitric acid medium, J. Mater. Environ. Sci. 10 (2019) 926–938.

[119] S. Ouattara, P. Niamien, E.A. Bilé, A. Trokourey, Ciprofloxacin hydrochloride as a potential inhibitor of copper corrosion in 1M HNO_3, Der Chemica Sinica 8 (2017) 398–412.

[120] L. Adardour, H. Lgaz, R. Salghi, M. Larouj, S. Jodeh, M. Zougagh, I. Warad, H. Oudda, Corrosion inhibition performance of sulfamethazine for mild steel in phosphoric acid solution: gravimetric, electrochemical and DFT studies, Der Pharm. Lett. 8 (2016) 126–137.

[121] L.Y. Hu, S.T. Zhang, X.H. Huang, H.H. Hu, Corrosion inhibition of tinidazole in KOH solution for zinc, Adv. Mater. Res., Trans. Tech. Publ. 317–319 (2011) 1852–1857.

[122] I.B. Obot, Synergistic effect of nizoral and iodide ions on the corrosion inhibition of mild steel in sulphuric acid solution, Port. Electrochim. Acta 27 (2009) 539–553.

[123] N.O. Eddy, E.E. Ebenso, U.J. Ibok, Adsorption, synergistic inhibitive effect and quantum chemical studies of ampicillin (AMP) and halides for the corrosion of mild steel in H2SO4, J. Appl. Electrochem. 40 (2010) 445–456.

[124] N. Vaszilcsin, V. Ordodi, A. Borza, Corrosion inhibitors from expired drugs, Int. J. Pharm. 431 (2012) 241–244.

[125] H.I. Al-Shafey, R.A. Hameed, F. Ali, A.-M. Ae-AS, M. Salah, Effect of expired drugs as corrosion inhibitors for carbon steel in 1M HCL solution, Int. J. Pharm. Sci. Rev. Res. 27 (2014) 146–152.

[126] N.K. Gupta, C. Gopal, V. Srivastava, M. Quraishi, Application of expired drugs in corrosion inhibition of mild steel, Int. J. Pharm. Chem. Anal 4 (2017) 8–12.

19

Natural polymers as green corrosion inhibitors

Polymers are materials or macromolecules made up of the repetition of a very large number of small molecules, called monomers. Depending upon their origin, they may be classified as synthetic and natural polymers. Synthetic polymers are derived chemically or isolated from petroleum resources, such as oil. Some common examples of synthetic polymers include polyethylene, nylon, epoxy, Teflon, and polyester. On the other hand, natural polymers are derived from natural resources. Silk, nucleic acids (DNA & RNA), wool, protein, cellulose, starch, chitosan (CH), etc., are common examples of natural polymers [1,2]. Polymers are widely used for different engineering applications because of their various salient features including high modulus to weight ratio, high mechanical strength, resistance to corrosive degradation, resilience, toughness, transparency, lack of heat and electrical conductance, ease of fabrication and processing, less permeable, and cost-effectivity. Properties of the polymers can be suitably designed by controlling their long-chain structures.

Literature investigation shows that natural polymers, especially carbohydrates and their derivatives are widely used as corrosion inhibitors for different metals and alloys in various electrolytes. Because of their polymeric nature, they provide excellent surface coverage and behave as effective corrosion inhibitors. More so, because of their biological and natural origin, they can be treated as environmental-friendly alternatives to be used for different biological and industrial applications. Natural polymers are associated with the following useful properties:

(i) Eco-friendly (biodegradable, nonbioaccumulative, and biotolerable).
(ii) Excellent corrosion inhibition potential (high surface area) at relatively low concentration.
(iii) Numerous polar functional groups (adsorption centers).
(iv) Abundant, natural, and commercial availability.
(v) Ease and cost-effective processing, isolation, and application.
(vi) Wide-range of anticorrosive activity (useful for different metal/electrolyte systems).
(vii) The low adverse effect of environment and living organisms.
(viii) May be suitably modified using chemical species (semisynthetic polymers).
(ix) Their solubility and corrosion inhibition potential can be enhanced by chemical modification.
(x) Nontoxic and noninflammatory.

However, use of polymers as corrosion inhibitors is also connected with some shortcomings such as:

(i) Limited solubility in polar electrolytes (mostly used in coating formulations).
(ii) Sometimes, isolation and purification of natural polymers utilize toxic and expensive chemicals/solvents.
(iii) For better corrosion inhibition potential, chemical modification is essentially required.
(iv) Pure natural polymers show reasonably less corrosion protection potential.
(v) Natural polymers are highly susceptible for heat and rarely used as high-temperature corrosion inhibitors.
(vi) They are also susceptible for hydrolysis, degradation, and/or rearrangement in highly aggressive electrolytes, especially at elevated temperature.

Nevertheless, natural polymers especially, carbohydrate-based natural and chemically modified polymers are extensively used as corrosion inhibitors. It is important to mention that most of the natural polymers exhibit very little or no solubility in the polar aqueous electrolytes. Therefore, natural polymers in their pure form are very rarely used as aqueous phase corrosion inhibitors. However, their solubility in such electrolytes can be enhanced by modifying them using surface-active chemicals containing polar functional groups. This type of functionalization not only increases their solubility in polar electrolytes but it also enhances their corrosion inhibition potential. For example, literatures on pure cellulose as an aqueous phase corrosion inhibitor are very less however the anticorrosive effect of methoxy-cellulose is widely reported. Similarly, chemically modified using different organic compounds are more widely investigated as compared to pure CH.

19.1 Literature survey: natural polymers as green corrosion inhibitors

Natural polymers are produced by plant and animal cells. They are established as one of the most environmental friendly class of organic compounds that possess different biological and industrial applications. More so, it is recall that most of the natural polymers including natural rubbers, polypeptides, nucleic acids (DNA & RNA), starch, CH, cellulose, lignin, etc., are biodegradable and nonbioaccumulative. In view of this, natural polymers are widely used for various applications including as corrosion inhibitors. Although polymers derived from various natural resources have been widely investigated as corrosion inhibitors, the present chapter mainly describes the inhibition effect of carbohydrate-based natural polymers and their chemically modified forms.

19.1.1 Chitosan as green corrosion inhibitors

One of the most frequently used carbohydrate polymers as corrosion inhibitors is CH. This is a leaner polymer that is associated with arbitrarily scattered D-glucosamine and N-acetyl-D-glucosamine units joined together through β-1-4-glycosidic linkage [3–5]. As presented in Fig. 19.1, CH is derived through the deacetylation of chitin, which is mainly presented in fungi cell walls, exoskeletons of crustaceans (such as shrimps and crabs), exoskeletons of insects, cephalopods beaks, mollusks's radulae, and the scales of fish [6–8]. CH with the

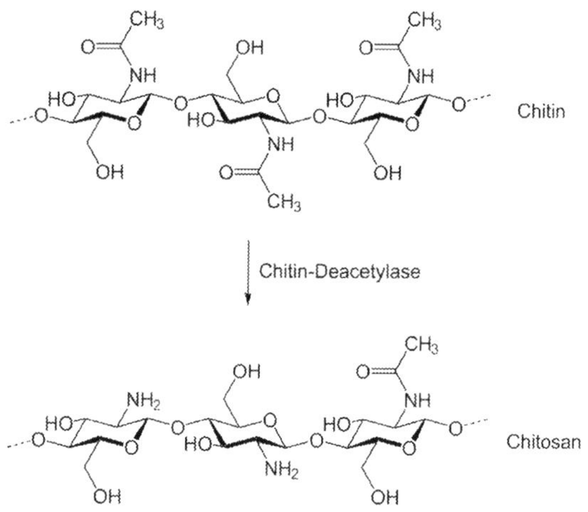

FIG. 19.1 Synthesis of chitosan from deacetylation of chitin.

degree of deacetylation (%DD) of 60–100 and having molecular weight of 3800–20,000 Daltons are commercially important [9,10]. The chemical structure of CH contains various polar functional groups including acetyl (–$COCH_3$), amide (–NHCO–), hydroxyl (–OH), hydroxymethyl (–CH_2OH), amino (–NH_2), and cyclic ether (–O–) that help in making CH and its derivatives soluble in the aqueous electrolytes. Functionalization of the CH using other surface-active molecules further enhances their corrosion inhibition effectiveness.

It is well-established that organic compounds, including CH, become effective corrosion inhibitors by adsorbing on the surface of the metallic surface using their electron-rich polar functional groups and π-bonds. Therefore, the above functional groups of CH also act as adsorption centers during their interaction with the metallic surface [11–14]. Due to its natural origin, CH and CH-based compounds can be treated as environmental friendly alternatives at the place of traditional toxic corrosion inhibitors [15–17]. Because of their high solubility and excellent tendency to adsorb on the metallic surface, CH, and its derivatives are widely used as corrosion inhibitors for different metals and alloys in various electrolytes.

Recently, Umoren and coworkers [18] described the inhibition effect of CH for mild steel (MS) corrosion in 1M HCl. Chemical, electrochemical and surface investigations were used to demonstrate the corrosion inhibition effect of CH in 1M HCl. Results showed that CH exhibits corrosion inhibition potential of 93% and 96% at 70°C and 60°C, respectively. Analyses also showed that CH becomes effective by adsorbing on the metallic surface following through the Langmuir adsorption isotherm model. Potentiodynamic polarization (PDP) study showed that CH becomes effective by retarding both anodic and cathodic Tafel reactions. CH was classified as a mixed-type corrosion inhibitor based on the outcomes of PDP analyses. Electrochemical impedance spectroscope study shows that CH becomes effective by adsorbing at the interface of metal and electrolyte (1M HCl). Adsorption of CH at the interface increased the values of change transfer resistance for the corrosion process. Thereby, CH behaved as an interface-type corrosion inhibitor. Corrosion inhibition power of CH for MS corrosion in 1M HCl system has also been reported in another study [19].

Recently, our research team reported the corrosion inhibition potential of CH for MS corrosion in sulfamic acid solution [20]. Various experimental methods were used to describe the corrosion inhibition power of CH in the studied electrolyte. The effect of synergism using potassium iodide (KI) was also studied. Results showed that CH exhibits the highest inhibition efficiencies of 73.8% and 90% at 200 ppm concentration in the absence and presence of 5 ppm concentration of KI, respectively. PDP study reveals that CH with and without KI inhibits both anodic and cathodic Tafel reactions and behaved as mixed-type corrosion inhibitors. PDP study further showed that the presence of CH in the corrosive medium decreases the values of corrosion current density (i_{corr}). This observation revealed that CH becomes effective by blocking the active sites, responsible for corrosion, through its adsorption. Adsorption of the CH on the metallic surface was supported by atomic force microscope (AFM) and scanning electron microscope (SEM) studies. A significant smoothness in the surface morphology of the metallic specimens was observed in the presence of CH. Improvement in the surface morphology of the protected metallic surface is attributed due to the adsorption and formation of protective film by CH. It is important to mention that the corrosion inhibition potential of organic corrosion inhibitors can be increased by adding inorganic salts (such as $ZnCl_2$ and KI). This is known as the synergism of synergistic effect. Obviously, inorganic slats help organic corrosion inhibitors in getting them adsorbed on the metallic surface. The phenomenon of synergism using inorganic salts is widely reported in the field of corrosion inhibition.

Not only steel alloys, CH has also been studied as corrosion inhibitors for corrosion in acidic [21] and neutral sodium chloride (NaCl) [22,23] solutions. Harmami et al. [24] reported the corrosion inhibition of water-soluble CH (WSC) for tinplate in sedum chloride solution. WSC utilized in the study was derived from mussel shells. The corrosion inhibition potential of WSC was determined at its various concentrations 10–1500 mg L^{-1}. Results showed that an increase in WSC concentration was associated with a subsequent increase in its corrosion inhibition potential. Analyses further showed that WSC collected from shrimp shells was more effective as compared to the CH derived from mussel shells. Weight loss and PDP studies suggest that WSC derived from shrimp shells showed the highest protection potential of 72.73% and 91.41%, respectively. On the other hand, WSC derived from mussel shells exhibited the highest protection inhibition potential of 54.55% at 1300 mg L^{-1} concentration. PDP study revealed that WSC derived from shrimp and mussel shells become effective by retarding both anodic and cathodic reactions and behaves as mixed-type corrosion inhibitors.

Literatures on enhanced solubility of CH using its functionalization are widely reported [25–33]. Corrosion inhibition potential and related salient features of functionalized CH of some common reports are given in Table 19.1. Schiff's bases (SBs) derived from the condensation of aromatic aldehydes and CH are extensively used as corrosion inhibitors for different metals and alloys in various electrolytes [29]. SBs show remarkably high corrosion inhibition potential at relatively lower concentrations. It is important to mention that the functionalization of CH using aromatic aldehydes increases the number of active sites (adsorption centers) and therefore protection efficiency.

Recently, Haque and coworkers [34] described the synthesis, characterization, and corrosion inhibition evaluation of three CH-based SBs (CSBs) from benzaldehyde (CSB-1), 4-(dimethylamino) benzaldehyde (CSB-2) and 4-hydroxy-3-methoxybenzeldehyde (CSB-3) for MS corrosion in 1M HCl. Chemical, electrochemical, surface, and computational examinations were performed to demonstrate the corrosion inhibition property of CH-based SBs.

19.1 Literature survey: natural polymers as green corrosion inhibitors 211

TABLE 19.1 Summary of some common reports on modified chitosan as corrosion inhibitors for different metals and alloys in various electrolytes.

S. No	Moiety attached from chitosan & abbreviation	Electrolyte and metal	Nature of adsorption	IE% and Conc.	References
1	Chitosan- cinnamaldehyd (Cinn-Cht)	15% HCl/carbon steel	Langmuir isotherm & mixed-type inhibitor	87.72% at 600 ppm & 92.67% at 600 ppm+10 mM KI	[30]
2	Carboxymethyl-Chitosan-benzaldehyde (CMChi-B) & Carboxymethyl-chitosan-urea-glutaric acid (CMChi-UGLU)	2% NaCl and 1-3M HCl/steel	—	CMChi-B (80.82%) > CMChi-UGLU (80.62%)	[29]
3	CH-benzaldehyde (CSB-1), CH-4(dimethylamino) benzaldehyde (CSB-2) & 4-hydroxy-3-methoxy benzaldehyde (CSB-3)	1M HCl/mild steel	Langmuir isotherm & slight cathodic-type inhibitor	CSB-3 (91.43%) > CSB-2 (89.87%) CSB-1 (88.63%) at 100 ppm	[34]
4	Chitosan-thiosemicarbazide (TSFCS) & chitosan-thiocarbohydrazide (TCFCS)	2% acetic acid/304 steel	Mixed-type inhibitors	TCFCS (92%) at 60 mg L^{-1}	[37]
5	Chitosan-Polyethylene glycol (Cht-PEG)	1M sulfamic acid/mild steel	Langmuir isotherm & slight cathodic-type inhibitor	93.9% at 200 ppm	[39]
6	Chitosan-methyl acrylate-ethylene diamine (CS-MAA-EN) & chitosan-methyl acrylate-triethylene tetramine (CS-MAA-TN)	5% HCl/carbon steel	—	CS-MAA-EN (88.06%) > CS (84.22%) > CS-MAA-TN (69.46%) at 0.3%	[31]
7	Chitosan –Vanillin (Van-Cht)	15% HCl/carbon steel	Langmuir isotherm & mixed-type inhibitor	92.72 % at 500 mg L^{-1}	[35]
8	Chitosan- salicylaldehyde (CHSA)	1M HCl/mild steel	Temkin adsorption isotherm/ mixed-type inhibitor	70.08% at 1500 ppm	[36]
9	Chitosan-Poly (vinyl butyral) (PVB-Ch)	0.3M salt solution/carbon steel	—	—	[38]
10	Chitosan-Polyethylene glycol (CS-PEG)	1M HCl/mild steel	Langmuir isotherm & slight cathodic-type inhibitor	93.9% at 200 mg L^{-1}	[41]
11	Chitosan-polyaniline (PANI/CTS)	0.5 M HCl/Q235 steel	Mixed-type inhibitor	79.02% at 200 ppm	[42]
12	Carboxymethyl chitosan (CMC)	3.5% NaCl/1020 carbon steel	Langmuir adsorption isotherm/mixed-type inhibitor	85.57% at 80 ppm	[33]
13	Chitosan-polyaspartic acid (PASP/CS)	3.5% NaCl/carbon steel	Anodic-type inhibitor	87.56% at 20 ppm	[32]

Among the tested SBs, CSB-3 showed the highest protection power of 90.65% at 50 ppm concentration. Weight loss study suggests that increase in CSB-1 to CSB-3 concentration causes improvement in their inhibition performance. PDP study suggests that CSB-1, CSB-2, and CSB-3 become effective by retarding both anodic as well as cathodic Tafel half-cell reaction. They behaved as mixed-type corrosion inhibitors. The presence of CSBs causes a significant reduction in the values of corrosion current density. This observation suggests that CSBs become effective by forming the corrosion protective surface film. Adsorption of CSBs and formation of anticorrosive film were further supported by energy dispersive X-ray (EDX), SEM, and Ft-IR analyses. Change in the composition of elements present on the metallic surface was observed through EDX analysis which validates the adsorption nature of CSBs on the MS surface. Improvement in the surface morphology of the protected MS surface further validated the adsorption mechanism of corrosion inhibition using CSBs. FT-IR spectral analysis was also conducted to demonstrate the adsorption of CSBs on the metallic surface. DFT-based computational analysis was conducted to support the adsorption mechanism of studied CSBs on the MS surface. Results showed that studied CSBs interacted with the metallic surface using donor–acceptor interactions in which polar functional groups act as adsorption centers. MD simulations analyses showed that studied CSBs spontaneously adsorbed on Fe (100) surface. Corrosion inhibition potentials of some other SCBs have also been reported [35,36]. In this report, investigated CSBs mostly act as mixed-type corrosion inhibitors as their presence adversely affects the anodic as well as cathodic Tafel reactions. Electrochemical impedance spectroscope study suggests that most of the investigated CSBs become effective by adsorbing at the metal and electrolyte interface. This type of adsorption results in the subsequent increase in the magnitude of charge transfer resistance.

Research on corrosion inhibition potential of cross-linked CH is relatively more recent. Therefore, cross-linking of CH using organic compounds, called as linkers, and their subsequent use as corrosion inhibitors is gaining immense attention. Obviously, this type of cross-linking results in the joining of two or more CH–polymer chains. It is important to mention that the corrosion inhibition effect, as well as solubility of CH, can be enhanced using this type of cross-linking [37,38]. Chauhan et al. [39] recently described the corrosion inhibition potential of CH cross-linked with polyethylene glycol (PEG), designated as Cht-PEG, for MS in 1M sulfamic acid solution. Results showed that Cht-PEG exhibited the highest protection effectiveness of 93.9% at 200 mg L^{-1} concentration. It was also derived that Cht-PEG becomes effective by adsorbing at the metallic surface following through Langmuir adsorption isotherm. PDP measurement showed that Cht-PEG act as mixed-type corrosion inhibitor and its presence adversely affect anodic as well as cathodic Tafel reactions. Decrease in the magnitude of corrosion current density indicated the Cht-PEG molecules inhibit MS corrosion in sulfamic acid solution by blocking the active sites, responsible for corrosion, through their adsorption.

Corrosion inhibition potential of CH cross-linked with thiocarbohydrazide, designated as (TC-Cht), was investigated for stainless steel (SS) in 3.5% sodium chloride solution [40]. Results showed that TC-Cht acts as an efficient corrosion inhibitor for SS in 3.5% NaCl and showed highest protection power of greater than 94% at 500 mg L^{-1} concentration. TC-Cht inhibits SS corrosion by adsorbing on its surface following through Langmuir adsorption isotherm model. PDP study showed that TC-Cht behaves as a mixed type but predominantly cathodic-type corrosion inhibitor. Adsorption mechanism of corrosion inhibition using TC-Cht was further supported by SEM and EDX analyses. EDX analysis showed elemental composition present on the metallic surface was greatly affect in the presence of TC-Cht. SEM analysis showed that surface of the SS corroded in the presence of TC-Cht was much smoother as compared to SS surface without TC-Cht.

Besides, chemically modified and cross-linked CH, composites of CH have also been widely investigated as corrosion inhibitors for different metals and alloys in various electrolytes. It is expected that CH-based composites would be relatively more effective corrosion inhibitors as compared to the CH itself. Kong and coworkers described the synthesis, characterization and corrosion inhibition effect of a CH-polyaniline (PANI/CTS) for Q235 steel in acidic solution [42]. Through PDP investigation, it was observed that PANI/CTS molecules decrease the corrosion current density in their presence. This observation suggests that PANI/CTS molecules become effective by blocking the active sites through their adsorption. PANI/CTS molecules interact with the metallic surface using donor–acceptor mechanism.

SEM analyses showed that the presence of PANI/CTS molecules in the corrosive electrolyte of 0.5M HCl causes significant improvement in the SEM surface morphology of protected metallic specimen. From the SEM images, it can be clearly seen that surface of polished steel was very smooth as compared to the metallic surfaces corroded in the absence and presence of PANI/CTS. However, the morphology of steel surface morphology of PANI/CTs protected metallic surface was much smoother than that of the nonprotected metallic surface. This observation suggests that PANI/CTS adsorb on the metallic surface avoids the aggressive attack of electrolyte molecules.

Corrosion inhibition potential of other CH-based composites including CH-Zn nanoparticle [43–46], carboxymethyl CH grafted poly(2-methyl-1-vinylimidazole) [47,48], CH-Cu, Ni, CH-nanocomposites [49–51], sulfonated CH [52,53], Au, and F [54–56], CH-polymer [38,42,57,58], CH/TiO_2 [59–61], CH-Ag nanoparticle [62,63], CH-boron nitrile [64], CH-polyamines [31], CH- hydroxyapatite [65–69], CH-drug [70], CH-polymer blends [38,71], etc., have also been investigated. Careful observation of literature shows that most of the CH composites behave as a superior corrosion inhibitor as compared to pure CH. Most of the derivatives behave as mixed-type corrosion inhibitors and they become effective by retarding both anodic and cathodic reactions. They become effective by adsorbing on the metallic surface following mostly through the Langmuir adsorption isotherm model.

19.1.2 Cellulose as green corrosion inhibitors

Cellulose is a highly industrially and biologically important carbohydrate polymer which present mainly in cotton (90%), wood (40%–50%), and help (57%) [72–75]. Cellulose is a linear polymeric chain of D-glucose units connected together by β→1-4-glycosidic linkage [76,77]. It is a structural component of plant cell walls and widely present in oomycetes and algae. Its molecular formula id $(C_5H_{10}O_5)_n$ because of their natural and commercial availability, cellulose and its derivatives are extensively used for versatile biological and industrial applications. Literature study suggests that cellulose and its derivatives are also widely used as corrosion inhibitors for different metal-electrolyte systems [78]. Again it is important to mention that cellulose derivatives are more effective corrosion inhibitors as compared to cellulose itself. Hydroxymethyl cellulose and carboxymethyl cellulose (CMC) are the two most frequently utilized cellulose-derived corrosion inhibitors. A collection of some major reports on the corrosion inhibition effect of cellulose derivatives is presented in Table 19.2 [79–95].

TABLE 19.2 Summary of some common reports on modified cellulose as corrosion inhibitors for different metals and alloys in various electrolytes.

S. No.	Moiety attached from chitosan & abbreviation	Electrolyte and metal	Nature of adsorption	IE% and Conc	References
1	Carboxymethyl cellulose (CMC)	2M H_2SO_4/mild steel	Langmuir adsorption isotherm	CMC (65% at 0.5 g L^{-1})	[79]
2	Hydroxyethylcellulose (HEC)	3.5 NaCl/1018 c-steel	Langmuir isotherm & mixed-type inhibitor	95.5% at 0.5 mM	[81]
3	Hydroxyethyl Cellulose (HEC)	0.5M H_2SO_4/mild steel	Freundlich isotherm & mixed-type inhibitor	70.35% at 2000 mg L^{-1} (333K)	[82]
4	Hydroxyethyl cellulose (HEC) 2017	1M HCl/A1020 carbon steel	Langmuir isotherm & mixed-type inhibitor	91.62% at 500 ppm	[83]
5	Ethyl hydroxyethyl cellulose (EHEC)	1M H_2SO_4/mild steel	Langmuir isotherm & Slightly cathodic-type inhibitor	68.19% (EHEC) and 91.05% (EHEC+KI) at 2.5 g L^{-1}	[84]
6	Hydroxyethyl cellulose (HEC)	2% NH_4Cl, zinc-carbon battery	Langmuir isotherm & mixed-type inhibitor	92.07% at 300 ppm	[85]
7	Hydroxypropyl cellulose (HPC)	0.5M HCl and 2M H_2SO_4/aluminum	Langmuir isotherm & mixed-type inhibitor	92.54% (H_2SO_4) & 80.33% (HCl) at 5 g L^{-1}	[86]
8	hydroxyethyl cellulose (HEC) & hydroxypropyl methylcellulose (HPMC)	1M HCl/Aluminum	Slightly cathodic-type inhibitors	HEC (83.25%) and HPMC (84.68%) at 2000 mg L^{-1} (1 day)	[87]
9	Hydroxyethyl cellulose (HEC)	0.5M H_2SO_4/Mild steel & aluminum	Langmuir isotherm & mixed-type inhibitor	93.61% (mild steel) 64.18% (Al) at 2000 mg L^{-1}	[88]
10	Cellulose acetate	0.5, 1, 2, & 3M HCl/aluminum	–	55.71%	[89]
11	Sodium carboxymethyl cellulose (Na-CMC) 2020	1M HCl/aluminum	Freundlich adsorption isotherm	86.0% at 1 g L^{-1} (at 35°C)	[90]
12	Hydroxyethyl cellulose (HEC)	0.5M HCl/mild steel & aluminum	Langmuir isotherm & mixed-type inhibitor	67.94% at 2.5×10^{-3}M	[91]
13	Carboxymethyl cellulose/AgNPs composite (CMC/AgNPs)	15% H_2SO_4/St37 Steel	Langmuir isotherm & mixed-type inhibitor	96.37% at 1000 ppm (at 60°C)	[80]
14	Chitosan (CH) and carboxymethyl cellulose (CMC)	3.5% NaCl+CO_2/API 5L X60 pipeline steel	Langmuir isotherm & mixed-type inhibitors	88% (Comme-rcial inh), 45% (CH) and 39% (CMC) at 100 ppm	[92]
15	Hydroxyethyl Cellulose (HEC)	1M HCl and 0.5M H_2SO_4/copper	Mixed-type Inhibitor	95% at 2000 mg L^{-1}	[93]
16	Sodium carboxymethyl cellulose (Na-CMC)	Simulated water (NaCl)/copper	Langmuir isotherm & Slightly cathodic-type inhibitor	83.34% at 5 mg L^{-1} (at 20°C)	[94]
17	NEC, NMCC & NCMC	3.5% NaCl/copper	Mixed-type inhibitors	94.7% (NEC), 33.2% (NMCC) & 83.4% (NCMC) at 100 ppm	[95]

Umoren and coworkers [79] described the inhibition effect of CMC for MS corrosion in 2M H_2SO_4 using hydrogen evolution and weight loss (gravimetric) methods. Weight loss study showed that CMC becomes effective by adsorbing on the metallic surface following through Langmuir adsorption isotherm. Results showed that adsorption of CMC obeyed the Langmuir adsorption isotherm model. The effect of synergism was also observed in the presence of potassium halide salts (KI, KBr, and KCl). It was observed that the presence of a small amount (5 mM) of potassium halide salts significantly improve the corrosion inhibition performance of CMC. In the presence of halide salts synergism order follows the sequence: KCl < KBr < KI. CMC (0.5 g L^{-1})/KI (5 mM), CMC (0.5 g L^{-1})/KBr (5 mM), CMC (0.5 g L^{-1})/KCl (5 mM) and CMC (0.5 g L^{-1}) showed optimum inhibition efficiency of 85%, 63%, 48%, and 56%, respectively.

In another study, this group of authors demonstrated the synthesis, characterization, and inhibition evaluation of carboxymethyl cellulose (CMC) and silver nanoparticles (AgNPs) nanocomposite (CMC/AgNPs) for St37 steel in 15% H_2SO_4 [80]. The inhibition effect of CMC/AgNPs was evaluated using electrochemical, chemical, and surface measurements. Adsorption of CMC/AgNPs nanocomposite on the metallic surface was supported by various EDX, AFM, SEM, and FT-IR methods. SEM and EDX spectra of St37 steel surface corroded in 15% H_2SO_4 solution with and without CMC/AgNPs are presented in Fig. 19.2. Investigation of the SEM images showed that the presence of CMC/AgNPs composite in corrosive medium causes significant improvement in the surface morphology. Changes in the composition of the surface elements were observed using EDX analysis. Cellulose derivatives are also used as corrosion inhibitors for MS in acidic [4,6,13,14] and neutral sodium chloride (NaCl) [3,14] electrolytes. Cellulose derivatives are also evaluated as corrosion inhibitors for aluminum [7,8,11,12,15,18] and copper [5,10,16]. Enhancement in the corrosion inhibition effect of the cellulose derivatives by synergism is also investigated [8,11]. A summary of some major reports on cellulose derivatives as corrosion inhibitors is presented in Table 19.2. Most of the cellulose derivatives become effective by adsorbing on the metallic surface using The Langmuir adsorption isotherm model. Using PDP study it is observed that most of the cellulose derivatives behave as mixed-type corrosion inhibitors and become effective corrosion inhibitors by retarding the anodic as well cathodic reactions.

19.1.3 Carbohydrates other than cellulose and chitosan as green corrosion inhibitors

Along with cellulose, CH and their derivatives, other carbohydrate-based polymers are also evaluated as corrosion inhibitors for different metals and alloys in various electrolytic media. In most cases, these compounds become effective on adsorbing on the metallic surface using their electron-rich polar functional groups. Through their adsorption, these natural polymers form corrosion protective film over the metallic surface. One of such carbohydrate polymers is starch which is consists of linear amylose and branched-chain amylopectin. Literature study suggests that starch has also been investigated as corrosion inhibitors for different metals and alloys however its application as an aqueous phase corrosion inhibitor is limited because of its limited solubility. Therefore, in the majority of cases,

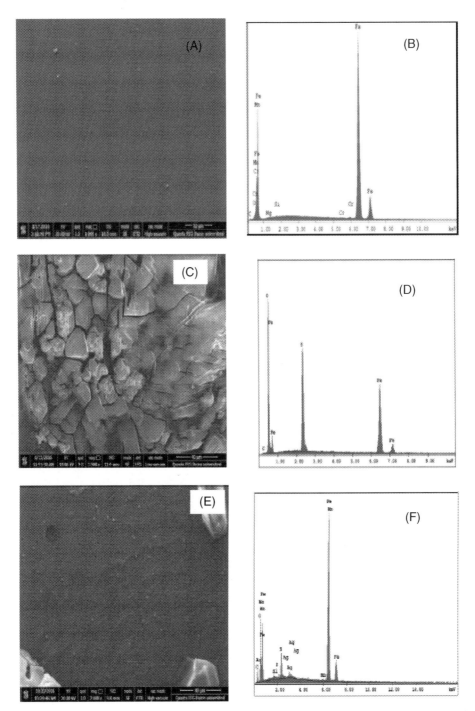

FIG. 19.2 SEM and EDX spectra of (A and B) abraded St37 surface, (C and D) corroded St37 surface in 15% H_2SO_4 without CMC/AgNPs and (E and F) corroded St37 surface in 15% H_2SO_4 with 1000 ppm of CMC/AgNPs [80].

chemically modified CH has been used as corrosion inhibitors. Brindha and coworkers [96] modified starch using 2,6-diphenyl-3-methylpiperidin-4-one (DPMP) and tested it as a corrosion inhibitor for MS in an acidic solution. Corrosion inhibition potential of DPMP was determined using various chemical and electrochemical methods. It was observed that the corrosion inhibition effect of modified starch is greatly depends upon the immersion time and experimental temperature. Obviously, an increase in temperature causes a subsequent increase in the corrosion rate. Further, the corrosion inhibition potential of starch modified with sodium dodecyl sulfate (DS) and cetyltrimethyl ammonium bromide has also been investigated elsewhere [97]. Corrosion inhibition potential of starch [98–100], algenates [101,102], dextrin and cyclodextrin [103–108], pectin [109–114], pectate [115,116], and exudate gums [117–122] are widely investigated widely. Corrosion inhibition performance of these carbohydrate polymers in the presence of halide ions (synergism) has also been widely reported [123–126]. These polymers behave as mixed-type corrosion inhibitors as they become effective by retarding both anodic as well as cathodic reactions. They inhibit metallic corrosion by adsorbing on the metallic surface following mostly through the Langmuir adsorption isotherm. Adsorption of these polymers on the metallic surface was supported by various surface investigation methods including SEM, AFM, EDX, XRD, XPS, FT-IR, and UV-vis. In few reports, computational analyses were carried out to support the experimental results. Carbohydrate polymers interact with the metallic surface using donor–acceptor mechanism.

19.2 Summary

Natural polymers are environmental-friendly, biodegradable, nonbioaccumulative, and biotolerable polymers derived from cells of living organisms. These compounds are widely used corrosion inhibitors for different metals and alloys in various electrolytes. Because of their high surface area, due to polymeric nature, natural polymers show excellent corrosion inhibition potential at relatively lower concentrations. These polymers contain various polar functional groups that enhance their solubility in the aqueous phase electrolytes ad also act as adsorption centers during metal–inhibitor interactions. Because of their natural availability, these compounds are abundantly and commercially available and can be used for wide range of anticorrosive activity (useful for different metal/electrolyte systems). They inhibit metallic corrosion by forming surface protective film through their adsorption. Adsorption of these compounds mostly follows the Langmuir adsorption isotherm model.

19.3 Useful links

https://www.cmu.edu/gelfand/lgc-educational-media/polymers/natural-synthetic-polymers/index.html
https://byjus.com/chemistry/natural-polymers/
https://www.sciencedirect.com/topics/materials-science/natural-polymer
https://www.sciencedirect.com/topics/chemical-engineering/natural-polymer

References

[1] A. Sionkowska, Current research on the blends of natural and synthetic polymers as new biomaterials, Prog. Polym. Sci. 36 (2011) 1254–1276.

[2] V.I. Lozinsky, Cryogels on the basis of natural and synthetic polymers: preparation, properties and application, Russ. Chem. Rev. 71 (2002) 489–511.

[3] T. Philibert, B.H. Lee, N. Fabien, Current status and new perspectives on chitin and chitosan as functional biopolymers, Appl. Biochem. Biotechnol. 181 (2017) 1314–1337.

[4] D. Raafat, H.G. Sahl, Chitosan and its antimicrobial potential–a critical literature survey, Microb. Biotechnol. 2 (2009) 186–201.

[5] P. Suresh, Enzymatic technologies of chitin and chitosan, Enzymatic Technologies for Marine Polysaccharides, CRC Press, Tailor and Francis (2019) 449.

[6] I. Hamed, F. Özogul, J.M. Regenstein, Industrial applications of crustacean by-products (chitin, chitosan, and chitooligosaccharides): a review, Trends Food Sci. Technol. 48 (2016) 40–50.

[7] P.K. Dutta, J. Dutta, V. Tripathi, Chitin and chitosan: chemistry, properties and applications, J. Sci. Ind. Res. 63 (2004) 20–31.

[8] S.-K. Kim, N. Rajapakse, Enzymatic production and biological activities of chitosan oligosaccharides (COS): a review, Carbohydr. Polym. 62 (2005) 357–368.

[9] T. Imai, S. Shiraishi, H. Saitô, M. Otagiri, Interaction of indomethacin with low molecular weight chitosan, and improvements of some pharmaceutical properties of indomethacin by low molecular weight chitosans, Int. J. Pharm. 67 (1991) 11–20.

[10] A. Singla, M. Chawla, Chitosan: some pharmaceutical and biological aspects-an update, J. Pharm. Pharmacol. 53 (2001) 1047–1067.

[11] R.R. Mohamed, A. Fekry, Antimicrobial and anticorrosive activity of adsorbents based on chitosan Schiff's base, Int. J. Electrochem. Sci. 6 (2011) 2488–2508.

[12] M.M. Solomon, H. Gerengi, T. Kaya, S.A. Umoren, Enhanced corrosion inhibition effect of chitosan for St37 in 15% H_2SO_4 environment by silver nanoparticles, Int. J. Biol. Macromol. 104 (2017) 638–649.

[13] J. Carneiro, J. Tedim, M. Ferreira, Chitosan as a smart coating for corrosion protection of aluminum alloy 2024: a review, Prog. Org. Coat. 89 (2015) 348–356.

[14] E. Avcu, F.E. Baştan, H.Z. Abdullah, M.A.U. Rehman, Y.Y. Avcu, A.R. Boccaccini, Electrophoretic deposition of chitosan-based composite coatings for biomedical applications: A review, Prog. Mater Sci. 103 (2019) 69–108.

[15] Z. Guo, R. Xing, S. Liu, Z. Zhong, X. Ji, L. Wang, P. Li, Antifungal properties of Schiff bases of chitosan, N-substituted chitosan and quaternized chitosan, Carbohydr. Res. 342 (2007) 1329–1332.

[16] E. Portes, C. Gardrat, A. Castellan, V. Coma, Environmentally friendly films based on chitosan and tetrahydrocurcuminoid derivatives exhibiting antibacterial and antioxidative properties, Carbohydr. Polym. 76 (2009) 578–584.

[17] M.F. Goosen, Applications of Chitan and Chitosan, CRC Press, Boca Raton, Florida, USA, 1996.

[18] S.A. Umoren, M.J. Banera, T. Alonso-Garcia, C.A. Gervasi, M.V. Mirífico, Inhibition of mild steel corrosion in HCl solution using chitosan, Cellulose 20 (2013) 2529–2545.

[19] T. Rabizadeh, S. Khameneh Asl, Chitosan as a green inhibitor for mild steel corrosion: thermodynamic and electrochemical evaluations, Mater. Corros. 70 (2019) 738–748.

[20] N.K. Gupta, P. Joshi, V. Srivastava, M. Quraishi, Chitosan: a macromolecule as green corrosion inhibitor for mild steel in sulfamic acid useful for sugar industry, Int. J. Biol. Macromol. 106 (2018) 704–711.

[21] A. Jmiai, B. El Ibrahimi, A. Tara, R. Oukhrib, S. El Issami, O. Jbara, L. Bazzi, M. Hilali, Chitosan as an eco-friendly inhibitor for copper corrosion in acidic medium: protocol and characterization, Cellulose 24 (2017) 3843–3867.

[22] K. El Mouaden, B. El Ibrahimi, R. Oukhrib, L. Bazzi, B. Hammouti, O. Jbara, A. Tara, D.S. Chauhan, M.A. Quraishi, Chitosan polymer as a green corrosion inhibitor for copper in sulfide-containing synthetic seawater, Int. J. Biol. Macromol. 119 (2018) 1311–1323.

[23] Y. Brou, N. Coulibaly, D. N'GYS, J. Creus, A. Trokourey, Chitosan biopolymer effect on copper corrosion in 3.5 wt.% NaCl solution: electrochemical and quantum chemical studies, Int. J. Corros. Scale Inhib. 9 (2020) 182–200.

[24] H. Harmami, I. Ulfin, A.H. Sakinah, Y.L. Ni'mah, Water-soluble chitosan from shrimp and mussel shells as corrosion inhibitor on tinplate in 2% NaCl, Malaysian Journal of Fundamental and Applied Sciences 15 (2019) 212–217.

[25] Q. Zhao, J. Guo, G. Cui, T. Han, Y. Wu, Chitosan derivatives as green corrosion inhibitors for P110 steel in a carbon dioxide environment, Colloids Surf. B 194 (2020) 111150.

[26] G.A. El-Mahdy, A.M. Atta, H.A. Al-Lohedan, A.O. Ezzat, Influence of green corrosion inhibitor based on chitosan ionic liquid on the steel corrodibility in chloride solution, Int. J. Electrochem. Sci. 10 (2015) 5812–5826.

[27] A.M. Alsabagh, M.Z. Elsabee, Y.M. Moustafa, A. Elfky, R.E. Morsi, Corrosion inhibition efficiency of some hydrophobically modified chitosan surfactants in relation to their surface active properties, Egypt. J. Pet. 23 (2014) 349–359.

[28] S.M. El-Sawy, Y.M. Abu-Ayana, F.A. Abdel-Mohdy, Some chitin/chitosan derivatives for corrosion protection and waste water treatments, Anti-Corros. Method M. 48 (2001) 227–235.

[29] D. Suyanto, H. Darmokoesemo, R. RURIYANTI, L.S. ANGGARA, Application chitosan derivatives as inhibitor corrosion on steel with fluidization method, Journal of Chemical and Pharmaceutical Research 2 (2015) 260–267.

[30] D.S. Chauhan, M.J. Mazumder, M. Quraishi, K. Ansari, Chitosan-cinnamaldehyde Schiff base: a bioinspired macromolecule as corrosion inhibitor for oil and gas industry, Int. J. Biol. Macromol. 158 (2020) 127–138.

[31] H. Li, H. Li, Y. Liu, X. Huang, Synthesis of polyamine grafted chitosan copolymer and evaluation of its corrosion inhibition performance, J. Korean Chem. Soc. 59 (2015) 142–147.

[32] T. Chen, D. Zeng, S. Zhou, Study of polyaspartic acid and chitosan complex corrosion inhibition and mechanisms, Pol. J. Environ. Stud. 27 (2018) 1441–1448.

[33] R.G.M. de Araújo Macedo, N. do Nascimento Marques, J. Tonholo, R. de Carvalho Balaban, Water-soluble carboxymethylchitosan used as corrosion inhibitor for carbon steel in saline medium, Carbohydr. Polym. 205 (2019) 371–376.

[34] J. Haque, V. Srivastava, D.S. Chauhan, H. Lgaz, M.A. Quraishi, Microwave-induced synthesis of chitosan Schiff bases and their application as novel and green corrosion inhibitors: experimental and theoretical approach, ACS Omega 3 (2018) 5654–5668.

[35] M. Quraishi, K. Ansari, D.S. Chauhan, S.A. Umoren, M. Mazumder, Vanillin modified chitosan as a new bioinspired corrosion inhibitor for carbon steel in oil-well acidizing relevant to petroleum industry, Cellulose 27 (2020) 6425–6443.

[36] R. Menaka, S. Subhashini, Chitosan Schiff base as eco-friendly inhibitor for mild steel corrosion in 1 M HCl, J. Adhes. Sci. Technol. 30 (2016) 1622–1640.

[37] M. Li, J. Xu, R. Li, D. Wang, T. Li, M. Yuan, J. Wang, Simple preparation of aminothiourea-modified chitosan as corrosion inhibitor and heavy metal ion adsorbent, J. Colloid Interface Sci. 417 (2014) 131–136.

[38] G.E. Luckachan, V. Mittal, Anti-corrosion behavior of layer by layer coatings of cross-linked chitosan and poly (vinyl butyral) on carbon steel, Cellulose 22 (2015) 3275–3290.

[39] D. Chauhan, V. Srivastava, P. Joshi, M. Quraishi, PEG cross-linked chitosan: a biomacromolecule as corrosion inhibitor for sugar industry, Int. J. Ind. Chem. 9 (2018) 363–377.

[40] K.E. Mouaden, D. Chauhan, M. Quraishi, L. Bazzi, Thiocarbohydrazide-crosslinked chitosan as a bioinspired corrosion inhibitor for protection of stainless steel in 3.5% NaCl, Sustain. Chem. Pharm. 15 (2020) 100213.

[41] V. Srivastava, D.S. Chauhan, P.G. Joshi, V. Maruthapandian, A.A. Sorour, M.A. Quraishi, PEG-functionalized chitosan: a biological macromolecule as a novel corrosion inhibitor, ChemistrySelect 3 (2018) 1990–1998.

[42] P. Kong, H. Feng, N. Chen, Y. Lu, S. Li, P. Wang, Polyaniline/chitosan as a corrosion inhibitor for mild steel in acidic medium, RSC Adv. 9 (2019) 9211–9217.

[43] A. Sanmugam, D. Vikraman, K. Karuppasamy, J.Y. Lee, H.-S. Kim, Evaluation of the corrosion resistance properties of electroplated chitosan-zn1− xcuxo composite thin films, Nanomaterials 7 (2017) 432.

[44] S. John, A. Joseph, A.J. Jose, B. Narayana, Enhancement of corrosion protection of mild steel by chitosan/ZnO nanoparticle composite membranes, Prog. Org. Coat. 84 (2015) 28–34.

[45] P.A. Rasheed, K.A. Jabbar, K. Rasool, R.P. Pandey, M.H. Sliem, M. Helal, A. Samara, A.M. Abdullah, K.A. Mahmoud, Controlling the biocorrosion of sulfate-reducing bacteria (SRB) on carbon steel using ZnO/chitosan nanocomposite as an eco-friendly biocide, Corros. Sci. 148 (2019) 397–406.

[46] Y. Liu, C. Zou, X. Yan, R. Xiao, T. Wang, M. Li, β-Cyclodextrin modified natural chitosan as a green inhibitor for carbon steel in acid solutions, Ind. Eng. Chem. Res. 54 (2015) 5664–5672.

[47] U. Eduok, E. Ohaeri, J. Szpunar, Conversion of Imidazole to N-(3-Aminopropyl) imidazole toward Enhanced Corrosion Protection of Steel in Combination with Carboxymethyl Chitosan Grafted Poly (2-methyl-1-vinylimidazole), Ind. Eng. Chem. Res. 58 (2019) 7179–7192.

[48] U. Eduok, J. Szpunar, Biocorrosion reduction of pipeline steel in desulfovibrio ferrophilus culture in the presence of carboxymethyl chitosan grafted poly (2-methyl-1-vinylimidazole)/cerium molybdate nanocomposite, ECS Meeting Abstracts, IOP Publishing, 2019, pp. 853.

[49] M. Srivastava, S. Srivastava, G. Ji, R. Prakash, Chitosan based new nanocomposites for corrosion protection of mild steel in aggressive chloride media, Int. J. Biol. Macromol. 140 (2019) 177–187.

[50] H. Ashassi-Sorkhabi, R. Bagheri, B. Rezaei-moghadam, Sonoelectrochemical synthesis of ppy-MWCNTs-chitosan nanocomposite coatings: Characterization and corrosion behavior, J. Mater. Eng. Perform. 24 (2015) 385–392.

[51] M. Barman, S. Mahmood, R. Augustine, A. Hasan, S. Thomas, K. Ghosal, Natural halloysite nanotubes/chitosan based bio-nanocomposite for delivering norfloxacin, an anti-microbial agent in sustained release manner, Int. J. Biol. Macromol. 162 (2020) 1849–1861.

[52] A. Farhadian, M.A. Varfolomeev, A. Shaabani, S. Nasiri, I. Vakhitov, Y.F. Zaripova, V.V. Yarkovoi, A.V. Sukhov, Sulfonated chitosan as green and high cloud point kinetic methane hydrate and corrosion inhibitor: experimental and theoretical studies, Carbohydr. Polym. 236 (2020) 116035.

[53] R. Farahati, A. Ghaffarinejad, H.J. Rezania, S.M. Mousavi-Khoshdel, H. Behzadi, Sulfonated aromatic polyamide as water-soluble polymeric corrosion inhibitor of copper in HCl, Colloids Surf. A 578 (2019) 123626.

[54] E. Tabesh, H. Salimijazi, M. Kharaziha, M. Mahmoudi, M. Hejazi, Development of an in-situ chitosan-copper nanoparticle coating by electrophoretic deposition, Surf. Coat. Technol. 364 (2019) 239–247.

[55] S. Karimi, E. Salahinejad, E. Sharifi, A. Nourian, L. Tayebi, Bioperformance of chitosan/fluoride-doped diopside nanocomposite coatings deposited on medical stainless steel, Carbohydr. Polym. 202 (2018) 600–610.

[56] X. Zhai, Y. Ren, N. Wang, F. Guan, M. Agievich, J. Duan, B. Hou, Microbial corrosion resistance and antibacterial property of electrodeposited Zn–Ni–chitosan coatings, Molecules 24 (2019) 1974.

[57] P. Sambyal, G. Ruhi, S. Dhawan, B. Bisht, S. Gairola, Enhanced anticorrosive properties of tailored poly (aniline-anisidine)/chitosan/SiO2 composite for protection of mild steel in aggressive marine conditions, Prog. Org. Coat. 119 (2018) 203–213.

[58] D. Junying, L. Weihua, W. Maotao, Z. Xia, H. Baorong, Synthesis and inhibition studies of carboxymethl chitosan doped polianiline for mild steel in hydrochloric acid, Acta Polym. Sin. 10 (2010) 588–593.

[59] S. John, A. Salam, A.M. Baby, A. Joseph, Corrosion inhibition of mild steel using chitosan/TiO2 nanocomposite coatings, Prog. Org. Coat. 129 (2019) 254–259.

[60] L. Cordero-Arias, S. Cabanas-Polo, H. Gao, J. Gilabert, E. Sanchez, J. Roether, D. Schubert, S. Virtanen, A.R. Boccaccini, Electrophoretic deposition of nanostructured-TiO 2/chitosan composite coatings on stainless steel, RSC Adv. 3 (2013) 11247–11254.

[61] P. Ledwig, M. Kot, T. Moskalewicz, B. Dubiel, Electrophoretic deposition of nc-TiO2/chitosan composite coatings on X2CrNiMo17-12-2 stainless steel, Arch. Metall. Mater. 62 (2017) 405–410.

[62] G. Jena, B. Anandkumar, S. Vanithakumari, R. George, J. Philip, G. Amarendra, Graphene oxide-chitosan-silver composite coating on Cu-Ni alloy with enhanced anticorrosive and antibacterial properties suitable for marine applications, Prog. Org. Coat. 139 (2020) 105444.

[63] M.M. Solomon, H. Gerengi, T. Kaya, S.A. Umoren, Performance evaluation of a chitosan/silver nanoparticles composite on St37 steel corrosion in a 15% HCl solution, ACS Sustain. Chem. Eng. 5 (2017) 809–820.

[64] S.P. Damari, L. Cullari, D. Laredo, R. Nadiv, E. Ruse, R. Sripada, O. Regev, Graphene and boron nitride nanoplatelets for improving vapor barrier properties in epoxy nanocomposites, Prog. Org. Coat. 136 (2019) 105207.

[65] B.-D. Hahn, D.-S. Park, J.-J. Choi, J. Ryu, W.-H. Yoon, J.-H. Choi, H.-E. Kim, S.-G. Kim, Aerosol deposition of hydroxyapatite–chitosan composite coatings on biodegradable magnesium alloy, Surf. Coat. Technol. 205 (2011) 3112–3118.

[66] S. Sutha, K. Kavitha, G. Karunakaran, V. Rajendran, In-vitro bioactivity, biocorrosion and antibacterial activity of silicon integrated hydroxyapatite/chitosan composite coating on 316 L stainless steel implants, Mater. Sci. Eng. C 33 (2013) 4046–4054.

[67] A. Molaei, M. Yari, M.R. Afshar, Investigation of halloysite nanotube content on electrophoretic deposition (EPD) of chitosan-bioglass-hydroxyapatite-halloysite nanotube nanocomposites films in surface engineering, Appl. Clay Sci. 135 (2017) 75–81.

[68] L.L.d. Sousa, V.P. Ricci, D.G. Prado, R.C. Apolinario, L.C.d.O. Vercik, E.C.d.S. Rigo, M.C.d.S. Fernandes, N.A. Mariano, Titanium coating with hydroxyapatite and chitosan doped with silver nitrate, Mater. Res. 20 (2017) 863–868.

[69] M. Stevanović, M. Đošić, A. Janković, V. Kojic, M. Vukasinovic-Sekulic, J. Stojanović, J. Odović, M. Crevar Sakač, K.Y. Rhee, V. Mišković-Stanković, Gentamicin-loaded bioactive hydroxyapatite/chitosan composite coating electrodeposited on titanium, ACS Biomater. Sci. Eng. 4 (2018) 3994–4007.

[70] F. Ordikhani, E. Tamjid, A. Simchi, Characterization and antibacterial performance of electrodeposited chitosan–vancomycin composite coatings for prevention of implant-associated infections, Mater. Sci. Eng. C 41 (2014) 240–248.

[71] G. Ruhi, O. Modi, S. Dhawan, Chitosan-polypyrrole-SiO2 composite coatings with advanced anticorrosive properties, Synth. Met. 200 (2015) 24–39.

[72] S. Kalia, B. Kaith, I. Kaur, Cellulose fibers: bio-and nano-polymer composites: green chemistry and technology, Springer Science & Business Media, Springer-Verlag, Berlin, Heidelberg, 2011.

[73] S. Gawande, Cellulose: A Natural Polymer on the Earth, Int. J. Polym. Sci. Eng. 3 (2017) 32–37.

[74] L. Walker, D. Wilson, Enzymatic hydrolysis of cellulose: an overview, Bioresour. Technol. 36 (1991) 3–14.

[75] A.C. O'sullivan, Cellulose: the structure slowly unravels, Cellulose 4 (1997) 173–207.

[76] M.W. Bauer, L.E. Driskill, W. Callen, M.A. Snead, E.J. Mathur, R.M. Kelly, An endoglucanase, EglA, from the hyperthermophilic ArchaeonPyrococcus furiosus hydrolyzes β-1, 4 bonds in mixed-linkage (1→ 3),(1→ 4)-β-D-glucans and cellulose, J. Bacteriol. 181 (1999) 284–290.

[77] A. Akaracharanya, T. Taprig, J. Sitdhipol, S. Tanasupawat, Characterization of cellulase producing Bacillus and Paenibacillus strains from Thai soils, J. App. Pharm. Sci. 4 (2014) 6.

[78] S.A. Umoren, U.M. Eduok, Application of carbohydrate polymers as corrosion inhibitors for metal substrates in different media: a review, Carbohydr. Polym. 140 (2016) 314–341.

[79] S. Umoren, M. Solomon, I. Udosoro, A. Udoh, Synergistic and antagonistic effects between halide ions and carboxymethyl cellulose for the corrosion inhibition of mild steel in sulphuric acid solution, Cellulose 17 (2010) 635–648.

[80] M.M. Solomon, H. Gerengi, S.A. Umoren, Carboxymethyl cellulose/silver nanoparticles composite: synthesis, characterization and application as a benign corrosion inhibitor for St37 steel in 15% H2SO4 medium, ACS Appl. Mater. Interfaces 9 (2017) 6376–6389.

[81] M.N. El-Haddad, Hydroxyethylcellulose used as an eco-friendly inhibitor for 1018 c-steel corrosion in 3.5% NaCl solution, Carbohydr. Polym. 112 (2014) 595–602.

[82] I. Arukalam, I. Madufor, O. Ogbobe, E. Oguzie, Inhibition of mild steel corrosion in sulfuric acid medium by hydroxyethyl cellulose, Chem. Eng. Commun. 202 (2015) 112–122.

[83] M. Mobin, M. Rizvi, Adsorption and corrosion inhibition behavior of hydroxyethyl cellulose and synergistic surfactants additives for carbon steel in 1 M HCl, Carbohydr. Polym. 156 (2017) 202–214.

[84] I. Arukalam, I. Madu, N. Ijomah, C. Ewulonu, G. Onyeagoro, Acid corrosion inhibition and adsorption behaviour of ethyl hydroxyethyl cellulose on mild steel corrosion, J. Mater. 2014 (2014) 1–11.

[85] M. Deyab, Hydroxyethyl cellulose as efficient organic inhibitor of zinc–carbon battery corrosion in ammonium chloride solution: Electrochemical and surface morphology studies, J. Power Sources 280 (2015) 190–194.

[86] S. Nwanonenyi, H. Obasi, I. Eze, Hydroxypropyl Cellulose as an Efficient Corrosion Inhibitor for Aluminium in Acidic Environments: Experimental and Theoretical Approach, Chemistry Africa 2 (2019) 471–482.

[87] I. Arukalam, I. Madufor, O. Ogbobe, E. Oguzie, Cellulosic polymers for corrosion protection of aluminium, Int. J. Eng. Tech. Res. 3 (2015) 2321 0869.

[88] I. Arukalam, I. Madufor, O. Ogbobe, E. Oguzie, Experimental and theoretical studies of hydroxyethyl cellulose as inhibitor for acid corrosion inhibition of mild steel and aluminium, The open corrosion Journal 6 (2014).

[89] K. Andarany, A. Sagir, A. Ahmad, S. Deni, W. Gunawan, Cellulose acetate layer effect toward aluminium corrosion rate in hydrochloric acid media, IOP Conference Series: Materials Science and Engineering, 237, Institute of Physics Publishing, 2017, p. 012042.

[90] O. Egbuhuzor, I. Madufor, S. Nwanonenyi, J. Bokolo, Adsorption behavior and corrosion rate model of sodium carboxymethyl cellulose (NA-CMC) polymer on aluminium in hcl solutioN, Niger. J. Technol. 39 (2020) 369–378.

[91] I.O. Arukalam, I.K. Nleme, A.E. Anyanwu, Comparative inhibitive effect of hydroxyethylcellulose on mild steel and aluminium corrosion in 0.5 M HCl solution, Acad. Res. Int. 1 (2011) 492.

[92] S.A. Umoren, A.A. AlAhmary, Z.M. Gasem, M.M. Solomon, Evaluation of chitosan and carboxymethyl cellulose as ecofriendly corrosion inhibitors for steel, Int. J. Biol. Macromol. 117 (2018) 1017–1028.

[93] I. Arukalam, I. Madufor, O. Ogbobe, E. Oguzie, Acidic corrosion inhibition of copper by hydroxyethyl cellulose, Curr. J. Appl. Sci. Technol. 4 (2014) 1445–1460.
[94] M.-M. Li, Q.-J. Xu, J. Han, H. Yun, Y. Min, Inhibition action and adsorption behavior of green inhibitor sodium carboxymethyl cellulose on copper, Int. J. Electrochem. Sci. 10 (2015) 9028–9041.
[95] M.S. Hasanin, S.A. Al Kiey, Environmentally benign corrosion inhibitors based on cellulose niacin nanocomposite for corrosion of copper in sodium chloride solutions, Int. J. Biol. Macromol. 161 (2020) 345–354.
[96] T. Brinda, J. Mallika, V. Sathyanarayana Moorthy, Synergistic effect between starch and substituted piperidin-4-one on the corrosion inhibition of mild steel in acidic medium, J. Mater. Environ. Sci 6 (2015) 120–191.
[97] M. Mobin, M. Khan, M. Parveen, Inhibition of mild steel corrosion in acidic medium using starch and surfactants additives, J. Appl. Polym. Sci. 121 (2011) 1558–1565.
[98] M. Bello, N. Ochoa, V. Balsamo, F. López-Carrasquero, S. Coll, A. Monsalve, G. González, Modified cassava starches as corrosion inhibitors of carbon steel: An electrochemical and morphological approach, Carbohydr. Polym. 82 (2010) 561–568.
[99] R. Rosliza, W.W. Nik, Improvement of corrosion resistance of AA6061 alloy by tapioca starch in seawater, Curr. Appl Phys. 10 (2010) 221–229.
[100] X. Li, S. Deng, Cassava starch graft copolymer as an eco-friendly corrosion inhibitor for steel in H_2SO_4 solution, Korean J. Chem. Eng. 32 (2015) 2347–2354.
[101] S.M. Shaban, I. Aiad, A.H. Moustafa, O.H. Aljoboury, Some alginates polymeric cationic surfactants; surface study and their evaluation as biocide and corrosion inhibitors, J. Mol. Liq. 273 (2019) 164–176.
[102] R. Hassan, I. Zaafarany, A. Gobouri, H. Takagi, A revisit to the corrosion inhibition of aluminum in aqueous alkaline solutions by water-soluble alginates and pectates as anionic polyelectrolyte inhibitors, Int. J. Corros. 2013 (2013) 1–8.
[103] A. Biswas, D. Das, H. Lgaz, S. Pal, U.G. Nair, Biopolymer dextrin and poly (vinyl acetate) based graft copolymer as an efficient corrosion inhibitor for mild steel in hydrochloric acid: electrochemical, surface morphological and theoretical studies, J. Mol. Liq. 275 (2019) 867–878.
[104] G.N. Devi, C.B.N. Unnisa, S.M. Roopan, V. Hemapriya, S. Chitra, I.-M. Chung, S.-H. Kim, M. Prabakaran, Floxacins: as mediators in enhancing the corrosion inhibition efficiency of natural polymer dextrin, Macromol. Res. 28 (2020) 1–9.
[105] A. Altin, M. Rohwerder, A. Erbe, Cyclodextrins as carriers for organic corrosion inhibitors in organic coatings, J. Electrochem. Soc. 164 (2017) C128.
[106] S. Amiri, A. Rahimi, Anticorrosion behavior of cyclodextrins/inhibitor nanocapsule-based self-healing coatings, J. Coat. Technol. Res. 13 (2016) 1095–1102.
[107] A. Khramov, N. Voevodin, V. Balbyshev, R. Mantz, Sol–gel-derived corrosion-protective coatings with controllable release of incorporated organic corrosion inhibitors, Thin Solid Films 483 (2005) 191–196.
[108] A. Khramov, N. Voevodin, V. Balbyshev, M. Donley, Hybrid organo-ceramic corrosion protection coatings with encapsulated organic corrosion inhibitors, Thin Solid Films 447 (2004) 549–557.
[109] M.V. Fiori-Bimbi, P.E. Alvarez, H. Vaca, C.A. Gervasi, Corrosion inhibition of mild steel in HCL solution by pectin, Corros. Sci. 92 (2015) 192–199.
[110] S.A. Umoren, I.B. Obot, A. Madhankumar, Z.M. Gasem, Performance evaluation of pectin as ecofriendly corrosion inhibitor for X60 pipeline steel in acid medium: Experimental and theoretical approaches, Carbohydr. Polym. 124 (2015) 280–291.
[111] A.N. Grassino, J. Halambek, S. Djaković, S.R. Brnčić, M. Dent, Z. Grabarić, Utilization of tomato peel waste from canning factory as a potential source for pectin production and application as tin corrosion inhibitor, Food Hydrocolloids. 52 (2016) 265–274.
[112] M.M. Fares, A. Maayta, M.M. Al-Qudah, Pectin as promising green corrosion inhibitor of aluminum in hydrochloric acid solution, Corros. Sci. 60 (2012) 112–117.
[113] R. Geethanjali, A. Sabirneeza, S. Subhashini, Water-soluble and biodegradable pectin-grafted polyacrylamide and pectin-grafted polyacrylic acid: electrochemical investigation of corrosion-inhibition behaviour on mild steel in 3.5% NaCl media, Indian J. Mater. Sci. 2014 (2014) 1–9.
[114] B. Charitha, P. Rao, Pectin as a potential green inhibitor for corrosion control of 6061Al–15%(V) SiC (P) composite in acid medium: electrochemical and surface studies, J. Fail. Anal. Prev. 20 (2020) 1–13.
[115] R.M. Hassan, I.A. Zaafarany, Kinetics of corrosion inhibition of aluminum in acidic media by water-soluble natural polymeric pectates as anionic polyelectrolyte inhibitors, Materials 6 (2013) 2436–2451.

[116] I. Zaafarany, Corrosion inhibition of aluminum in aqueous alkaline solutions by alginate and pectate water-soluble natural polymer anionic polyelectrolytes, Portugaliae Electrochimica Acta 30 (2012) 419–426.

[117] S. Umoren, E. Ebenso, Studies of the anti-corrosive effect of Raphia hookeri exudate gum-halide mixtures for aluminium corrosion in acidic medium, Pigm. Resin Technol. 37 (2008) 173–182.

[118] S. Umoren, I. Obot, E. Ebenso, P. Okafor, Eco-friendly inhibitors from naturally occurring exudate gums for aluminium corrosion inhibition in acidic medium, Portugaliae Electrochimica Acta 26 (2008) 267–282.

[119] J. Buchweishaija, G. Mhinzi, Natural products as a source of environmentally friendly corrosion inhibitors: the case of gum exudate from Acacia seyal var. seyal, Portugaliae Electrochimica Acta 26 (2008) 257–265.

[120] S. Umoren, I. Obot, E. Ebenso, N. Obi-Egbedi, The Inhibition of aluminium corrosion in hydrochloric acid solution by exudate gum from Raphia hookeri, Desalination 247 (2009) 561–572.

[121] S. Umoren, I. Obot, E. Ebenso, Corrosion inhibition of aluminium using exudate gum from Pachylobus edulis in the presence of halide ions in HCl, E-J. Chem. 5 (2008) 1–10.

[122] J. Buchweishaija, Plants as a source of green corrosion inhibitors: the case of gum exudates from Acacia species (A. drepanolobium and A. senegal), Tanz. J. Sci. 35 (2009) 93–106.

[123] F. Waanders, S. Vorster, A. Geldenhuys, Biopolymer corrosion inhibition of mild steel: electrochemical/mössbauer results, Hyperfine Interact. 139 (2002) 133–139.

[124] S.A. Umoren, M.M. Solomon, Synergistic corrosion inhibition effect of metal cations and mixtures of organic compounds: a review, J. Environ. Chem. Eng. 5 (2017) 246–273.

[125] D.E. Arthur, A. Jonathan, P.O. Ameh, C. Anya, A review on the assessment of polymeric materials used as corrosion inhibitor of metals and alloys, Int. J. Ind. Chem. 4 (2013) 2.

[126] A. Peter, I. Obot, S.K. Sharma, Use of natural gums as green corrosion inhibitors: an overview, Int. J. Ind. Chem. 6 (2015) 153–164.

Carbohydrates as green corrosion inhibitors

The previous chapter describes the corrosion inhibition effect of natural polymers including carbohydrates polymers and their derivatives. Carbohydrates are compounds that are mainly composed of carbon hydrogen and oxygen. They are regarded are polyhydroxy aldehydes or ketones. The present chapter aims to describe the corrosion inhibition effect of non-polymeric carbohydrates and their derivatives. Literature investigation shows that recently the use of non-polymeric carbohydrates including glucose and glucosamine and their derivatives are extensively used as green corrosion inhibitors for different metals and alloys in various electrolytes are gaining particular attention. Application of these compounds as corrosion inhibitors is associated with the following benefits:

1. Eco-friendly and sustainable alternatives (biodegradable, nonbioaccumulative, and biotolerable) for traditional toxic corrosion inhibitors.
2. High corrosion inhibition potential because of the presence of a large number of polar functional groups (adsorption centers).
3. High solubility in polar electrolytes.
4. Huge, economic and commercial availability.
5. Cost-effective and ease of purification and isolation.
6. Effective for different metal/electrolyte systems.
7. Low-toxicity and minimized negative impact for living organisms and the environment.
8. Their corrosion inhibition potential and solubility can be suitably modified using organic (heterocyclic) compounds.

Although, nonpolymeric carbohydrates and their derivatives are widely used as corrosion however, their use is associated with various shortcomings.

1. Some of the carbohydrates and their derivatives show limited solubility, especially in polar electrolytes.
2. Purification, processing and isolation of these compounds can't treat as cost-effective protocol.
3. Chemical modification requires the use of toxic and expensive chemicals, solvents and catalysts.
4. These compounds show reasonably low protection efficiency.

5. They undergo degradation at high temperatures.
6. They undergo hydrolysis, fragmentation or rearrangement, especially at high temperature.

20.1 Carbohydrates as green corrosion inhibitors: literature survey

Although, literatures on pure non-polymeric carbohydrates are relatively lesser however their derivatives are widely used as corrosion inhibitors. Verma et al. [1] developed three glucose derivatives using barbituric acid and aniline derivatives (Scheme 20.1), designated as GPHs, and used as corrosion inhibitors for mild steel/1M HCl system using computational and experimental techniques. In this synthesis, ethanol and PTSA (p-Toluenesulfonic acid) were used as solvent and catalyst, respectively. Experimental and computational investigations show that compounds Corrosion inhibition studies suggest that compounds containing electron releasing hydroxyl (–OH; GPH-2) and methoxy (–OMe; GPH-3) substituents show relatively high inhibition potential that of non-substituted compound (GPH-1; –H). Corrosion inhibition potential of the studied compounds followed the sequence: GPH-1 < GPH-2 < GPH-3.

Weight loss study showed that the protection power of GPHs increases with their concentration and GPH-3, GPH-2 and GPH-1 exhibited the highest protection potential of 97.82%, 95.21%, and 93.91%, respectively at the optimum concentration of 10.15×10^{-5} mol L^{-1}. Because of their electron releasing nature, –OH and –OMe substituents increase electron density at the center(s) responsible for interaction with the metallic surface (adsorption centers). The higher corrosion protection potential of GPH-3 as compared to GPH-2 is resulted due to higher electron donating ability of the –OMe as compared to the –OH. Weight loss studied carried out at different temperature suggest that increase in temperature causes a subsequent decrease in the corrosion potential which can be resulted due to increased kinetic energy, fragmentation, rearrangement and/or degradation of inhibitor molecules at high temperatures. Presence of inhibitor molecules increase the value of activation energy (E_a) for the corrosion process and this increase in the values of E_a was consistent with the order of their protection potential. It was also observed that GPHs become effective through adsorbing on the metallic surface. Adsorption of GPHs on the metallic surface occurs through physiochemisorption mechanism following through Langmuir adsorption isotherm.

Potentiodynamic polarization (PDP) study suggests GPHs inhibit both anodic oxidation and cathodic hydrogen evolution reactions. It was observed that an increase in GPHs

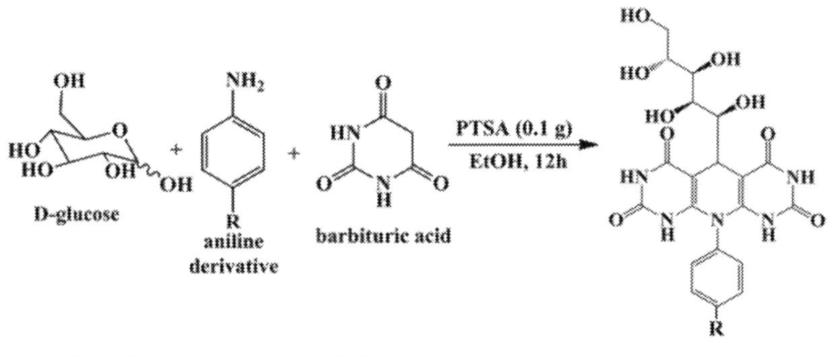

SCHEME 20.1 Scheme for the synthesis of GPHs.

concentrations causes a corresponding decrease in the corrosion current density (icorr). This observation revealed that GPHs become effective by blocking the active sites responsible for corrosion through their adsorption. PDP study also suggests that investigated GPHs act as mixed-type corrosion inhibitors as displacement in the values of corrosion potential (E_{corr}) between inhibited and non-inhibited Tafel curves were less than −85 mV. However, displacements in the values of cathodic Tafel constant (β_c) were greater than that of the displacements in the values of anodic Tafel constant (β_a). On this basis, it was concluded that GPHs mainly act as cathodic-type corrosion inhibitors. Electrochemical impedance spectroscope (EIS) study shows that the presence of GPHs in the corrosive electrolyte causes increase in the values of charge transfer resistance (R_{ct}) for corrosion process. This observation suggests that GPHs become effective by adsorbing at the interface of metal and electrolyte (1M HCl). Therefore, GPHs categorized as interface-type corrosion inhibitors. Increase in the concentration of GPHs causes corresponding increase in the diameter of Nyquist curves. Bode phase angle plots analysis suggests that the morphology of protected metallic surfaces was much smoother as compared to the surface morphology of non-protected metallic specimen. This observation further suggests that GPHs become effective by adsorbing on the metallic surface.

Adsorption mechanism of corrosion inhibition using GPHs was also supported by SEM and EDX analyses. SEM study shows that the morphology of the nonprotected metallic surface was much corroded and damaged as compared to the morphology of protected metallic species. This outcome suggests that GPHs inhibit corrosion by adsorbing on the metallic surface. Change in the composition of elements present on the metallic surface using EDX analyses also confirms the same conclusion. Results of SEM-EDX analyses and adsorption of GPHs on the metallic surface was further supported by AFM study where a significant improvement in the surface morphology of protected metallic specimens was observed. Average surface roughnesses (nm) of protected metallic surfaces were much lower as compared to the surface roughness of non-protected metallic surface. Lastly, corrosion inhibition potential of GPHs was measured using computational simulations. Density functional theory (DFT) based quantum chemical calculations (QCCs) show that GPHs interact with metallic surface using donor-acceptor mechanism. The presence of electron donating –OH and –OMe substituents enhances the effectiveness of metal-inhibitor interactions. Various DFT based parameters were consistent with the order of inhibition potential derived experimentally. Monte Carlo (MC) study shows that magnitudes of adsorption energy (E_{ads}) enhance in the presence of –OH and –OMe substituents.

In another study [2], Haque and coworkers developed ethylenediamine, tetramethylenediamine and hexamethylenediamine modified glucose, designated respectively as EMG, TMG and HMG, and used as corrosion inhibitors for mild steel/1M HCl system. In this synthesis, two units of glucose were joined together using diamine spacers. Methanol was used as a solvent. Synthesis of EMG, TMG and HMG is presented in Scheme 20.2.

Electrochemical studies showed that inhibition efficiencies of the investigated compounds followed the order: HMG (95.42%) > TMG (92.81%) > EMG (74.66%) that is, increase in the length of spacer results in a subsequent increase in their corrosion inhibition potential. Through these analyses, it was also observed that all chemically modified glucose derivatives act as mixed- and interface-type corrosion inhibitors. EMG, TMG and HMG become effective by adsorbing on the mild steel surface following through Langmuir adsorption isotherm. Electrochemical results and order of protection efficiency of studied compounds was also

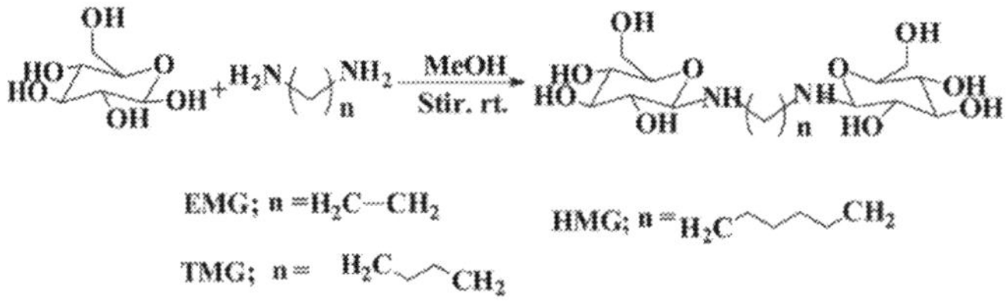

SCHEME 20.2 Scheme for the synthesis of EMG, TMG and HMG.

supported using surface investigation carried out through AFM, XPS and contact angle (CA) measurements. DFT based computational investigations were carried out for neutral as well as the protonated form of EMG, TMG and HMG to support the experimental results and to describe their interactions with the metallic surface. It was observed that all studied compounds acted as efficient corrosion inhibitors. They interact through the donor-acceptor mechanism. It was also reported that protonated form of EMG, TMG and HMG interacted with metallic surface more strongly as compared to their neutral form. Experimental findings were well supported by DFT based computational studies.

In the continuation of similar reports, Chauhan et al. [3] developed another glucose derivative (DHA) by its reaction with 1, 6-hexamethylene diamine and tested it as a corrosion inhibitor for copper in 3.5% NaCl solution. DHA exhibits the highest inhibition potential of 95.2% at 0.27 mmol L^{-1} concentration. A potentiodynamic polarization study showed that DHA inhibits corrosion by blocking the active sites present over the metallic surface through their adsorption. Both anodic and cathodic Tafel reaction rates were adversely affected in the presence of DHA without any significant change in values of corrosion potential (E_{corr}). DHA acts as a mixed-type but predominantly cathodic-type corrosion inhibitor. The adsorption of DHA on metallic surface obeyed Langmuir adsorption isotherm. Adsorption mechanism of corrosion inhibition was supported by SECM and FTIR-ATR techniques. DFT study provides good support to the experimental studies. Literature survey reveals that glucose derivatives modified with epoxy resins are widely used as corrosion inhibitors [4–7]. Most of such derivatives behave as mixed-type corrosion inhibitors and their adsorption mostly follows the Langmuir adsorption isotherm.

Recently, Zhang et al. [8] studied the inhibition potential of two glucosamine-based derivatives (GAs) derived by the reaction of glucosamine and phenyl isothiocyanate using a mixture of water and ethanol as solvent. Both investigated compounds were synthesized using an eco-friendly approach. Among the tested compounds, GA-2 showed acts as better corrosion inhibitor (97.7% at 0.64 mM) as compared to the GA-1. Both studied GAs become effective by blocking the active sites present over the metallic surface through their adsorption. They inversely affect the rate of anodic as well as cathodic Tafel reactions without causing any appreciable change in the values of corrosion potential (E_{corr}). DFT and MD simulations based computational analyses were carried out to support the experimental findings and elucidate the adsorption/interaction behavior of GAs on/with the metallic surface.

Verma et al. [9] reported the synthesis of four Glucosamine-Based, Pyrimidine-Fused Heterocycles (CARBs) by the reaction of glucosamine, barbituric acid and substituted aromatic aldehydes using ethanol as a solvent and PTSA as a catalyst. Synthesis of CARBs is presented in Scheme 20.3. Synthesized CARBs were characterized using spectral techniques and evaluated as potential inhibitors for mild steel/1M HCl system. Experimental and computational analyses suggest that inhibition efficiencies of CARBs greatly depend upon the

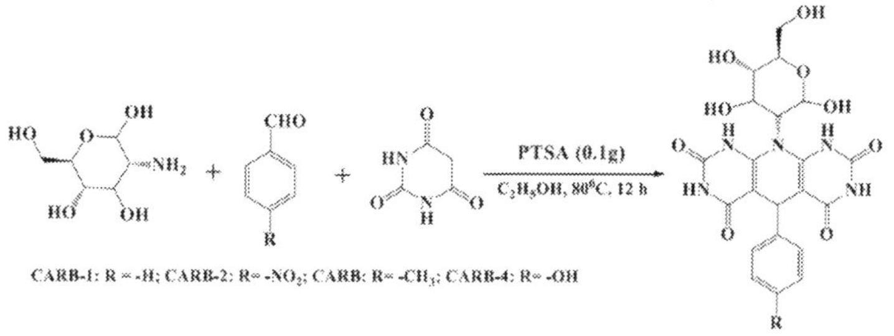

SCHEME 20.3 Scheme for the synthesis of CARBs.

nature of substituents present in their molecule structures. In summary, the presence of electron withdrawing nitro (–NO$_2$) substituent decreases the inhibition performance whereas electron donating methyl (–Me) and hydroxyl (–OH) substituents increase inhibition performance. The inhibition efficiencies of tested compounds followed the sequence: CARB-1 (–H) < CARB-2 (–NO$_2$) CARB-3 (–Me) < CARB-4 (–OH). Experimental studies showed that studied CARBs behave as mixed but predominantly cathodic-type corrosion inhibitors and they become effective by suppressing the rate of anodic and cathodic Tafel reactions. Adsorption of CARBs on metallic surface obeyed the Langmuir adsorption isotherm. Surface analyses confirmed that CARBs become effective by adsorbing on the metallic surface. This type of adsorption resulted in the formation of corrosion inhibitive film.

DFT based computational studies were conducted to support eh experimental observations and to demonstrate the effect of substituents on the corrosion inhibition potential of CARBs. Optimized, HOMO and LUMO (collectively called as FMOs) pictures of investigated CARBs are shown in Fig. 20.1. It can be clearly seen that the distribution of HOMO and

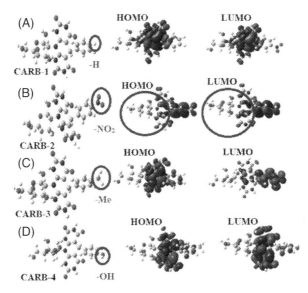

FIG. 20.1 FMOs of (A) CARB-1-H, (B) CARB-2-NO$_2$, (C) CARB-3-Me, and (D) CARB-4-OH [9]. (Reproduced with Permission @ Copyright the American Chemical Society).

LUMO varies depending upon the nature of substituents. As compare to non-substituted compound (CARB-1), in nitro (–NO$_2$) substituted CARB-2 contribution of HOMO and LUMO is greatly reduced. This observation suggests that because of its electron withdrawing nature, –NO$_2$ substituent decreases the electron density at the donor sites and therefore decreases the corrosion inhibition potential of CARB-2. Reduction in the contribution of HOMO and LUMO in the presence of –NO$_2$ substituent is presented by a red circle. Unlike to this, –CH$_3$ and –OH substituted compounds (CARB-3 & CARB-4, respectively) contribution of HOMO and/or LUMO increased as compared to the non-substituted, CARB-1.

Most stable configurations or orientations of investigated CARBs on Fe (110) surface derived through MC simulations are presented in Fig. 20.2. It can be seen that in CARB-1, CARB-3 and CARB-4 containing –H, –Me and –OH substituent, respectively, acquire the planar or horizontal (with the metal surface) orientations. Whereas, CARB-2, having –NO$_2$ substituent acquires the vertical orientation. This observation suggests that because of their horizontal orientations CARB-1, CARB-3 and CARB-4 cover larger metallic surface and behave as superior corrosion inhibitors as compared to the CARB-2 that acquire the vertical orientation. It can be clearly seen that nitrophenyl moiety is perpendicular with the plane of metal surface Negative values of adsorption energy (E_{ads}) derived for the adsorption of investigated CARBs suggests that they interact spontaneously with the metallic surface using their electron rich adsorption centers.

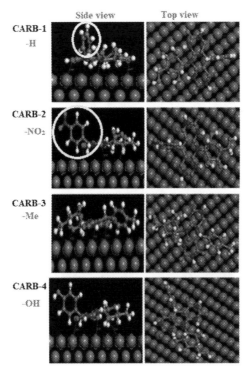

FIG. 20.2 Side (left-hand side) and top (right-hand side) views of (A) CARB-1-H, (B) CARB-2-NO$_2$, (C) CARB-3-Me, and (D) CARB-4-OH on Fe (110) surface [9]. (Reproduced with Permission @ Copyright the American Chemical Society).

20.2 Summary

Non-polymeric carbohydrates such as glucose and glucosamine and their derivatives are evaluated as corrosion inhibitors for various metal/electrolyte systems. Their corrosion inhibition performances are tested using various experimental and computational techniques. They serve as effective and environmental friendly alternatives for traditional toxic corrosion inhibitors. Potentiodynamic polarization measurement reveals that most of such compounds become effective by suppressing both anodic and cathodic Tafel reactions without causing any significant variation in the value of corrosion potential, E_{corr}. They act as mostly mixed-type corrosion inhibitors however cathodic predominance has also been reported in few reports. They behave as interface-type corrosion inhibitors as they increase the value of charge transfer resistance in their presence. Adsorption of these compounds on the metallic surface mostly follows the Langmuir adsorption isotherm. Adsorption mechanism of corrosion inhibition using non-polymeric carbohydrates has also been supported by surface investigations, especially through SEM, AFM, XPS, FT-IR, contact angle and EDX analyses. DFT and MD (or MC) simulation studies are widely used to demonstrate the nature and effectiveness of interactions between non-polymeric carbohydrates and metallic surface. These analyses show that most of the non-polymeric carbohydrates and their derivatives spontaneously interacted through donor-acceptor interactions. Substituents greatly affect the orientation and effectiveness of these molecules adsorption on the metallic surface.

References

[1] C. Verma, M.A. Quraishi, K. Kluza, M. Makowska-Janusik, L.O. Olasunkanmi, E.E. Ebenso, Corrosion inhibition of mild steel in 1M HCl by D-glucose derivatives of dihydropyrido [2, 3-d: 6, 5-d'] dipyrimidine-2, 4, 6, 8 (1H, 3H, 5H, 7H)-tetraone, Sci. Rep. 7 (2017) 1–17.

[2] J. Haque, V. Srivastava, D.S. Chauhan, M. Quraishi, A.M. Kumar, H. Lgaz, Electrochemical and surface studies on chemically modified glucose derivatives as environmentally benign corrosion inhibitors, Sustain. Chem. Pharm. 16 (2020) 100260.

[3] D.S. Chauhan, A.M. Kumar, M. Quraishi, Hexamethylenediamine functionalized glucose as a new and environmentally benign corrosion inhibitor for copper, Chem. Eng. Res. Des. 150 (2019) 99–115.

[4] M. Rbaa, P. Dohare, A. Berisha, O. Dagdag, L. Lakhrissi, M. Galai, B. Lakhrissi, M.E. Touhami, I. Warad, A. Zarrouk, New epoxy sugar based glucose derivatives as eco friendly corrosion inhibitors for the carbon steel in 1.0 M HCl: experimental and theoretical investigations, J. Alloys Compd. 833 (2020) 154949.

[5] M. Rbaa, F. Benhiba, P. Dohare, L. Lakhrissi, R. Touir, B. Lakhrissi, A. Zarrouk, Y. Lakhrissi, Synthesis of new epoxy glucose derivatives as a non-toxic corrosion inhibitors for carbon steel in molar HCl: experimental, DFT and MD simulation, Chem. Data Collect. 27 (2020) 100394.

[6] A. Koulou, F. Benhiba, M. Rbaa, N. Errahmany, Y. Lakhrissi, R. Touir, B. Lakhrissi, A. Zarrouk, M. Elyoubi, Synthesis of new epoxy glucose derivatives as inhibitor for mild steel corrosion in 1.0 M HCl: DMol3 theory and molecular dynamics simulation study: part-2, Mor. J. Chem. 8 (8-1) (2020) 2157–2166.

[7] A. Koulou, M. Rbaa, N. Errahmany, F. Benhiba, Y. Lakhrissi, R. Touir, B. Lakhrissi, A. Zarrouk, M. Elyoubi, Synthesis of new epoxy glucose derivatives as inhibitor for mild steel corrosion in 1.0 M HCl, experimental study: part-1, Mor. J. Chem. 8 (8-4) (2020) 2775–2787.

[8] Q. Zhang, B. Hou, Y. Li, G. Zhu, H. Liu, G. Zhang, Effective corrosion inhibition of mild steel by eco-friendly thiourea functionalized glucosamine derivatives in acidic solution, J. Colloid. Interface. Sci. 585 (2020) 355–367.

[9] C. Verma, L.O. Olasunkanmi, E.E. Ebenso, M.A. Quraishi, I.B. Obot, Adsorption behavior of glucosamine-based, pyrimidine-fused heterocycles as green corrosion inhibitors for mild steel: experimental and theoretical studies, J. Phys. Chem. C 120 (2016) 11598–11611.

CHAPTER 21

Amino acids as green corrosion inhibitors

Amino acids (AAs) are a highly important biologically active class of compounds that contain at least one carboxylic acid (–COOH) and one amino (–NH$_2$) substituents attached to the same carbon. Fig. 21.1 represents the basic structure of AAs. AAs are the building blocks (monomers) of various proteins (polymers). In the polymeric chain two amino acid units joined together by a peptide (–CO-NH–) bond. Most of the biologically important AAs are α-amino acids in which amino group present at the alpha-position with respect to the carboxylic carbon. Because of the presence of both basic (–NH$_2$) and acidic (–COOH) substituents, AAs and their derivatives possess some unique properties. They exhibit remarkably high solubility in the polar media. They can be classified into different categories based on their various properties including polarity, solubility, resources and acidic or basic nature. Because of their natural and biological origin, AAs and their derivatives are widely used as alternative environmental sustainable materials for various biological and industrial applications. They have also been used as effective corrosion inhibitors for different metal/electrolytes. The se of AAs and their derivatives are associated with the following advantages:

1. Eco-friendly (biodegradable, nonbioaccumulative, and biotolerable)
2. High solubility and excellent corrosion inhibition potential
3. Various polar functional groups, for example, –COOH and –NH$_2$ that can serve as adsorption centers during binding with the metallic surface
4. Natural and commercial availability
5. Ease and economic isolation and application
6. Useful for various metal/electrolyte systems
7. Minimized adverse effect on living organisms
8. Nontoxic to environmental
9. Corrosion inhibition potential can be suitably tuned using proper functionalization
10. A big library of corrosion inhibitors can be made using twenty AAs
11. Their solubility can be enhanced by chemical modification
12. Some of the AAs, especially high molecular weight AAs can be used as effective corrosion inhibitors in their pure form

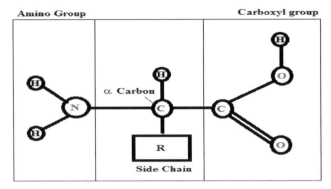

FIG. 21.1 Basic structure of AAs. *AAs*, amino acids.

In spite of their various advantages, application of AAs and their derivatives as corrosion inhibitors is associated with the following shortcomings:

1. Isolation and purification of AAs from natural resources utilize expensive and toxic chemicals/solvents.
2. AAs and their derivatives are susceptible to degradation at high temperatures.
3. Chemical modifications are essentially required for high corrosion inhibition performance.
4. In most of the cases, pure AAs are relatively less effective.
5. They undergo hydrolysis, rearrangement and degradation in highly aggressive electrolytes, especially at high temperatures.

21.1 Literature survey: amino acids as green corrosion inhibitors

Literature study suggests that AAs and their derivatives are extensively utilized as corrosion inhibitors for different metal/electrolyte systems [1]. It is important to mention that AAs inhibit corrosion by forming a protective surface film on the metal surface through their adsorption. AAs acquire both positive as well as negative charges because of their zwitterionic form. Therefore, when the metallic surface acquires a negative charge (with respect to point of zero charge; PZC) the cationic end of amino acid gets participated in the adsorption process and vice versa. AAs may adsorb physically or chemically depending upon the charges present on AAs and the metallic surface. Literature study suggests that in most cases, adsorption of AAs and their derivatives on metallic surface mostly follows the physiochemisorption mechanism. Physical or electrostatic adsorption/interaction is resulted through electrostatic force of attraction between charge metallic surface and charge AA molecules. On the other hand, chemisorption occurs through charge sharing (donor-acceptor binding) between them.

21.1.1 Amino acids as corrosion inhibitors for copper

Copper and its alloys are associated with numerous benefits such as high corrosion resistance, excellent thermal and chemical stability, high mechanical strength and electrical conductivity.

Therefore, these materials are extensively used constructional and building materials for different household and industrial applications [2,3]. They are used as electric wires, heat exchanges in cooling towers, durable metal pipelines and sheets, electronic and the marine equipments. It is well-known that copper acquires the ability to form surface protective oxide (cuprous and cupric oxides) layers that isolate the metal from aggressive atmospheres and protect it from corrosion. Therefore, copper and its alloys can be classified as noble materials [4,5]. However, these materials may undergo corrosive damage in highly aggressive industrial based electrolytes and marine environments. Therefore, proper corrosion monitoring practices have been developed to mitigate the corrosion of copper in such conditions. The use of organic compounds has been developed as one of the most common, economic and cost-effective methods because of their ease of synthesis and high inhibition potential. Nevertheless, most of these compounds are highly toxic therefore their recent application is highly limited because of increasing ecological awareness and strict environmental regulations.

Barouni and coworkers [5] reported the corrosion inhibition effectiveness of five amino acids namely, Lysine (Lys), Arginine (Arg), Cysteine (Cys), Glycine (Gly) and valine (Val) for copper corrosion in 1M HNO_3 solution. It was observed that Lysine (Lys), Arginine (Arg) and Cysteine (Cys) suppress the corrosion rate while Glycine (Gly) and valine (Val) enhance the corrosion rate. Among the evaluated amino acids, cysteine acts as the best corrosion inhibitor and it exhibits the highest protection power of 61% at 10^{-3}M concentration. The inhibition efficiencies of tested amino acids followed the order: Cys (61%) > Lys (54%) > Arg (38%) > Gly (−4%) > Val (−15%). Experimental results were supported by MNDO and AM1 theoretical studies. The corrosion inhibition potential of eight AAs was evaluated for copper corrosion in 1M HNO_3 [6]. It was observed that methionine (Met) exhibited the best inhibition effectiveness of 93.98%.

In another study [7], corrosion inhibition of four amino acids namely glycine (I), threonine (II), phenylalanine (III) and glutamic acid (IV) was determined for M3 copper and St. 3 low carbon steel alloy 0.5 M HCl solution. Analysis showed that inhibition effectiveness of the tested AAs for copper corrosion followed the order: glutamic acid (IV) > glycine (I) > threonine (II) > phenylalanine (III). On the other hand, the inhibition efficiency order was slightly different for St. 3 low-carbon steel corrosion. Recently, Kumar et al. [8] reported the inhibition potential of four AAs and their glutathione derivatives for copper corrosion using computational methods. Various computational based descriptors were calculated for the interaction of tested AAs with Cu (111), Cu (110) and Cu (100) surfaces. Through optimized structures it was observed that investigated AAs form bonding with Cu (111) surface using their O, N and S heteroatoms.

21.1.2 Amino acids as corrosion inhibitors for steel alloys

Steel-based alloys, especially carbon steel and mild steel, are widely used as constructional materials for different industrial as well as household applications. However, these materials are highly prone to corrosion when allowing in contact with aggressive environments. The problem of corrosion becomes more severe during industrial cleaning processes, including acid pickling, descaling, cleaning, and oil-well acidification. These processes utilize the consumption of highly aggressive acidic solutions to wash out the surface impurities such as metal oxides, carbonates, sulfides etc. However, because of their aggressiveness, these solutions also dissolve metallic components along with the impurities. Therefore, the cleaning processes require the use of some external additives called corrosion inhibitors, to mitigate the metallic dissolution. Literature investigation shows that AAs are extensively utilized as corrosion inhibitors for various metal/electrolyte systems.

Dehdab et al. [9] described the inhibition potential of three amino acids namely, serine (A), tryptophan (B) and tyrosine (C) for carbon steel corrosion using density functional theory (DFT) method. The DFT studies were carried out in gas as well as aqueous phases. Numerous DFT based computational parameters including total energy (TE) and molecular volume (MV) were derived using B3LYP level of theory with 6-311++G** basis set. Investigation showed that inhibition efficiencies of investigated amino acids followed the sequence: Serine (A) < tyrosine (C) < tryptophan (B). Local reactivity parameters for studied amino acids were determined using Fukui indices to determine the sites responsible for nucleophilic and electrolytic attacks. In the aqueous phase, a DFT study was also conducted for neutral as well as the protonated form of the amino acids. Results show that FMOs (HOMO and LUMO) were distributed over the entire part of the molecules indicating that the entire part of the amino acids involved in bonding (charge sharing) with the metallic surface. Order of corrosion inhibition efficiency determined was corroborated with experimental order of inhibition efficiency determined previously using experimentally.

The inhibition potential of four amino acids, namely, aspartic acid, asparagine, glutamine and glutamic acid was determined for mild steel corrosion in HCl solution using theoretical studies [10]. Different theoretical parameters were derived to describe the corrosion inhibition potential of investigated amino acids. The inhibition performances of the studied amino acids followed the sequence: glutamic acid < aspartic acid < asparagine < glutamine. Local reactivity parameters for studied amino acids were also determined using Fukui indices to determine the sites responsible for nucleophilic and electrolytic attacks. Corrosion inhibition of mild steel, Fe (100) surface using different L-amino acids has also been reported theoretically [11]. The inhibition efficiencies of evaluated amino acids followed the order: Arg (basic) > Trp (nonpolar) > Gln (polar) > Glu (acidic).

Hluchan et al. [12] described the corrosion inhibition of twenty-two amino acids and their two derivatives for iron in 1M hydrochloric acid solution. The corrosion inhibition potential of amino acids and their derivatives investigated in the study was tested using potentiodynamic polarization methods. Analysis showed that investigated compounds showed the highest protection effectiveness at 10 mM concentration. Among the tested compounds, 3,5-diiodotyrosine exhibits the best inhibition potential of 87% at the optimum concentration (10 mM). Among the studied amino acids, tryptophan exhibits the best performance, and it showed the highest protection of 80%. Hydroxyproline, Cystine and Cysteine act as corrosion accelerators that is, these amino acids accelerate the corrosion rate. Corrosion inhibition performance of the amino acids depends upon the molecular structure of the amino acids. In general, inhibition performance of amino acids increases on increasing the hydrocarbon chain lengths that is, increase in hydrophobicity results in an increase in the inhibition performance. The presence of additional electron releasing groups increases the inhibition performance of amino acids.

Along with pure amino acids, amino acid derivatives have also tested as green corrosion inhibitors for different meta;/electrolyte systems. Olivares-Xometl and coworkers [13] reported the corrosion inhibition potential of alkylamides derived from amino acids for carbon steel in hydrochloric acid solution. The inhibition potential of the alkylamides was determined using chemical and electrochemical methods. These compounds exhibit different inhibition efficiencies depending upon their concentrations and testing temperatures. These compounds exhibit their best inhibition performance at 100 ppm concentration. Tyr C-12, Gly C-12, Tyr C-8, and Gly C-8 showed the highest inhibition efficiencies of 97%, 94%, 81% and 72%, respectively at 25°C and 100 ppm concentration. All studied amino acid derivatives behaved as mixed-type corrosion inhibitors as they suppressed the rate of anodic as well as cathodic Tafel reactions without any significant shift in the values of corrosion potential, E_{corr}. SEM analysis suggests that in the presence of amino acid derivatives metal surface morphology was significantly improved due to the adsorption of Tyr C-12.

Recently, Aslam et al. [14] described the corrosion inhibition effect of three amino acid ester salts based ionic liquids (ILs), namely, namely L-Phenyl Alanine methyl ester saccharinate, L-Leucine methyl ester saccharinate and L-Alanine methyl ester saccharinate, designated respectively as [PheME][Sac], [LeuME][Sac], and [AlaME][Sac] for mild steel in 1M HCl. Synthesis of [PheME][Sac], [LeuME][Sac], and [AlaME][Sac] is illustrated in Scheme 21.1. various chemical, electrochemical, surface morphological and computational investigations were used to demonstrate the corrosion inhibition performance of studied amino acid derivatives.

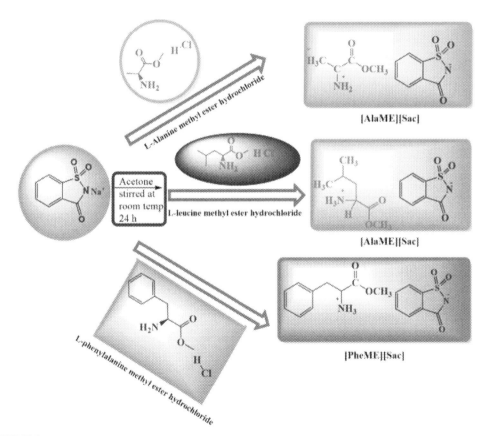

SCHEME 21.1 Synthesis of [PheME][Sac], [LeuME][Sac], and [AlaME][Sac] [14]. (Reproduced with permission@ Copyright Elsevier).

Weight loss or gravimetry measurement shows that [PheME][Sac], [LeuME][Sac], and [AlaME][Sac] act as effective inhibitors for mild steel corrosion in 1M HCl and their inhibition performance increases with their concentrations. [AlaME][Sac], [LeuME][Sac], and [PheME][Sac] showed highest protection effectiveness of 71.99%, 76.99% and 83.99%, respectively. Potentiodynamic polarization study shows that [PheME][Sac], [LeuME][Sac], and [AlaME][Sac] become effective by adsorbing at the active sites responsible for corrosion. Magnitudes of corrosion current density for mild steel corrosion in acidic solution were greatly reduced in the presence of investigated ionic liquids derivatives. It was also observed that [PheME][Sac], [LeuME][Sac], and [AlaME][Sac] acted as mixed-type corrosion inhibitors as they inhibit both anodic and cathodic half-cell reactions without any appreciable change in the magnitudes of E_{corr}. EIS study suggests that these compounds increase the value of charge transfer resistance for the corrosion process. This observation suggests that investigated amino acid derivatives adsorb at the interface of mild steel and 1M HCl solution. Adsorption of these compounds on metallic surface obeyed the Langmuir adsorption isotherm model.

Adsorption of [PheME][Sac], [LeuME][Sac], and [AlaME][Sac] on metallic surface was tested by SEM-EDX and FT-IR methods. Using SEM analysis, it was observed that the surface of mild steel specimen corroded in 1M HCl in the absence of investigated compounds was much damaged and corroded as compared to the surface morphology of protected metallic specimens. This observation suggests that investigated amino acid derivatives become effective by adsorbing on the metallic surface thereby forming corrosion protective surface films. Change in the elemental composition derived from EDX and FT-IR analyses validate this conclusion. Experimental studies were also supported by DFT and Monte Carlo (MC) simulation-based studies. FMOs (optimized, HOMO and LUMO) pictures of studied amino acid derivatives are presented in Fig. 21.2. It can be clearly seen that HOMO and LUMO are

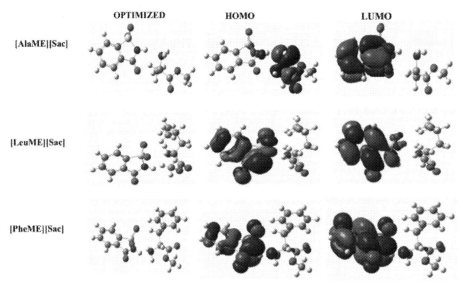

FIG. 21.2 FMOs (optimized, HOMO and LUMO) pictures of studied amino acid derivatives derived through e PM6/ZDO//DFT- B3LYP//631G (d) level of theory [14]. (Reproduced with permission@ Copyright Elsevier).

distributed over the major fraction of [PheME][Sac], [LeuME][Sac], and [AlaME][Sac] molecules. This observation suggests that studied molecules strongly participate in the charge sharing process with the metallic surface. Careful observation of this figure also shows that distribution of HOMO, as well as LUMO, is greatest in the case of [PheME][Sac], which indicates that [PheME][Sac] is one of the most effective corrosion inhibitors. Values of DFT parameters were consistent with the experimental order of inhibition efficiency. Negative values of adsorption energy derived through MC simulation studies indicated that investigated amino acid derivatives spontaneously and strongly adsorbed over the metallic surface. Corrosion inhibition potential of amino acid derivatives for steel alloys has also been investigated elsewhere [15–18].

21.1.3 Amino acids as corrosion inhibitors for aluminum and alloys

Aluminum and its alloys are extensively used as constructional materials for different industries because of their numerous advantageous properties. Similar to copper, aluminum also acquires oxide film forming ability. Therefore, in the presence of oxygen, aluminum shows remarkable resistivity against its corrosion. However, in aggressive solutions including acidic and neutral sodium chloride solutions, aluminum and its alloys readily undergo corrosive damage. Organic compounds, especially heterocyclic compounds are most frequently used as corrosion inhibitors in such electrolytes. However, traditional organic compounds are toxic and nonenvironmental friendly that limits their current anticorrosive applications. Because of increasing ecological awareness and strict environmental regulations, the use of these materials is highly restricted. In view of this, various new environmental friendly alternatives have been developed to replace the traditional toxic corrosion inhibitors. Amino acids and derivatives are also frequently utilized for the inhibition of aluminum corrosion in various electrolytes.

Ayuba et al. [19] investigated the corrosion inhibition potential of two amino acids, namely, aspartic acid and glutamic acid using DFT based quantum chemical calculations and molecular dynamics (MD) simulation methods. Various DFT and MDS based parameters were derived to describe the corrosion inhibition effectiveness of aspartic acid and glutamic acid. It was observed that glutamic acid interacts with the metallic surface more strongly as compared to aspartic acid. Ashassi-Sorkhabi et al. [20] reported the inhibition potential of some amino acids for aluminum corrosion in a mixture of 1M HCl+1M H_2SO_4 solution. The inhibition potential of the amino acids was determined using gravimetry, linear polarization and SEM investigations. Weight loss and polarization studies suggest that all studied amino acids act as effective anticorrosive materials for aluminum corrosion and their inhibition performance was concentration dependent. An increase in their concentration resulted in a subsequent increase in their corrosion protection potential. Weight loss study was conducted at 0.1, 0.01, 0.001, and 0.0001 M concentrations and tested amino acids exhibited the highest protection efficiency at 0.1M concentration. SEM analyses show that an increase in the concentration of electrolyte increases the roughness of the metallic surface while converse order is observed in the presence of amino acids. All studied amino acids behaved as mixed-type corrosion inhibitors. Their adsorption on metallic surface obeyed the Langmuir adsorption isotherm.

In another study, Bereket and Yurt [21] reported the pitting corrosion inhibition effect of some amino acids and hydroxyl-carboxylic acids for 7075 aluminum alloy in 0.05M sodium chloride solution at different pH. SEM analysis showed that in the absence of amino acids significant pitting was observed at pH 4 and 8. However, in 0.05M NaCl solution having a small amount of $K_2Cr_2O_7$ pitting was significantly reduced but pitting depth was increased. More so, in 0.05 M NaCl solution containing glycolic acid (an amino acid), the number and size of pits were greatly reduced at pH 4 and 8. Further, in 0.05M NaCl solution containing $NaNO_3$ and glycolic acid (at pH 8), intergranular corrosion was observed instead of pitting corrosion. SEM images of 7075 aluminum alloy corroded in 0.05 M NaCl solution with and without glycolic acid, $K_2Cr_2O_7$ and $NaNO_3$ at different pH are shown in Fig. 21.3. Literature

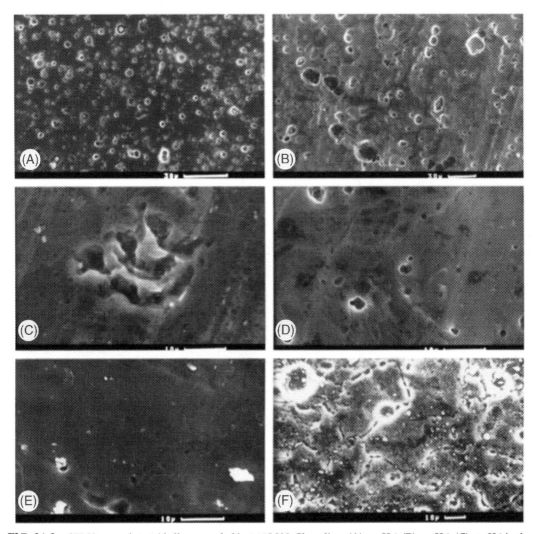

FIG. 21.3 SEM image of 7075 Al alloy corroded in 0.05 M NaCl medium (A) at pH 4, (B) at pH 8, (C) at pH 8 in the presence of 10^{-2} M $K_2Cr_2O_7$, (D) at pH 8 in the presence of 10^{-2} M glycolic acid, (E) at pH 4 in the presence of 10^{-2} M glycine, and (F) at pH 8 in the presence of 10^{-2} M glycolic acid and $NaNO_3$ [21]. (Reproduced with permission@ Copyright Elsevier).

study suggests that amino acids and their derivatives have also been investigated as green corrosion for aluminum in other numerous reports. In most of the studies, these compounds behave as mixed-type corrosion inhibitors as their presence affects the anodic as well as cathodic Tafel reactions without causing any significant change in corrosion potential. EIS study suggests that most of these compounds become effective by adsorbing at the interface of metal and electrolyte. Adsorption of these compounds on the metallic surface mostly follows the Langmuir adsorption isotherm model. Adsorption of these compounds on metallic surface is supported by various surface investigations including SEM, EDX, AFM and FT-IR methods.

21.2 Summary

Amino acids and their derivatives are extensively investigated as green corrosion inhibitors for different metal and electrolyte systems. Their corrosion inhibition performances are tested using various experimental and computational techniques. They serve as effective and environmental friendly alternatives for traditional toxic corrosion inhibitors. Potentiodynamic polarization measurement reveals that most of the amino acids and their derivatives become effective by suppressing both anodic and cathodic Tafel reactions without causing any significant variation in the value of corrosion potential, E_{corr}. They act as mostly mixed-type corrosion inhibitors. Amino acids and their derivatives behave as interface-type corrosion inhibitors as they increase the values of charge transfer resistance (R_{ct}) for the corrosion process. The adsorption of amino acids and their derivatives mostly follows the Langmuir adsorption isotherm. Adsorption mechanism of corrosion inhibition using amino acids and their derivatives has also been supported by surface investigations. DFT and MD (or MC) simulation studies are widely used to demonstrate the nature and effectiveness of interactions between these compounds and metallic surface. These analyses show that most of the amino acids and their derivatives spontaneously interacted through donor-acceptor interactions.

21.3 Useful websites

https://en.wikipedia.org/wiki/Amino_acid
https://medlineplus.gov/ency/article/002222.htm#:~:text=Amino%20acids%20are%20organic%20compounds,Break%20down%20food
https://www.healthline.com/nutrition/essential-amino-acids

References

[1] B. El Ibrahimi, A. Jmiai, L. Bazzi, S. El Issami, Amino acids and their derivatives as corrosion inhibitors for metals and alloys, Arab. J. Chem. 13 (2020) 740–771.
[2] J. Konieczny, Z. Rdzawski, Antibacterial properties of copper and its alloys, Arch. Mater. Sci. Eng. 56 (2012) 53–60.
[3] M. Ebrahimi, M. Par, Twenty-year uninterrupted endeavor of friction stir processing by focusing on copper and its alloys, J. Alloys Compd. 781 (2019) 1074–1090.

[4] A. Forty, Corrosion micromorphology of noble metal alloys and depletion gilding, Nature 282 (1979) 597–598.
[5] A. Epstein, J. Smallfield, H. Guan, M. Fahlman, Corrosion protection of aluminum and aluminum alloys by polyanilines: a potentiodynamic and photoelectron spectroscopy study, Synth. Met. 102 (1999) 1374–1376.
[6] K. Barouni, A. Kassale, A. Albourine, O. Jbara, B. Hammouti, L. Bazzi, Amino acids as corrosion inhibitors for copper in nitric acid medium: experimental and theoretical study, J. Mater. Environ. Sci. 5 (2014) 456–463.
[7] N. Makarenko, U. Kharchenko, L. Zemnukhova, Effect of amino acids on corrosion of copper and steel in acid medium, Russ. J. Appl. Chem. 84 (2011) 1362–1365.
[8] D. Kumar, N. Jain, V. Jain, B. Rai, Amino acids as copper corrosion inhibitors: a density functional theory approach, Appl. Surf. Sci. 514 (2020) 145905.
[9] M. Dehdab, M. Shahraki, S.M. Habibi-Khorassani, Theoretical study of inhibition efficiencies of some amino acids on corrosion of carbon steel in acidic media: green corrosion inhibitors, Amino Acids 48 (2016) 291–306.
[10] N.O. Eddy, Part 3. Theoretical study on some amino acids and their potential activity as corrosion inhibitors for mild steel in HCl, Mol. Simul. 36 (2010) 354–363.
[11] A. Kasprzhitskii, G. Lazorenko, T. Nazdracheva, V. Yavna, Comparative computational study of L-amino acids as green corrosion inhibitors for mild steel, Computation 9 (2021) 1.
[12] V. Hluchan, B. Wheeler, N. Hackerman, Amino acids as corrosion inhibitors in hydrochloric acid solutions, Mater. Corros. 39 (1988) 512–517.
[13] O. Olivares-Xometl, N. Likhanova, M. Domínguez-Aguilar, E. Arce, H. Dorantes, P. Arellanes-Lozada, Synthesis and corrosion inhibition of α-amino acids alkylamides for mild steel in acidic environment, Mater. Chem. Phys. 110 (2008) 344–351.
[14] R. Aslam, M. Mobin, I.B. Obot, A.H. Alamri, Ionic liquids derived from α-amino acid ester salts as potent green corrosion inhibitors for mild steel in 1M HCl, J. Mol. Liq. 318 (2020) 113982.
[15] H.M. Abd El-Lateef, M. Ismael, I.M. Mohamed, Novel Schiff base amino acid as corrosion inhibitors for carbon steel in CO2-saturated 3.5% NaCl solution: experimental and computational study, Corros. Rev. 33 (2015) 77–97.
[16] N. Negm, M. Zaki, Synthesis and characterization of some amino acid derived Schiff bases bearing nonionic species as corrosion inhibitors for carbon steel in 2N HCl, J. Dispers. Sci. Technol. 30 (2009) 649–655.
[17] L. Goni, M.J. Mazumder, S. Ali, M. Nazal, H. Al-Muallem, Biogenic amino acid methionine-based corrosion inhibitors of mild steel in acidic media, Int. J. Miner. Metall. Mater. 26 (2019) 467–482.
[18] M.M. Kabanda, I.B. Obot, E.E. Ebenso, Computational study of some amino acid derivatives as potential corrosion inhibitors for different metal surfaces and in different media, Int. J. Electrochem. Sci 8 (2013).
[19] A. Ayuba, A. Uzairu, H. Abba, G.A. Shallangwa, Theoretical study of aspartic and glutamic acids as corrosion inhibitors on aluminium metal surface, Moroc. J. Chem. 6 (2018) (2018) 2160–2172.
[20] H. Ashassi-Sorkhabi, Z. Ghasemi, D. Seifzadeh, The inhibition effect of some amino acids towards the corrosion of aluminum in 1 M HCl+ 1 M H2SO4 solution, Appl. Surf. Sci. 249 (2005) 408–418.
[21] G. Bereket, A. Yurt, The inhibition effect of amino acids and hydroxy carboxylic acids on pitting corrosion of aluminum alloy 7075, Corros. Sci. 43 (2001) 1179–1195.

CHAPTER 22

Oleochemicals as corrosion inhibitors

Chemicals derived from animal oils and fats and vegetable oils are called oleochemicals from (Latin: oleum "olive oil") [1,2]. The study of oleochemicals is known as oleochemistry [3,4]. Along with fats and oils, fatty acids, fatty amines, fatty alcohols and fatty acid methyl esters (FAMEs) are the common examples of oleochemicals. One of the major products of the oleochemical industry is soap, roughly 8.9×10^6 tons of which were produced in 1990. Various intermediates such as monoacylglycerols (MAGs), sugar esters, alcohol sulfates, quaternary ammonium salts, diacylglycerols (DAGs), alcohol ether sulfates, structured triacylglycerols (TAGs), alcohol ethoxylates also form during oleochemicals processing. Because of their natural and biological origin, oleochemicals can be regarded as environmental friendly alternatives for expensive, toxic and non-environmental friendly petrochemicals. Among the various modes of oleochemicals synthesis, transesterification and hydrolysis of triglycerides are important processes [5]. Hydrolysis and trans-esterification of triglycerides are presented in Scheme 22.1 and 22.2, respectively. Generally, hydrolysis is performed in an aqueous medium at 250°C. Hydrolysis in the presence of base is called as saponification, which produces soap and fatty acids as hydrolysis products. Triglycerides also react with methanol and form corresponding fatty acid esters of alcohol and glycerol. This process is known as transesterification. Methanol is the most frequently used alcohol for transesterification.

22.1 Oleochemicals as corrosion inhibitors: literature survey

Because of their cost-effectivity and environmental friendly behavior, oleochemicals are widely used, especially for the synthesis of detergents and soaps. They are also widely used in pharmaceutical, painting and food industries. They have also been established as effective environmental friendly inhibitors for metallic corrosion [6–9]. Suitably modified fatty acids, containing polar heteroatoms such as nitrogen and oxygen are extensively utilized as corrosion inhibitors for various metal/electrolyte systems. The application of oleochemicals as corrosion inhibitors are associated with various advantages, including:

1. Cost-effective and huge availability (natural and biological origin)
2. Ease of processing and fabrication
3. Environmental friendly (bio-originated, bio-degradable, biotolerable, and nonbioaccumulative)

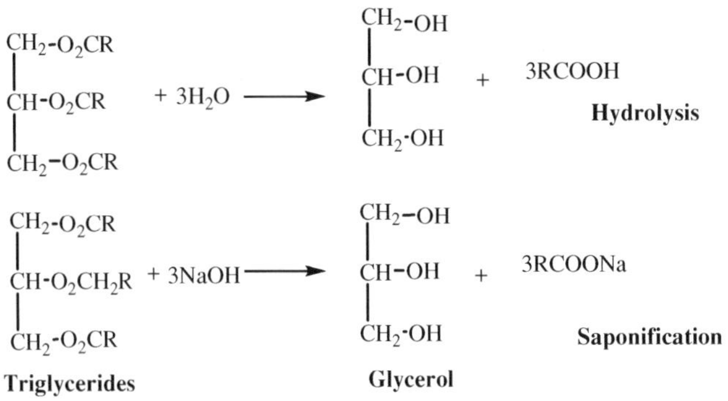

SCHEME 22.1 Hydrolysis (upper) and saponification (lower) of a triglyceride.

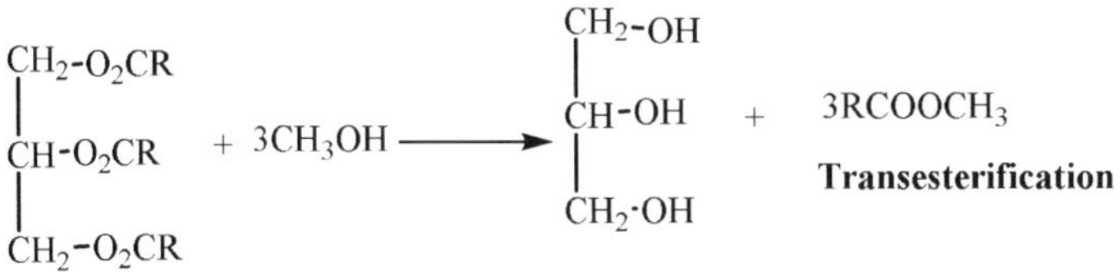

SCHEME 22.2 Transesterification of a triglyceride.

4. Excellent combination of hydrophilicity and hydrophobicity that is, surfactant-type properties
5. Reasonably soluble in the polar electrolytes
6. Useful for different metal/electrolyte systems (wide range of anticorrosive activity)
7. Minimized environmental risks
8. Corrosion inhibition potential and solubility in the polar electrolytes can be enhanced through proper modifications
9. Can be used in anticorrosive coatings and paintings

Tripathy et al. [10] synthesized palmitic acid imidazole (PI) by the reaction of palmitic acid and diethylenetriamine under microwave irradiation. Scheme 22.3 illustrates the synthesis of PI. PI was used as corrosion inhibitors for mild steel in 1M H_2SO_4 solution experimental and computational methods. PDP study suggests that PI exhibits the highest inhibition

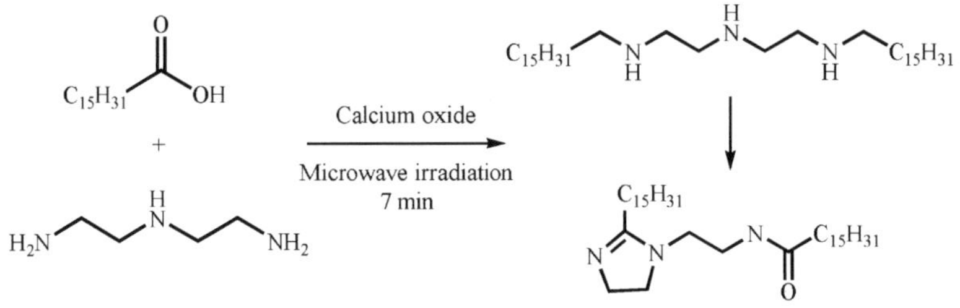

SCHEME 22.3 Synthetic route of palmitic acid imidazole (PI) [10].

effectiveness of 90.10% at 1×10^{-3} M concentration. It was also derived that the presence of the small amount of potassium iodide (6×10^{-3} M) synergistically enhances the protection effectiveness of PI. Thermodynamic investigations suggest the adsorption of PI with and without KI followed the Langmuir adsorption isotherm model. PDP study suggests that PI in the absence and presence of KI acts as mixed-type corrosion inhibitors.

Donor-acceptor mechanism (DFT) based quantum chemical analyses suggest that FMOs (HOMO and LUMO), especially in neutral form, were mainly distributed over the imidazole moiety. This finding suggests that imidazole moiety is principally participated in interaction that is, charge sharing with the metallic surface and long alkyl chains play the role of hydrophobic moieties. Through DFT analyses it was also derived that both protonated as neutral forms of PI interact strongly with the iron surface. MD simulations were also carried for neutral as well as protonated forms of PI and it was derived that both the forms of PI adsorb spontaneously on Fe (110) surface using their flat orientations. Top and side views of the most stable configurations of neutral PI in various conditions are presented in Fig. 22.1. It can be clearly seen that PI acquires the flat or horizontal orientations.

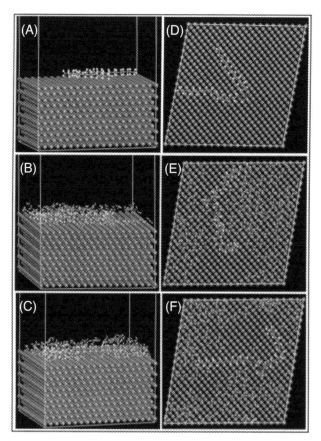

FIG. 22.1 Top (right hand-side) and side (heft hand-side) views of most stable configurations of PI (A and D) in vaccum, (B and E) in a system containing H_2O, H_3O^+, and SO_4^{2-}, and (C and F) in a system containing H_2O, H_3O^+, and SO_4^{2-} and I^- on Fe (110) surface [10]. (Reproduced with permission@ Copyright Elsevier).

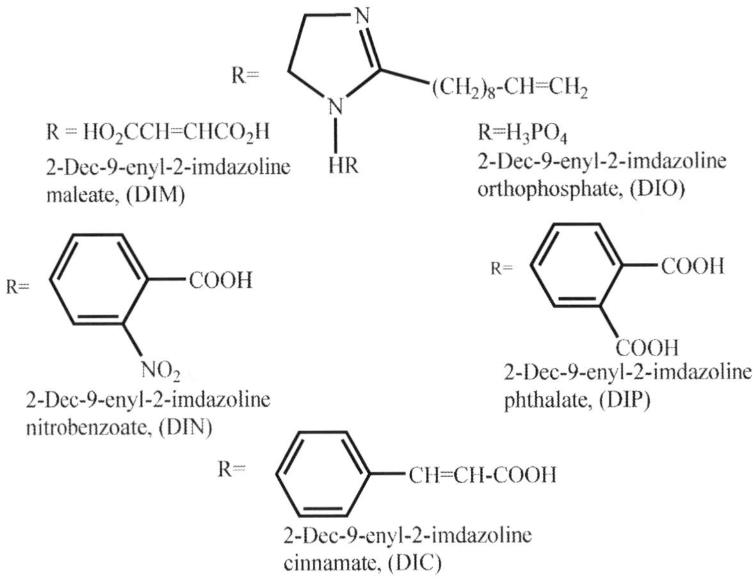

SCHEME 22.4 Structure of some oleochemical based volatile corrosion inhibitors (*VCIs*) [11].

In another study [11], Quraishi and Jamal developed five volatile compounds (VCIs) using 2-Dec-9-enyl- 2-imidazoline and used them as corrosion inhibitors for ferrous (mild steel) and nonferrous (brass & copper) alloys. Chemical structures of 2-Dec-9-enyl- 2-imidazoline based VCIs are presented in Scheme 22.4. Weight loss study suggests that inhibition efficiencies of investigated 2-Dec-9-enyl- 2-imidazoline based VCIs for ferrous and non-ferrous alloys followed the order: DIC > DIP > DIN > DIO > DIM.

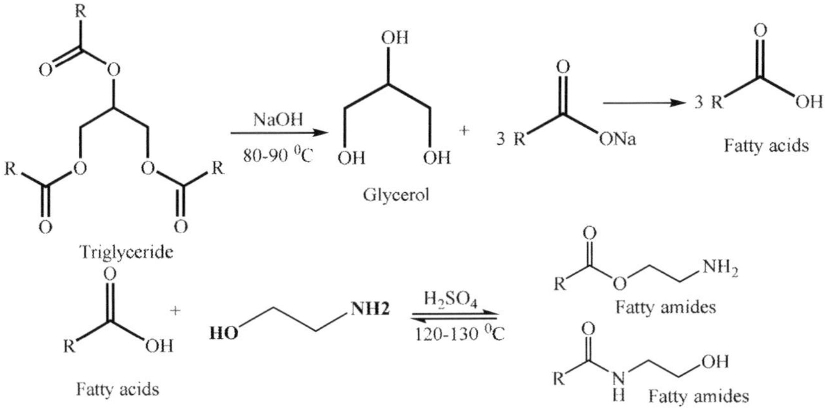

SCHEME 22.4A Synthesis of fatty acids and their derivatization to form a mixture of amides and esters.

Imidazoline-based fatty acid derivatives as investigated as effective corrosion inhibitors in other studies [12–15] Scheme 22.4a. The Imidazoline-based fatty acid derivatives mostly become effective by adsorbing at the interface metal and electrolyte and behaving as interface-type corrosion inhibitors. The presence of these chemicals in the corrosive electrolytes increases the diameter of Nyquist curves and values of charge transfer resistance. Through their adsorption, they block the active sites responsible for corrosion. They adversely affect the Tafel reactions without causing any appreciable change in the magnitude of corrosion potential that is, they act as mixed-type corrosion inhibitors. Rafiquee et al. [14] reported that a proper combination of hydrophobicity and hydrophilicity is essential for effective anticorrosive activities. Either too high or too low hydrophobic character adversely affects the inhibition effectiveness. Inhibition efficiencies of evaluated imidazoline based fatty acid derivatives followed the order: UDI [–$(CH_2)_{10}$–CH_3; 97.43%] > NI [–$(CH_2)_8$–CH_3; 95.16%] > PDI [–$(CH_2)_{14}$–CH_3; 91.87%] > HDI [–$(CH_2)_{10}$–CH_3; 87.20%].

Corrosion inhibition effectiveness of fatty amide type [16], sulfonated fatty acid diethanolamide, sulfated fatty acid- diethanolamine complexes (DC) [17], fatty acids based surfactants [18], fatty acid from palm oil [19], carboxylic acid from palm kernel oil (PKO) [20,21], 13-Docosenoic acid based amide derivatives of aliphatic and aromatic amines [22], fatty acid amide (KSFA) from the seed oil [23] and oleic acid hydrazide (OAH) [24] has also been tested for various metal/electrolyte systems. These compounds mostly behaved as mixed- and interface-type corrosion inhibitors. Their adsorption on metallic surface mainly followed the Langmuir adsorption isotherm model. Ali and coworkers [25] developed products (esters and amides) of palm oil and ethanolamine and tested them as corrosion inhibitors for carbon steel in 0.5 M HCl solution. Synthesis of fatty acids and their reactions with ethanolamine is presented in Scheme 22.5. The mixture of esters and amides exhibits the highest protection efficiency of 88.86% at 80 ppm concentration. Adsorption of the products on the metallic surface followed the Langmuir adsorption isotherm model. Potentiodynamic polarization studies suggest that a mixture of amides and esters become effective by blocking the active sites of the metallic surface. The mixture inhibits both anodic and cathodic Tafel reactions without any significant shift in the values of corrosion potential. This observation reveals that investigated mixture of esters and amides behaved as mixed-type inhibitors. Kinetic and thermodynamic studies suggest that magnitude of activation energy was greatly enhanced in the presence of fatty acids and amides mixture.

Sudheer and coworkers [26] described the corrosion inhibition potential of oleic hydrazide benzoate (OHB) and oleic hydrazide salicylate (OHS) as environmental friendly VCIs for mild steel using chemical and electrochemical methods. Syntheses of the OHB and OHS are illustrated in Scheme 22.5. Using weight loss study, it was derived that OHB and OHS exhibit

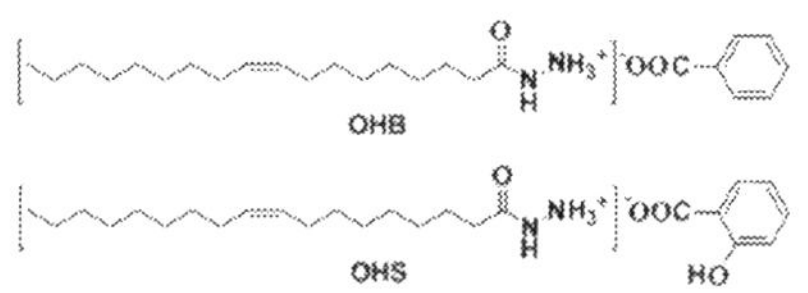

SCHEME 22.5 Scheme for the synthesis of OHB and OHS.

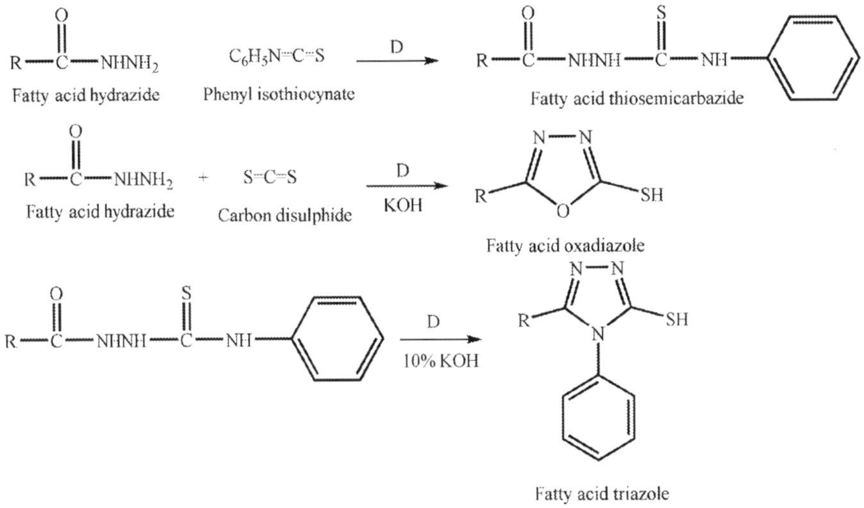

SCHEME 22.6 Schemes for the synthesis of hydrazides, thiosemicarbazides, oxadiazoles, triazoles based oleochemicals.

the same %IE of 89.97% at 100 ppm concentration. Adsorption of OHB and OHS on mild steel surface follows Langmuir adsorption isotherm model. Potentiodynamic polarization study suggests that OHB and OHS act as mixed-type corrosion inhibitors as they become effective by slowing down both anodic and Tafel reactions through their adsorption. An increase in the diameter of Nyquist curves and charge transfer resistance in the presence OHB and OHS suggested that they become effective by adsorbing at the interface of metal and electrolyte. In other studies, few relatively new hydrazides and thiosemicarbazides of long-chain fatty acids were developed and tested as corrosion inhibitors [27–30].

Quraishi and coworkers synthesized hydrazide, thiosemicarbazide, oxadiazole and triazole derivatives of lauric, undecenoic acids [31–33] (Scheme 22.6) and evaluated them as corrosion inhibitors for mild steel. A summary of some major oleochemicals as corrosion inhibitors is presented in Table 22.1. It can be clearly seen that the adsorption behavior of most of the oleochemical followed the Langmuir adsorption isotherm model. Using potentiodynamic polarization studies it clear that most of the oleochemicals become effective by adsorbing at the active sites of the metallic surface. Through their adsorption they inhibit corrosion by retarding both anodic as well as cathodic Tafel reactions. EIS based electrochemical studies suggest that oleochemicals mostly act as interface-type corrosion inhibitors as they increase the values of charge transfer resistance and diameter of Nyquist curves in their presence.

TABLE 22.1 A comparison of literature on the performance of oleochemicals as corrosion inhibitors.

S/N	Oleochemical(s)	Metal/alloy	Electrolyte	%IE and concentration	Isotherm nature	Electrochemical nature	References
1	Palmitic acid imidazole (PI)	Mild steel	H_2SO_4	$90/1 \times 10^{-3}$ mol/L	Langmuir	Mixed-type	[10]
2	Stearic acid imidazole (SI)	Mild steel	15% HCl	89.83%/400 mg/L + 1 mM KI	–	Mixed-type	[13]
3	2-Undecyl-1, 3-Imidazoline (UDI) Imidazoline derivatives, R = $-(CH_2)_8 \cdot CH_3$, $-(CH_2)_{10}CH_3$, $-(CH_2)_{14}CH_3$, $-(CH_2)_{16}CH_3$	Mild steel	$0.5\ H_2SO_4$	96.2%/500 ppm	Langmuir	Mixed-type	[14]
4	Fatty amide of crude rice bran oil (CRBO)	API X70 steel	3.5% NaCl + CO_2	95%/100 ppm	–	Anodic-type	[16]
5	Sulfated fatty acid-ethylamine complex	Mild steel	1% NaCl solution + CO_2	99.72%/150 ppm	Langmuir	–	[34]
6	$R-(CH_3)_7-CH_2-(CH_2)_7-CO-N(CH_2-CH-OH)_2$ $O-SO_2-N(CH_2-CH_2-OH)_2$	Mild steel	1% NaCl solution + CO_2	99.9%/100 ppm	Langmuir	Mixed-type	[35]

(continued)

TABLE 22.1 (Cont'd)

S/N	Oleochemical(s)	Metal/alloy	Electrolyte	%IE and concentration	Isotherm nature	Electrochemical nature	References
7	$R-(CH_2)_8-CH-(CH_2)_7-COONH_2^+(CH_2CH_2-OH)_2$ $\quad\quad\quad\quad\quad\ \ |$ $\quad\quad\quad\quad\quad O-SO_3^-NH_2^+(CH_2-CH_2-OH)_2$ SFADC, sulfated fatty acid	Mild steel	CO_2/ oilfield water	96.22%/457 ppm	Langmuir	–	[17]
8	$R-(CH_2)_8-CH-(CH_2)_7-COONH_2^+-(CH_2-CH_2-OH)_2$ $\quad\quad\quad\quad\quad\ \ |$ $\quad\quad\quad\quad\quad O-SO_3^-NH_3^+(CH_2-CH_2-OH)_2$ Sulfated fatty acid diethanolamine complexes	Mild steel	1% NaCl	99.95%/100 ppm	Langmuir	–	[18]
9	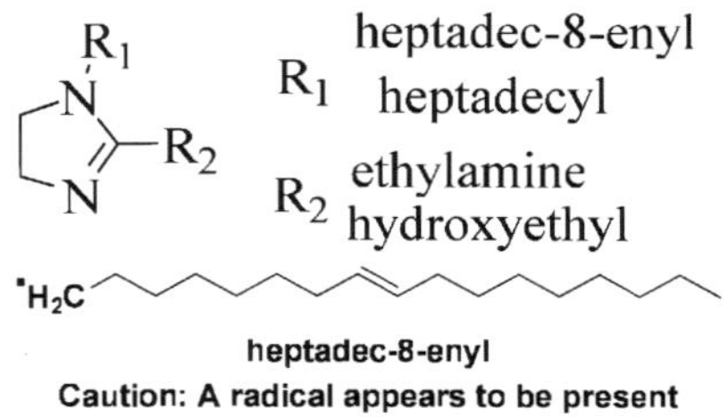 R_1 = heptadec-8-enyl heptadecyl R_2 = ethylamine hydroxyethyl heptadec-8-enyl Caution: A radical appears to be present	Carbon steel	1% NaCl	(39.59%)/ 8 ppm	Langmuir	–	[15]
10	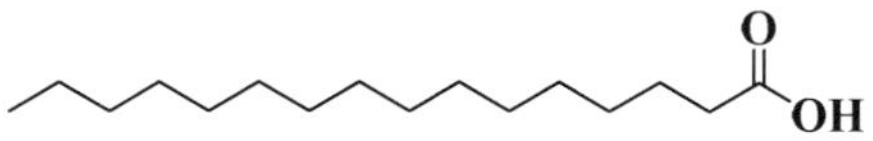 Palm oil	Mild steel/	1 M NaOH	–	–	–	[19]
11	Palm kernel oil	Carbon steel	1 M NaOH	96.67%/ 8 v/v%	–	Mixed-type	[20]
12	13-Docosenoic acid amide	Mild steel	1.0 HCl	96.8/500 ppm	Langmuir	–	[22]
13	Khaya senegalensis fatty hydroxylamide (KSFA)	Al	0.5 HCl	90.43/ × 10^{-6} g/L	–	–	[23]

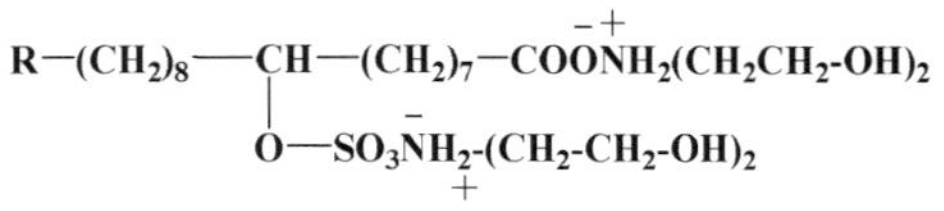

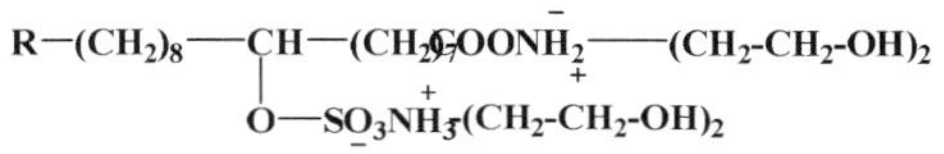

(continued)

S/N	Oleochemical(s)	Metal/alloy	Electrolyte	%IE and concentration	Isotherm nature	Electrochemical nature	References
14	Oleic acid hydrazide (OAH)	API X70 steel	Oilfield water	87.7/ 0.30 g/	–	–	[24]
15	N-(2-hydroxyethyl) fatty amides, 2-aminoethyl fatty ester	Mild steel	0.5 M HCl	80%/80 ppm	Langmuir	Mixed-type	[25]
16	9,12,15-octadecatrienoic acid; 9,12-octadecadienoic acid (C18:2), (C18:3) fatty acids	Mild steel	1 M HCl	95.1%/36 ppm	Langmuir	Mixed-type	[6]
17	PK_4, PR_4, PN_4	Mild steel	1 N HCl	–	Lanmguir	–	[36]

22.2 Summary

Because of their environmental friendly nature and huge availability at relatively low cost, oleochemicals are extensively used as corrosion inhibitors for numerous metal/electrolyte systems. Generally, these compounds become effective by adsorbing on the metallic surface following through the Langmuir adsorption isotherm model. Oleochemicals form a hydrophobic surface film at the interface of metal and electrolyte in which polar/hydrophilic site(s) oriented towards the metallic surface and non-polar/ hydrophobic site(s) directed towards electrolyte. A proper combination of hydrophilicity and hydrophobicity is essential for effective anticorrosive effectiveness. Obviously, a very high magnitude of hydrophobicity adversely affects the inhibition potential of oleochemicals by reducing their solubility in the polar electrolytes. Outcomes of electrochemical studies suggest the oleochemicals mostly act as mixed- and interface-type corrosion inhibitors and they become effective by adsorbing at the interface of metal and electrolyte. Various suitably modified oleochemicals have also been tested as corrosion inhibitors for different metal/electrolyte systems. Oleochemicals and their derivatives act as mixed-type corrosion inhibitors. Oleochemicals interact with the metallic surface using a DFT and acquire flat orientation (MDS or MCS).

22.3 Important websites

https://www.crodaindustrialchemicals.com/en-gb/products-and-applications/oleochemicals
https://www.twinriverstechnologies.com/blog/what-are-oleochemicals

References

[1] J. Salimon, N. Salih, E. Yousif, Industrial development and applications of plant oils and their biobased oleochemicals, Arab. J. Chem. 5 (2012) 135–145.
[2] K. Hill, Industrial development and application of biobased oleochemicals, Pure Appl. Chem. 79 (2007) 1999–2011.
[3] A. Rybak, P.A. Fokou, M.A. Meier, Metathesis as a versatile tool in oleochemistry, Eur. J. Lipid Sci. Technol. 110 (2008) 797–804.
[4] A.J. Vorholt, A. Behr, Oleochemistry, Applied Homogeneous Catalysis with Organometallic Compounds: A Comprehensive Handbook in Four Volumes, Wiley, USA, 2017, pp. 1579–1600.
[5] J.O. Metzger, U. Bornscheuer, Lipids as renewable resources: current state of chemical and biotechnological conversion and diversification, Appl. Microbiol. Biotechnol. 71 (2006) 13–22.
[6] A. Khanra, M. Srivastava, M.P. Rai, R. Prakash, Application of unsaturated fatty acid molecules derived from microalgae toward mild steel corrosion inhibition in HCl solution: a novel approach for metal–inhibitor association, ACS Omega 3 (2018) 12369–12382.
[7] S. Toliwal, K. Jadav, Inhibition of Corrosion of Mild Steel by Phenyl Thiosemicarbazides of Nontraditional Oils, Journal of Scientific and Industrial Research (JSIR), 68, CSIR, 2009, pp. 235–241.
[8] S. Toliwal, K. Jadav, T. Pavagadhi, Corrosion Inhibition Study of a New Synthetic Schiff Base Derived From Nontraditional Oils on Mild Steel in 1N HCl Solution, Journal of Scientific and Industrial Research (JSIR), 69, CSIR, 2010, pp. 43–47.
[9] J. Mayne, E. Ramshaw, Inhibitors of the corrosion of iron. II. Efficiency of the sodium, calcium and lead salts of long chain fatty acids, J. Appl. Chem. 10 (1960) 419–422.

[10] D.B. Tripathy, M. Murmu, P. Banerjee, M.A. Quraishi, Palmitic acid based environmentally benign corrosion inhibiting formulation useful during acid cleansing process in MSF desalination plants, Desalination 472 (2019) 114128.

[11] M. Quraishi, D. Jamal, Synthesis and evaluation of some organic vapour phase corrosion inhibitors, Indian J. Chem. Technol. 11 (2004) 459–464.

[12] M.M. Solomon, S.A. Umoren, M.A. Quraishi, M. Salman, Myristic acid based imidazoline derivative as effective corrosion inhibitor for steel in 15% HCl medium, J. Colloid Interface. Sci. 551 (2019) 47–60.

[13] M.M. Solomon, S.A. Umoren, M.A. Quraishi, D.B. Tripathy, E. Abai, Effect of akyl chain length, flow, and temperature on the corrosion inhibition of carbon steel in a simulated acidizing environment by an imidazoline-based inhibitor, J. Petrol. Sci. Eng. 187 (2020) 106801.

[14] M. Rafiquee, S. Khan, N. Saxena, M. Quraishi, Investigation of some oleochemicals as green inhibitors on mild steel corrosion in sulfuric acid, J. Appl. Electrochem. 39 (2009) 1409–1417.

[15] D. Wahyuningrum, S. Achmad, Y.M. Syah, B. Buchari, B. Ariwahjoedi, The synthesis of imidazoline derivative compounds as corrosion inhibitor towards carbon steel in 1% NaCl solution, J. Math. Fundam. Sci. 40 (2008) 33–48.

[16] E. Reyes-Dorantes, J. Zuñiga-Díaz, A. Quinto-Hernandez, J. Porcayo-Calderon, J. Gonzalez-Rodriguez, L. Martinez-Gomez, Fatty amides from crude rice bran oil as green corrosion inhibitors, J. Chem. 2017 (2017) 1–14.

[17] H.M. Abd El-Lateef, I. Ismayilov, V. Abbasov, E. Efremenko, L. Aliyeva, E. Qasimov, Green surfactants from the type of fatty acids as effective corrosion inhibitors for mild steel in CO_2-saturated NaCl solution, Am. J. Phys. Chem. 2 (2013) 16–23.

[18] H.M. Abd El-Lateef, V. Abbasov, L. Aliyeva, E. Qasimov, I. Ismayilov, Inhibition of carbon steel corrosion in CO_2-saturated brine using some newly surfactants based on palm oil: experimental and theoretical investigations, Mater. Chem. Phys. 142 (2013) 502–512.

[19] A. Daniyan, O. Ogundare, B. AttahDaniel, B. Babatope, Effect of palm oil as corrosion inhibitor on ductile iron and mild steel, Pac. J. Sci. Technol. 12 (2011) 45–53.

[20] M.Y. Zulkafli, N.K. Othman, A.M. Lazim, A. Jalar, Effect of carboxylic acid from palm kernel oil for corrosion prevention, Int. J. Basic Appl. Sci. 13 (2013) 29–32.

[21] M. Zulkafli, N. Othman, A. Lazim, A. Jalar, Inhibitive effects of palm kernel oil on carbon steel corrosion by alkaline solution, AIP Conference Proceedings, American Institute of Physics, 2013, pp. 42–47.

[22] A.M. Elsharif, S.A. Abubshait, I. Abdulazeez, H.A. Abubshait, Synthesis of a new class of corrosion inhibitors derived from natural fatty acid: 13-Docosenoic acid amide derivatives for oil and gas industry, Arab. J. Chem. 13 (2020) 5363–5376.

[23] A. Adewuyi, R.A. Oderinde, Synthesis of hydroxylated fatty amide from underutilized seed oil of Khaya senegalensis: a potential green inhibitor of corrosion in aluminum, J. Anal. Sci. Technol. 9 (2018) 1–13.

[24] T. Ajmal, S.B. Arya, L. Thippeswamy, M. Quraishi, J.J.C.E. Haque, Science, technology, influence of green inhibitor on flow-accelerated corrosion of API X70 line pipe steel in synthetic oilfield water, Corros. Eng. Sci. Technol. 55 (2020) 487–496.

[25] M.M. Ali, T.T. Irawadi, N. Darmawan, M. Khotib, Z.A. Mas'ud, Reaction products of crude palm oil-based fatty acids and monoethanolamine as corrosion inhibitors of carbon steel, Makara J. Sci. 23 (2019) 155–161.

[26] M.Quraishi Sudheer, E.E. Ebenso, M. Natesan, Inhibition of atmospheric corrosion of mild steel by new green inhibitors under vapour phase condition, Int. J.Electrochem. Sci. 7 (2012) 7463–7475.

[27] M. Quaraishi, D. Jamal, M.Tariq Saeed, Fatty acid derivatives as corrosion inhibitors for mild steel and oil-well tubular steel in 15% boiling hydrochloric acid, J. Am. Oil Chem. Soc. 77 (2000) 265–268.

[28] M. Ajmal, D. Jamal, M. Quraishi, Fatty acid oxadiazoles as acid corrosion inhibitors for mild steel, Anti-Corrosion Methods and Materials 47 (2000) 77–82.

[29] M. Rafiquee, N. Saxena, S. Khan, M. Quraishi, Some fatty acid oxadiazoles for corrosion inhibition of mild steel in HCl, Indian J. Chem. Technol. 14 (2007) 576–583.

[30] M.A. Quraishi, D. Jamal, Corrosion inhibition by fatty acid oxadiazoles for oil well steel (N-80) and mild steel, Mater. Chem. Phys. 71 (2001) 202–205.
[31] M. Quraishi, F.A. Ansari, Fatty acid oxadiazoles as corrosion inhibitors for mild steel in formic acid, J. Appl. Electrochem. 36 (2006) 309–314.
[32] M. Quraishi, F. Ansari, Corrosion inhibition by fatty acid triazoles for mild steel in formic acid, J. Appl. Electrochem. 33 (2003) 233–238.
[33] M.A. Quraishi, D. Jamal, Fatty acid triazoles: novel corrosion inhibitors for oil well steel (N-80) and mild steel, J. Am. Oil Chem. Soc. 77 (2000) 1107–1111.
[34] V.M. Abbasov, H.M. Abd El-Lateef, L.I. Aliyeva, I.T. Ismayilov, E.E. Qasimov, M.M. Narmin, Efficient complex surfactants from the type of fatty acids as corrosion inhibitors for mild steel C1018 in CO_2-environments, J. Korean Chem. Soc. 57 (2013) 25–34.
[35] V. Abbasov, H.M. Abd El-Lateef, L. Aliyeva, E. Qasimov, I. Ismayilov, M.M. Khalaf, A study of the corrosion inhibition of mild steel C1018 in CO_2-saturated brine using some novel surfactants based on corn oil, Egypt. J. Pet. 22 (2013) 451–470.
[36] S. Toliwal, K. Jadav, Inhibition of corrosion of mild steel by phenyl thiosemicarbazides of nontraditional oils, J. Sci. Ind. Res. 68 (2009) 235–241.

CHAPTER 23

High temperature corrosion and corrosion inhibitors

High temperature corrosion principally encountered in the oil and gas industries that forced these industries to experience unique challenges. Obviously, high temperature corrosion which is also known as hot corrosion occurs in the presence of hot gases, especially in anhydrous conditions. In the oil and gas industries, huge economic and safety losses have been experienced because of the hot corrosion. High temperature corrosion that usually occurs above 250–450°C adversely affects the structures and lifespan assessment of engineering parts of metallic assents. Obviously, hot corrosion proceeds through the formation of various corrosion products, including sulfides, carbides, oxides, nitrides, etc., vary upon the nature of hot gases. The formation of these products at high temperature unfavorably affects the stability, lifespan, reliability and load bearing capacity of the metallic structures [1,2]. The undesirable diffusion of atmospheric gases including oxygen, nitrogen oxides, sulfur dioxide, carbon oxides (CO and CO_2) at high temperature makes the metallic structures relatively brittle that cut-short their durability. Therefore, similar to ordinary corrosion, high temperature corrosion deteriorates the metallic structures and exerts unfavorable direct or indirect effects on engineering structures as well as on the environment.

Application of corrosion inhibitors is the first line of defense against high temperature corrosion. However, designing of anticorrosive formulations is one of the biggest challenges for corrosion scientist and engineers. Literature study suggests that anticorrosive formulations for high temperature corrosion are relatively lesser. Therefore, there is an extreme demand of designing, synthesis and utilization of high temperature anticorrosive formulations [3,4]. In view of this, some high temperature anticorrosive formulations have been developed and consumed. One of the biggest problems using high temperature corrosion inhibitors is their temperature dependent sensitivities. Recently, few research reports [5–8] and patents [9,10] anticorrosive formulations for high temperature corrosion have been proposed. It is also important to mention that at high temperature most of the metals convert in their oxides and scales (corrosion products). Obviously, corrosion products provide little bit of protection from further corrosion because oxide layers don't allow the diffusion or penetration of corrosive gases and moisture. However, the protection potential of corrosion products greatly depends on their porosity. Obviously, a more compact corrosion products deposit would provide better corrosion protection as compared to a relatively porous corrosion products deposit.

Protective and corrosive nature of corrosion products deposits can be accessed through Pilling–Bedworth ratio which commonly illustrated as Md/nmD [11]. In the Pilling-Bedworth ratio, D and M are the density and molar mass of corrosion product (scale), respectively and d and m are the density and atomic mass of corroding metal, respectively. Finally, n is the atomic number of the atom in the molecular formula of the corrosion product. For examples, in Fe_2O_3, Fe_3O_4 and Al_2O_3 values of n would be 2, 3, and 2, respectively. The deposits of corrosion products would be porous and non-protective for the case, Md/nmD < 1. On the other hand, it would be compact, nonporous and protection for the case, Md/nmD > 1. Depending upon the nature of scale deposits, high temperature corrosion can be categorized as oxidation, sulfidation, chlorination, carburization, and nitridation. Obviously, oxidation, sulfidation, chlorination, carburization and nitridation processes involve the formation of metal oxides, sulfides, chlorides, nitrides, and carbides, respectively. Generally, carburization requires relatively high temperature (~700–850°C).

23.1 High temperature corrosion inhibitors: literature survey

Literature study suggests that various high temperature anticorrosive formulations have been developed and utilized in last decades [7,12]. Though, their application is limited because of their temperature dependent sensitivity. More so, increase kinetic energy of the corrosion inhibitors, especially at elevated temperatures, decreases their attraction ability between metallic surfaces which adversely affects the inhibition potential of inhibitors. Some common high temperature corrosion inhibitors are described below.

23.1.1 Well acidization inhibitors

Well acidization is a common practice in oil and gases industries where highly concentrated acidic solutions are used to liquefy the rocks and scale blockages in order to permit the oil and gas to reach near the oil well. Concentrated hydrochloric acid (HCl) and hydrofluoric acid (HF) and their mixtures are frequently used for the well acidification process. However, these solutions are highly corrosive and cause corrosive dissolution of metallic structures of downhole and tubing equipments. Generally, well acidization is carried out at high temperatures further enhance the possibility and rate of metallic dissolution. Therefore, in order to mitigate corrosion in such conditions, some external additives, called as corrosion inhibitors, are added in the acidizing solutions. Chen and coworkers [13] described the corrosion inhibition effect of a precursor amide and its imidazole derivatives for gas and oil industries. Relative inhibition effectiveness of amide and imidazoline was accessed. It was derived that both amide and imidazoline act as effective anticorrosive formulations, especially at a lower temperature. The amide and imidazoline exhibit more than 95% inhibition effectiveness at 150°F temperature. However, amide and imidazoline and exhibit only 72% and 38% inhibition efficiencies at 100 ppm concentration after 72 h immersion time. Inhibition effectiveness of imidazoline and amide increased from 60% to 90% and 60% to 90%, respectively on increasing their concentration from 400 to 1000 ppm. It was also observed that at higher temperatures, a higher inhibitor concentration is required for proper anticorrosive activity. Both amide and imidazoline show comparable inhibition performance however at relatively high temperature, amide exhibits relatively better corrosion inhibition potential. The corrosion rate in the absence of amide and imidazoline increases on increasing temperature however after long-time exposure, especially at high temperature, becomes slower due to the formation of protective film.

In another report, Abayarathna and coworkers [14] described the inhibition potential of some imidazoline derived compounds for 1018 carbon steel in 3.3% NaCl solution saturated with CO_2 (at 150 psi). The Inhibition potential investigated imidazoline-derived compounds at 135°C. Corrosion inhibition potential of imidazoline based compounds was tested by enhancing the aggressiveness of the 3.3% NaCl solution by mixing 30 psi H_2S and/or 10 g/L elemental sulfur. Inhibition efficiency of the formulation was tested was determined at 500 ppm concentration at 2000 rpm. 3-phenyl-2-propyl-1-ol has also been tested as an inhibitor for API J55 oilfield pipes in HCl acidification [15]. Quraishi and Jamal [16] showed that demonstrated the corrosion inhibition potential of three triazoles, namely, 3-undecane-4-aryl-5-mercapto-1,2,4-triazole (triazole 1), 3(heptadeca-8-ene)-4-aryl-5-mercapto-1,2,4-triazole (triazole 2), and 3(deca-9-ene)-4-aryl-5-mercapto-1,2-4 triazole (triazole 3) for oil well steel (N80) in 15% HCl at temperatures above 105°C. A suggest that triazole 1-3 exhibit excellent corrosion inhibition potential at 500 ppm concentration and their inhibition effectiveness followed the sequence: triazole 3 (96.2%) > triazoles 2 (90.5%) > triazoles 1 (60.2%) inhibition efficiency. All studied triazoles act as mixed-type corrosion inhibitors.

The same authors [17], demonstrated the protection potential of three long-chain fatty acid oxadiazole-based derivatives, namely 2-undecan-5-mercapto-1-oxa-3,4-diazole (UMOD), 2-heptadecene-5-mercapto-1-oxa-3,4-diazole (HMOD) and 2-decene-5-mercapto-1-oxa-3,4-diazole (DMOD) for mild steel in 15% HCl. Using weight loss method, corrosion inhibition testing was carried out at 105±2°C temperature. The inhibition potential of UMOD (at 5000 ppm) was also tested for N80 steel in 15% HCl. Analysis establishes that all studied compounds act as effective inhibitors and UMOD behaved the best among them. These authors also reported the inhibition potential of few hydrazide-based compounds, namely, 1-decene-4-phenyl-thiosemicarbazide (DPTS), 1-heptadecene- 4-phenyl-thiosemicarbazide (HPTS) and 1-undecane-4- phenyl-thiosemicarbazide (UPTS) for mild steel and oil-well steel (N-80) in boiling 15% HCl at 105°C [18]. These compounds exhibit reasonably good anticorrosive activity and their inhibition potential follow the sequence: DPTS (96.0%) > HPTS (86.1%) > UPTS (77.8%). Similar reports have also been published elsewhere [19,20]. Inorganic oxide including ZrO_2 [21], Y_2O_3 [22], and SnO_2 [12] have also been investigated as high temperature corrosion inhibitors.

23.1.2 Naphthenic acid inhibitors

Sometimes, the presence of naphthenic acid accelerates metallic corrosion, especially at high temperatures. Generally, naphthenic acid accelerated corrosion occurs in the temperature range of 200–400°C. Nature and rate of naphthenic acid accelerated corrosion depend upon various factors including the concentration of naphthenic acid, physical state, engineering structure of metallic equipments, flow velocity, availability of sulfur containing compounds, Since naphthenic corrosion occurs at high temperature, therefore traditional nitrogenous based corrosion inhibitors are ineffective for this purpose as they are highly sensitive for degradation at high temperature. In general, naphthenic acid mediated corrosion occurs in the presence of H_2S. Therefore, corrosion inhibitor formulations for naphthenic acid corrosion mediated corrosion should contain inhibitors for naphthenic acid as well as H_2S. Some of the common anticorrosive formulations for naphthenic acid mediated corrosion are alkaline earth metal phenatesulfide-phosphonate and trialkyl-phosphate (1:1 and 1:5 mole), sulfur dichloride and monoalkyl-substituted phenol (3:4 mol), sulfur dichloride + alkyl phenol substituted (1:2 mol) and sulfur dichloride and alkyl phenol ratio (1:1) [21].

Besides these, mercaptotriazines [21], phosphite-based compounds [23] and sulphidizing agents [7] have also been evaluated as naphthenic acid corrosion inhibitors. Along with well acidization and naphthenic acid inhibitors, some non-specific inhibitors including oxotungstates [1], terephthalic acid [24], and amino amide and imidazoline-based compounds [9] are also tested as high temperature corrosion inhibitors. A summary of some common reports on high temperature corrosion inhibitors is presented in Table 23.1 and structures some corrosion high temperature corrosion inhibitors are presented in Fig. 23.1.

TABLE 23.1 A summary of some common reports on high temperature corrosion inhibitors.

S/N.	Inhibitor	Metal	Electrolyte/ temperature	Maximum protection	References
1.	Imidazoline derivative	Oil and gas steel	T = 148.8°C	90% at 25 ppm	[13]
2.	3-Undecane-4-aryl-5-mercapto-1,2,4-triazole (triazole 1)	N80 and cold rolled steel	15% HCl/T = 105°C	60.2% at 5,000 ppm	[16]
3.	3(Heptadeca-8-ene)-4-aryl-5-mercapto-1,2,4-triazole (triazole 2)	N80 and cold rolled steel	15% HCl/T = 105°C	90.5% at 5,000 ppm	[16]
4.	3(Deca-9-ene)-4-aryl-5-mercapto-1,2-4-triazole (triazole 3)	N80 and cold rolled steel	15% HCl/T = 105°C	96.2% at 5,000 ppm	[16]
5.	2-Undecane-5-mercapto-1-oxa-3,4-diazole (UMOD)	N80 steel	15% HCl/T = 105°C	98.9% at 5,000 ppm	[17]
6.	2-Heptadecene-5-mercapto-1-oxa-3,4-diazole (HMOD)	N80 steel	15% HCl/T = 105°C	69.1% at 5,000 ppm	[17]
7.	2-Decene-5-mercapto-1-oxa-3,4-diazole (DMOD)	N80 steel	15% HCl/T = 105°C	97.7% at 5,000 ppm	[17]
8.	Lauric acid hydrazie (LAH)	N80 and cold rolled steel	15% HCl/T = 105°C	71.3% at 5,000 ppm	[18]
9.	Oleic acid hydrazide (OAH)	N80 and cold rolled steel	15% HCl/T = 105°C	84.4% at 5,000 ppm	[18]
10.	Undecenoic acid hydrazide (UAH)	N80 and mild steel	15% HCl/T = 105°C	90.4% at ,5000 ppm	[29]
11.	1-Undecane-4-phenyl-thiosemicarbazide (UPTS)	15% HCl/ T = 105°C	15% HCl/T = 105°C	77.8% at 5,000 ppm	[18]
12.	1-Heptadecene-4-phenyl-thiosemicarbazide (HPTS)	15% HCl/ T = 105°C	15% HCl/T = 105°C	86.1% at 5,000 ppm	[18]
13.	1-Decene-4-phenyl-thiosemicarbazide (DPTS)	15% HCl/ T = 105°C	15% HCl/T = 105°C	96.0% at 5,000 ppm	[18]

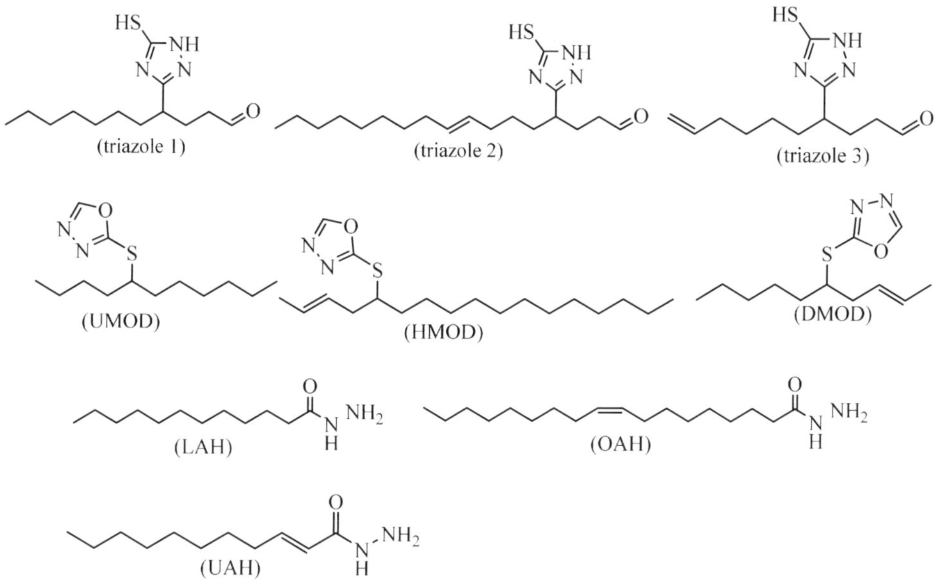

FIG. 23.1 Chemical structures some corrosion high temperature corrosion inhibitors.

23.2 Conclusion

From the ongoing discussion, it is clear that hot corrosion or high temperature acquires immense scientific and academic attention. Hot corrosion mostly takes place at the temperature of 250–450°C. However, carburization corrosion requires 700–800°C. Most of the metals and their alloys form metal oxides-based corrosion products that provide little bit of protection from further corrosion. Nevertheless, corrosive and protection nature of the scales depends upon the porosity of the corrosion products. Generally, the Pilling-Bedworth ratio, (Md/nmD), is used to describe the protective or corrosive behavior of the scales. Various specific corrosion inhibitors are effectively used as corrosion inhibitors for the well acidification process and naphthenic acid mediated corrosion. Oxidation, sulfidation, chlorination, carburization and nitridation are the common scale forming processes that involve the formation of metal oxides, sulfides, chlorides, nitrides, and carbides, respectively.

23.3 Important websites

https://www.materials.sandvik/en-in/materials-center/corrosion/high-temperature-corrosion/#:~:text=High%2Dtemperature%20corrosion%20refers%20to,of%20high%2Dtemperature%20corrosion%20are%3A&text=Flue%20gas%20and%20deposit%20corrosion,Nitridation

https://en.wikipedia.org/wiki/High-temperature_corrosion

References

[1] L.T. Popoola, A.S. Grema, G.K. Latinwo, B. Gutti, A.S. Balogun, Corrosion problems during oil and gas production and its mitigation, Int. J. Ind. Chem. 4 (2013) 1–15.

[2] W. Gao, Developments in High Temperature Corrosion and Protection of Materials, Elsevier, Woodhead Publishing Series, UK, 2008.

[3] S. Schauhoff, C. Kissel, New corrosion inhibitors for high temperature applications, Mater. Corros. 38 (2000) 51.

[4] R.C. John, A.D. Pelton, A.L. Young, W.T. Thompson, I.G. Wright, T.M. Besmann, Assessing corrosion in oil refining and petrochemical processing, Mater. Res. 7 (2004) 163–173.

[5] N. Hackerman, Use of inhibitors in corrosion control, Corrosion 4 (1948) 45–60.

[6] I. Gurappa, Protection of titanium alloy components against high temperature corrosion, Mater. Sci. Eng. A 356 (2003) 372–380.

[7] I. Obot, I.B. Onyeachu, S.A. Umoren, M.A. Quraishi, A.A. Sorour, T. Chen, N. Aljeaban, Q. Wang, High temperature sweet corrosion and inhibition in the oil and gas industry: Progress, challenges and future perspectives, J. Pet. Sci. Eng. 185 (2020) 106469.

[8] H.-W. Hsu, W.-T. Tsai, High temperature corrosion behavior of siliconized 310 stainless steel, Mater. Chem. Phys. 64 (2000) 147–155.

[9] D. Sullivan III, C. Strubelt, K. Becker, High Temperature Corrosion Inhibitor, US Patent 4028268, US patent, 1977.

[10] A. Groysman, N. Brodsky, J. Pener, D. Shmulevich, Low temperature naphthenic acid corrosion study, corrosion 2007, CORROSION Conference (2007) Paper Number: NACE-07569.

[11] C. Verma, E.E. Ebenso, M. Quraishi, Corrosion inhibitors for ferrous and non-ferrous metals and alloys in ionic sodium chloride solutions: a review, J. Mol. Liq. 248 (2017) 927–942.

[12] R.L. Jones, Corrosion Inhibition in High Temperature Environment, Google Patents, 1994.

[13] H.J. Chen, W.P. Jepson, T. Hong, High temperature corrosion inhibition performance of imidazoline and amide, corrosion 2000, CORROSION Conference 2000, Orlando, Florida (2000) Paper Number: NACE-00035.

[14] D. Abayarathna, A. Naraghi, N. Obeyesekere, Inhibition of Corrosion of Carbon Steel in the Presence of CO_2, H_2S and S, corrosion 2003, CORROSION Conference 2003, San Diego, California (2003) Paper Number: NACE-03340.

[15] M. Finšgar, J. Jackson, Application of corrosion inhibitors for steels in acidic media for the oil and gas industry: a review, Corros. Sci. 86 (2014) 17–41.

[16] M. Quraishi, D. Jamal, Fatty acid triazoles: novel corrosion inhibitors for oil well steel (N-80) and mild steel, J. Am. Oil Chem. Soc. 77 (2000) 1107–1111.

[17] M. Quraishi, D. Jamal, Corrosion inhibition by fatty acid oxadiazoles for oil well steel (N-80) and mild steel, Mater. Chem. Phys. 71 (2001) 202–205.

[18] M. Quraishi, V. Bhardwaj, J. Rawat, Prevention of metallic corrosion by lauric hydrazide and its salts under vapor phase conditions, J. Am. Oil Chem. Soc. 79 (2002) 603–609.

[19] M. Amosa, I. Mohammed, S. Yaro, O. Arinkoola, O. Ogunleye, Corrosion inhibition of oil well steel (N80) in simulated hydrogen sulphide environment by ferrous gluconate and synthetic magnetite, Nafta 61 (2010) 239–246.

[20] N. Abdel Ghany, M. Shehata, R. Saleh, A. El Hosary, Novel corrosion inhibitors for acidizing oil wells, Mater. Corros. 68 (2017) 355–360.

[21] V. Chukwuike, R.C. Barik, Inhibitors for high-temperature corrosion in oil and gas fields, Corrosion inhibitors in the oil and gas industry, Chapter 10 (2020) 271–288.

[22] S.H. Gitanjaly, S. Prakash, High temperature corrosion behavior of some Fe-, Co-and Ni-base superalloys in the presence of Y_2O_3 as inhibitor, Appl. Surf. Sci. 255 (2009) 7062–7069.

[23] I.D. Robertson, L.M. Dean, G.E. Rudebusch, N.R. Sottos, S.R. White, J.S. Moore, Alkyl phosphite inhibitors for frontal ring-opening metathesis polymerization greatly increase pot life, ACS Macro Lett. 6 (2017) 609–612.

[24] R. Zagidullin, U. Rysaev, M. Abdrashitov, D. Rysaev, Y. Kozyreva, I. Mazitova, P. Bulyukin, Method of preparing corrosion inhibitor by reaction of polyethylene-polyamine with terephathalic acid, RU2357007C2 Russia (2009) 2357007.

CHAPTER 24

Nanomaterials as corrosion inhibitors

Nanomaterials (NMs) represent a special class of compounds that are associated with various salient properties including high surface area. These materials are also called as nanoscale materials and characterized by a specific size of 1–100 nm. According to the European Commission held on October 18, 2011, NMs are defined as "natural, incidental or manufactured material containing particles, in an unbound state or as an aggregate or as an agglomerate and for 50% or more of the particles in the number size distribution, one or more external dimensions is in the size range 1–100 nm. Nanotechnology or nanotech is a field of research, science and engineering which is related with the synthesis or building of nanoscale materials and devices that is, atoms or molecules. NMs can be suitably modified for any specific engineering purposes. These materials are called engineered NMs. NMs are used for different industrial and biological applications. They are also used as corrosion inhibitors, especially in the coating phase for different metal/electrolyte systems. NMs are associated with various advantages properties that make them ideal candidates for different biological and industrial applications including as corrosion inhibitors:

1. Ease and cost-effective synthesis and scale up (nonrigorous processing).
2. Don't require high temperature, energy and pressure for synthesis. More so, their syntheses utilize environmental benign and safe chemicals, solvents and catalysts.
3. Large adsorption cross section area responsible for high surface coverage and excellent anticorrosive activities.
4. Their anticorrosion potential can be suitable magnified using proper surface modification by linking to molecular probes.
5. Useful for several of metal/electrolyte systems that is, wide range of applications.
6. Some of the NMs are less cytotoxic, biocompatible, nonbioaccumulative and biodegradable (environmental friendly).
7. Long-term protection that is, durable corrosion inhibitors.
8. Useful in highly aggressive electrolytes at high temperature and relatively low concentration.
9. Reduced adverse effects on living environment and living organisms.
10. Useful for coating and aqueous phase applications that is, proper combination of hydrophobicity and hydrophilicity.

However, use of NMs, especially as anticorrosive materials, is associated with following shortcomings:

1. Because of their limited solubility, mostly NMs are useful for anticorrosive coating formulations.
2. In an aqueous solution NMs may convert in their molecular size.
3. Some of the nanoparticles, especially metal oxides are toxic and nonenvironmental friendly in nature.
4. Some of the nanoparticles are highly toxic and their syntheses consume expensive chemicals, solvents and/or catalysts.
5. In the aqueous phase, the anticorrosive activity of these materials decreases with time because of agglomeration.

24.1 Nanomaterials as corrosion inhibitors: literature survey

Because of their various advantageous properties, NMs are widely employed as anticorrosive materials in aqueous as well as coating conditions. NMs synthesized using bottom-up methods including sol-gel, chemical vapor deposition and biosynthesis as well as top-down methods such as laser ablation and thermal decomposition are widely employed as corrosion inhibitors.

24.1.1 Metal/metal oxides as corrosion inhibitors

Metal oxides are crystalline solids of a metal cation and an oxide anion. Generally, metal oxides react with acids to form salts or with water to form bases. Metals, especially alkali and alkaline earth metals react with oxide anion to form various binary oxides (O^{2-}), peroxides (O_2^{2-}) and superoxides (O_2^-). It is important to mention that alkali metals having +1 oxidation state mostly form MO_2 (superoxides), M_2O (oxides) and M_2O_2 (peroxides), whereas alkaline earth metals having +2 oxidation state mostly form MO (oxides) and MO_2 (peroxides). Different metal oxides are used as additives in polymer-based anticorrosive coatings. Obviously, metal oxides mostly block micropores present in the surface of polymer-based coatings. This type of blockage retards the leakage of corrosive species such as water, oxide and aggressive ions through the surface micropores. Therefore, the presence of metal oxides in the polymer-based anticorrosive coatings enhances the corrosion resistance properties of the formulations. Literature study suggests that various metal and metal oxides such as Fe_2O_3 [1,2], Fe_3O_4 [3,4], TiO_2 [5,6], Cu_2O/CuO_2 [7], ZrO_2 [8], SiO_2 [9], ZnO [10,11], and NiO_2 [12] based composites are used as effective corrosion inhibitors for different metals and alloys in numerous electrolytic media. Literature studies suggest that pure metals and nano-crystal alloys are also widely used as corrosion inhibitors in different corroding systems.

24.1.2 Nanotube as corrosion inhibitors

Carbon nanotubes (CNTs) exist in two forms, single-walled CNTs (SWCNTs) and multi-walled (MWCNTs) and anticorrosive application of either form of CNTs is limited due to their limited solubility in aqueous based electrolytes [13]. Though, their dispersibility in such solutions can be enhanced through suitable functionalization and the addition of surface active polar functional groups [13,14]. Functionalization of SWCNTs and MWCNTs can be achieved using covalent and non-covalent methods [15–18]. Noncovalent functionalization occurs through π-π stacking, hydrogen bonding, electrostatic interactions and hydrophobic interactions. Due to their various advantageous properties such as nanosized, high mechanical and tensile strength, SWCNTs and MWCNTs and their derivatives, called as functionalized CNTs, are widely employed for various industrial applications. Functionalization of CNTs brought about the following physiochemical changes:

1. Decrease in hydrophobicity
2. Decrease in the agglomeration ability
3. Decrease in mechanical strength
4. Decrease in the electrical conductance Increase in the dispersion
5. Increase in the interfacial adhesion ability

Literature investigation suggests that CNTs mostly used as catalysts for various chemical transformations because of their high chemical and thermal stability, high surface-to-volume ratio, excellent mechanical and tensile strength etc. properties [19]. However, recently their use for electrochemical applications such as in corrosion inhibition and electrochemical sensors is gaining particular attention [20–22]. Obviously, CNTs are conductors of heat and electricity therefore they are expected to increase corrosion rate in their pure form. Therefore, most of the effective CNTs based corrosion inhibitors are their covalently functionalized species. Ionita and Pruna [23] reported the anticorrosive potential of (PPy)/polyaminobenzene sulfonic acid-functionalized single-walled CNT (CNT-PABS) and PPy/carboxylic acid-functionalized single-walled carbon nanotubes (CNT-CA) composites for carbon steel (OL 48–50) alloys in 3.5% NaCl. Results suggest that inhibition protection effectiveness of these species followed the order: CNT-PABS > CNT-CA > PPy. Corrosion inhibition potential of another PPy/MWCNTs nanocomposite for 304 SS in 3% NaCl has been investigated elsewhere [24]. Recently, the application of polyaniline/f-CNTs nanocomposites as anticorrosive coatings is gaining immense attention [25–28]. Anticorrosive applications of other CNTs based nanocomposites are summarized in Table 24.1.

24.1.3 Graphene and graphene oxide as corrosion inhibitors

Similar to CNTs, graphene (G), graphene oxide (GO) and their derivatives/composites are widely investigated as anticorrosive materials, especially as anticorrosive coating formulations. They can be also functionalized using covalent and non-covalent functionalization methods. Non-covalent functionalization of G and GO occurs through π-π stacking, hydrogen bonding, electrostatic interactions and hydrophobic interactions (Fig. 24.1). Literature study suggests that both single and multiwalled G, GO and their derivatives are evaluated as excellent corrosion inhibitors, especially in the coating systems. Similar to CNTs, G, GO and their derivatives behave as nanofillers that is, they uniformly distribute in the polymer matrixes and fill the micropores present in the coating structures. Because of this, they avoid the penetration of corrosive species such as water, oxygen and aggressive ions. Nevertheless, graphene based anticorrosive coatings experience numerous challenges and shortcomings. One of the biggest shortcomings of graphene-based materials using as anticorrosive formulations

TABLE 24.1 A summary of CNTs-based nanocomposites for different metals/alloys in various electrolytes.

S. No.	Carbon nanotubes based nanocomposites	Metal/ alloys	Electrolyte	References
1.	CNT-PABS and CNT-CA	Carbon steel (OL 48–50)	3.5% NaCl	[23]
2.	Ppy/MWCNT	304 stainless steel	3% NaCl	[24]
3.	PANI/f-CNT	Mild steel	3.5% NaCl	[25]
4.	PANI-MWCNT composite	Carbon steel	1M HCl	[26]
5.	c-PANI	Carbon steel	3.5% NaCl, acidic, and alkaline solution	[27]
6.	PANI–CNT nanocomposites	Mild steel	1M HCl	[28]
7.	(GFNACTL)	Copper	Simulated sea water	[29]
8.	MWNT/PU	Stainless steel	3% NaCl	[30]
9.	HA/f-MWCNT	316L stainless steel	Simulated body fluid (SBF)	[31]
10.	Ag-HA/f-MWCNT	316L stainless steel	Simulated body fluid (SBF)	[32]
11.	PoPDA@MWCNT	Steel	3.5% wt.% NaCl	[33]
12.	Ni-MWCNTs	Mild steel	3.5% NaCl	[34]

GFNACTL, graphitic filamentous nanocarbon-aligned carbon thin layer; HA, hydroxyapatite; PANI, polyaniline; PoPDA, poly(o-phenylenediamine); Ppy, polypyrrole; PU, polyurethane.

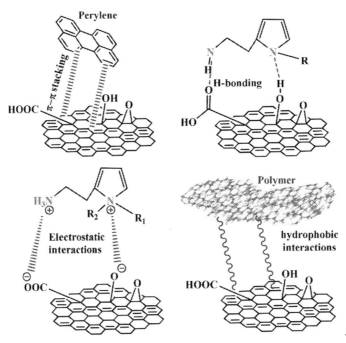

FIG. 24.1 Noncovalent functionalization of GO using (A) π-π stacking, (B) hydrogen bonding, (C) electrostatic interactions, and (D) hydrophobic interactions.

is their uncontrolled stacking in the polymer matrixes. This type of stacking adversely affects the corrosion inhibition potential of these materials because of their reduced dispersibility.

Literature investigation suggests the pure graphene and suitably modified graphene are widely used as anticorrosive coating formulations for different metals and alloys in various electrolytes. Anticorrosive coating effect of graphene derivatives for numerous metals/ electrolyte systems are illustrated in Table 24.2. It is important to mention that modified graphene exhibit superior anticorrosive activity. Through literature, it can also be observed that graphene oxide modified using organic compounds show outstanding dispersibility in the aqueous solutions therefore some of such derivatives have been tested as aqueous phase corrosion inhibitors. Mostly, these compounds become effective by adsorbing at the interface of metal surface and electrolyte. Their adsorption mostly followed the Langmuir adsorption isotherm. Using the PDP method, it is derived that these compounds mostly acted as mixed-type corrosion inhibitors as they become effective by blocking the anodic as well as cathodic Tafel reactions. Recently, our research team described the synthesis, characterization and

TABLE 24.2 Modified graphene as anticorrosive coatings for different metals and alloys in various electrolytes.

S. No	Modified graphene	Electrolyte	Metal	References
1.	Graphene@SiO_2	3.5% NaCl	Copper	[35]
2.	rGO@APTES	3.5% NaCl	Cu	[36]
3.	Fluorographene/epoxy (FG)	3.5% NaCl	Cu	[37]
4.	P2BA-G	3.5% NaCl	Q235 steel	[38]
5.	Graphene-reinforced ZRE	3.5% NaCl	Carbon steel	[39]
6.	Carboxylated aniline trimer stabilized grapheme	3.5% NaCl	Carbon steel	[40]
7.	EVOH/BA/GO	3.5% NaCl	Stainless steel	[41]
8.	AISi-GO	3.5% NaCl	Steel	[42]
9.	GO-Al_2O_3 hybrid (GO–Al_2O_3)	3.5% NaCl	Steel	[43]
10.	PS/modified-GO	3.5% NaCl	Steel	[44]
11.	PMMA embedded in graphene (HNPN)	3.5% NaCl	Steel	[45]
12.	GO-Fe_3O_4 hybrid@polydopamine + KH550	3.5% NaCl	Steel	[46]
13.	mGO-ODA/MAPP	3.5% NaCl	Steel	[47]
14.	Epoxy/SiO_2-GO nanohydride	3.5% NaCl	Steel	[48]
15.	Sulfonated oligoanilines /GO	3.5% NaCl	Steel	[49]
16.	GPTMS-GO	3.5% NaCl	Steel	[50]
17.	PEI-G	3.5% NaCl	Steel	[51]
18.	Graphene-tin oxide composite film	1M HCl	Al	[52]
19.	Graphene/Epoxy coating	3.5% NaCl	Al 2024-T3	[53]

AISi, amino and isocyanate silane; APTES, aminopropyl triethoxysilane; EVOH/BA, poly(vinyl alcohol-co-ethylene); GPTMS-GO, silane coupling agents/GO/epoxy; mGO-ODA/MAPP, octadecylamine/graphene oxide/maleic anhydride, grafted polypropylene; PEI-G, poly (ether imide)/graphene; PS, polystyrene; P2BA, poly(2-butylaniline); ZRE, zinc rich nanocomposite.

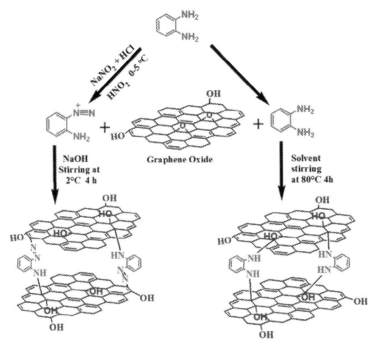

FIG. 24.2　Schemes for the synthesis of AAB-GO and DAB-GO [54]. *AAB-GO*, aminoazobenzene-graphene oxide; *DAB-GO*, diaminobenzene-graphene oxide.

corrosion inhibition potential of two graphene derivatives, namely aminoazobenzene-GO (AAB-GO) and diaminobenzene-GO (DAB-GO) [54]. The anticorrosive potential of AAB-GO and DAB-GO was tested for mild steel corrosion in 1M HCl solution. Schemes for the synthesis of AAB-GO and DAB-GO are illustrated in Fig. 24.2.

Characterization of AAB-GO and DAB-GO was achieved using XPS, XRD, TEM and FT-IR methods. Electrochemical studies suggest that both functionalized GO act as mixed- and interface-type corrosion inhibitors. Experimental results were also supported by DFT based computational analyses. Using the DFT method, it was derived that AAB-GO and DAB-GO react through each other following through charge sharing that is, donor-acceptor mechanism. A summary of graphene based aqueous phase corrosion inhibitors is presented in Table 24.3.

24.1.4　Nanofibers and nanocontainers as corrosion inhibitors

Nanofibers are the NMs that have fiber-like shape having diameter in nanoscale. Nanofibers mainly derived through various natural and synthetic polymers and they acquire special physiochemical properties. Generally, nanofibers acquire extremely high porosity, high surface area-to-volume ratio, high mechanical strength and flexibility. Literature study suggests the various nanofibers derived from natural and synthetic polymers are evaluated as anticorrosive materials, especially in coating condition, in various corroding systems [58–60]. Mostly, nanofibers are used in self-healing coatings. Similarly, nanocontainers are also used as corrosion inhibitors in different anticorrosive formulations [61–63].

TABLE 24.3 Chemically modified graphene oxide as aqueous phase corrosion inhibitors.

Nature of graphene oxide	Electrolyte	Metal	%IE of the best	Optimum conc	Electrochemical nature	References
AAB-GO and DAB-GO	1M HCl	Mild steel	96.80% (AAB-GO)	25 mgL^{-1}	Mixed type	[54]
DAMP-GO and DAZP-GO	1M HCl	Mild steel	96.73% (DAZP-GO)	25 mgL^{-1}	Cathodic type	[55]
p-Aminophenol-GO	1M HCl	Mild steel	92.86%	25 mgL^{-1}	Cathodic type	[56]
PEI-GO	15% HCl	Carbon steel	95.77%	+5 mM KI + at 50 mgL^{-1}	Cathodic type	[57]

AAB, aminoazobenzene; DAB, diaminobenzene; DAMP, diaminopyridine; DAZP, diazopyridine; PEI, polyethyleneimine.

24.2 Summary

NMs are established as one of the potential classes of anticorrosive materials. Obviously, these materials acquire various salient features including surface-area-to-volume ratio that make suitable candidates to replace the traditional corrosion inhibitors. The use of NMs as corrosion inhibitors is associated with various advantages. Some of the commonly used anticorrosive NMs are metal oxides and their composites, CNTs (SWCNTs and MWCNTs) and their covalent and non-covalent functionalized derivatives/composites, G, GO, and their covalent and non-covalent functionalized derivatives/composites, nanofibers and nanocontainers. Obviously, covalent and non-covalent functionalized derivatives exhibit reasonably high dispersibility in the aqueous solution. Nevertheless, most of the NMs are utilized as anticorrosive materials in anticorrosive formulations. Recently, few reports have been reported in which suitably modified graphene oxide using organic compounds such as aminoazobenzene, diaminobenzene, diaminopyridine, diazopyridine, and polyethyleneimine are tested as the excellent aqueous phase anticorrosive materials. In the aqueous solution, NMs behave as mixed- and interface-type corrosion inhibitors as they become effective by adsorbing at the interface of metal and electrolyte and retarding the anodic and cathodic Tafel reactions.

References

[1] T. Liu, Y. Liu, Y. Ye, J. Li, F. Yang, H. Zhao, L. Wang, Corrosion protective properties of epoxy coating containing tetraaniline modified nano-α-Fe2O3, Prog. Org. Coat. 132 (2019) 455–467.

[2] S. Umare, B. Shambharkar, Synthesis, characterization, and corrosion inhibition study of polyaniline-α-Fe2O3 nanocomposite, J. Appl. Polym. Sci. 127 (2013) 3349–3355.

[3] M. Izadi, T. Shahrabi, B. Ramezanzadeh, Synthesis and characterization of an advanced layer-by-layer assembled Fe3O4/polyaniline nanoreservoir filled with Nettle extract as a green corrosion protective system, J. Ind. Eng. Chem. 57 (2018) 263–274.

[4] A. Javidparvar, B. Ramezanzadeh, E. Ghasemi, The effect of surface morphology and treatment of Fe3O4 nanoparticles on the corrosion resistance of epoxy coating, J. Taiwan Inst. Chem. Eng. 61 (2016) 356–366.

[5] B. Bhuvaneshwari, S. Vivekananthan, G. Sathiyan, G. Palani, N.R. Iyer, P.K. Rai, K. Mondal, R.K. Gupta, Doping engineering of V-TiO2 for its use as corrosion inhibitor, J. Alloys Compd. 816 (2020) 152545.

[6] Y. Zhang, H. Zhu, C. Zhuang, S. Chen, L. Wang, L. Dong, Y. Yin, TiO2 coated multi-wall carbon nanotube as a corrosion inhibitor for improving the corrosion resistance of BTESPT coatings, Mater. Chem. Phys. 179 (2016) 80–91.

[7] A. Palit, S.O. Pehkonen, Copper corrosion in distribution systems: evaluation of a homogeneous Cu2O film and a natural corrosion scale as corrosion inhibitors, Corros. Sci. 42 (2000) 1801–1822.

[8] A. Ghafari, M. Yousefpour, A. Shanaghi, Corrosion protection determine of ZrO2/AA7057 nanocomposite coating with inhibitor using a mathematical ranking methods, Appl. Surf. Sci. 465 (2019) 427–439.

[9] E.V. Skorb, D. Fix, D.V. Andreeva, H. Möhwald, D.G. Shchukin, Surface-modified mesoporous SiO2 containers for corrosion protection, Adv. Funct. Mater. 19 (2009) 2373–2379.

[10] O.T. de Rincon, O. Perez, E. Paredes, Y. Caldera, C. Urdaneta, I. Sandoval, Long-term performance of ZnO as a rebar corrosion inhibitor, Cem. Concr. Compos. 24 (2002) 79–87.

[11] A. Popoola, O. Fayomi, ZnO as corrosion inhibitor for dissolution of zinc electrodeposited mild steel in varying HCl concentration, Int. J. Phys. Sci. 6 (2011) 2447–2454.

[12] M.Z. Ansari, M. Shoeb, P.S. Nayab, M. Mobin, I. Khan, W.A. Siddiqi, Honey mediated green synthesis of graphene based NiO2/Cu2O nanocomposite (Gr@ NiO2/Cu2O NCs): Catalyst for the synthesis of functionalized Schiff-base derivatives, J. Alloys Compd. 738 (2018) 56–71.

[13] N. Nakashima, Soluble carbon nanotubes: fundamentals and applications, Int. J. Nanosci. 4 (2005) 119–137.

[14] B.I. Kharisov, O.V. Kharissova, H. Leija Gutierrez, U. Ortiz Méndez, Recent advances on the soluble carbon nanotubes, Industr. Eng. Chem. Res. 48 (2009) 572–590.

[15] V. Georgakilas, M. Otyepka, A.B. Bourlinos, V. Chandra, N. Kim, K.C. Kemp, P. Hobza, R. Zboril, K.S. Kim, Functionalization of graphene: covalent and non-covalent approaches, derivatives and applications, Chem. Rev. 112 (2012) 6156–6214.

[16] M. Marcia, A. Hirsch, F. Hauke, Perylene-based non-covalent functionalization of 2D materials, FlatChem 1 (2017) 89–103.

[17] A. Hirsch, F. Hauke, Post-Graphene 2D Chemistry: The Emerging Field of Molybdenum Disulfide and Black Phosphorus Functionalization, Angew. Chem. Int. Ed. 57 (2018) 4338–4354.

[18] V.D. Punetha, S. Rana, H.J. Yoo, A. Chaurasia, J.T. McLeskey Jr, M.S. Ramasamy, N.G. Sahoo, J.W. Cho, Functionalization of carbon nanomaterials for advanced polymer nanocomposites: a comparison study between CNT and graphene, Prog. Polym. Sci. 67 (2017) 1–47.

[19] Y.G. Peng, J.L. Ji, Y.L. Zhang, H.X. Wan, D.J. Chen, Preparation of poly (m-phenylenediamine)/ZnO composites and their photocatalytic activities for degradation of CI acid red 249 under UV and visible light irradiations, Environ. Prog. Sustain. Energy 33 (2014) 123–130.

[20] E.N. Zare, M.M. Lakouraj, A. Ramezani, Efficient sorption of Pb (II) from an aqueous solution using a poly (aniline-co-3-aminobenzoic acid)-based magnetic core–shell nanocomposite, New J. Chem. 40 (2016) 2521–2529.

[21] M. Baghayeri, E.N. Zare, M.M. Lakouraj, Monitoring of hydrogen peroxide using a glassy carbon electrode modified with hemoglobin and a polypyrrole-based nanocomposite, Microchim. Acta 182 (2015) 771–779.

[22] A. Olad, A. Rashidzadeh, M. Amini, Preparation of polypyrrole nanocomposites with organophilic and hydrophilic montmorillonite and investigation of their corrosion protection on iron. Adv. Polym. Technol., 32, Wiley Periodicals, Inc., (2013) pp. 1–10.

[23] M. Ioniţă, A. Prună, Polypyrrole/carbon nanotube composites: molecular modeling and experimental investigation as anti-corrosive coating, Prog. Org. Coat. 72 (2011) 647–652.

[24] A. Ganash, Electrochemical synthesis and corrosion behaviour of polypyrrole and polypyrrole/carbon nanotube nanocomposite films, J. Compos. Mater. 48 (2014) 2215–2225.

[25] A.M. Kumar, Z.M. Gasem, In situ electrochemical synthesis of polyaniline/f-MWCNT nanocomposite coatings on mild steel for corrosion protection in 3.5% NaCl solution, Prog. Org. Coat. 78 (2015) 387–394.

[26] A.A. Farag, K.I. Kabel, E.M. Elnaggar, A.G. Al-Gamal, Influence of polyaniline/multiwalled carbon nanotube composites on alkyd coatings against the corrosion of carbon steel alloy, Corros. Rev. 35 (2017) 85–94.

[27] G. Qiu, A. Zhu, C. Zhang, Hierarchically structured carbon nanotube–polyaniline nanobrushes for corrosion protection over a wide pH range, RSC Adv. 7 (2017) 35330–35339.

[28] T. Rajyalakshmi, A. Pasha, S. Khasim, M. Lakshmi, M. Murugendrappa, N. Badi, Enhanced charge transport and corrosion protection properties of polyaniline–carbon nanotube composite coatings on mild steel, J. Electron. Mater. 49 (2020) 341–352.

[29] N. Jeong, E. Jwa, C. Kim, J.Y. Choi, J.-y. Nam, K.S. Hwang, J.-H. Han, H.-k. Kim, S.-C. Park, Y.S. Seo, One-pot large-area synthesis of graphitic filamentous nanocarbon-aligned carbon thin layer/carbon nanotube forest hybrid thin films and their corrosion behaviors in simulated seawater condition, Chem. Eng. J. 314 (2017) 69–79.
[30] H. Wei, D. Ding, S. Wei, Z. Guo, Anticorrosive conductive polyurethane multiwalled carbon nanotube nanocomposites, J. Mater. Chem. A 1 (2013) 10805–10813.
[31] D. Sivaraj, K. Vijayalakshmi, Novel synthesis of bioactive hydroxyapatite/f-multiwalled carbon nanotube composite coating on 316L SS implant for substantial corrosion resistance and antibacterial activity, J. Alloys Compd. 777 (2019) 1340–1346.
[32] D. Sivaraj, K. Vijayalakshmi, Enhanced antibacterial and corrosion resistance properties of Ag substituted hydroxyapatite/functionalized multiwall carbon nanotube nanocomposite coating on 316L stainless steel for biomedical application, Ultrason. Sonochem. 59 (2019) 104730.
[33] E.N. Zare, M.M. Lakouraj, S. Ghasemi, E. Moosavi, Emulsion polymerization for the fabrication of poly (o-phenylenediamine)@ multi-walled carbon nanotubes nanocomposites: characterization and their application in the corrosion protection of 316L SS, RSC Adv. 5 (2015) 68788–68795.
[34] R. Prasannakumar, V. Chukwuike, K. Bhakyaraj, S. Mohan, R. Barik, Electrochemical and hydrodynamic flow characterization of corrosion protection persistence of nickel/multiwalled carbon nanotubes composite coating, Appl. Surf. Sci. 507 (2020) 145073.
[35] W. Sun, L. Wang, T. Wu, Y. Pan, G. Liu, Inhibited corrosion-promotion activity of graphene encapsulated in nanosized silicon oxide, J. Mater. Chem. A 3 (2015) 16843–16848.
[36] W. Sun, L. Wang, T. Wu, M. Wang, Z. Yang, Y. Pan, G. Liu, Inhibiting the corrosion-promotion activity of graphene, Chem. Mater. 27 (2015) 2367–2373.
[37] Z. Yang, L. Wang, W. Sun, S. Li, T. Zhu, W. Liu, G. Liu, Superhydrophobic epoxy coating modified by fluorographene used for anti-corrosion and self-cleaning, Appl. Surf. Sci. 401 (2017) 146–155.
[38] C. Chen, S. Qiu, M. Cui, S. Qin, G. Yan, H. Zhao, L. Wang, Q. Xue, Achieving high performance corrosion and wear resistant epoxy coatings via incorporation of noncovalent functionalized graphene, Carbon 114 (2017) 356–366.
[39] H. Hayatdavoudi, M. Rahsepar, A mechanistic study of the enhanced cathodic protection performance of graphene-reinforced zinc rich nanocomposite coating for corrosion protection of carbon steel substrate, J. Alloys Compd. 727 (2017) 1148–1156.
[40] L. Gu, S. Liu, H. Zhao, H. Yu, Facile preparation of water-dispersible graphene sheets stabilized by carboxylated oligoanilines and their anticorrosion coatings, ACS Appl. Mater. Interf. 7 (2015) 17641–17648.
[41] X. Li, P. Bandyopadhyay, M. Guo, N.H. Kim, J.H. Lee, Enhanced gas barrier and anticorrosion performance of boric acid induced cross-linked poly (vinyl alcohol-co-ethylene)/graphene oxide film, Carbon 133 (2018) 150–161.
[42] N. Parhizkar, B. Ramezanzadeh, T. Shahrabi, Corrosion protection and adhesion properties of the epoxy coating applied on the steel substrate pre-treated by a sol-gel based silane coating filled with amino and isocyanate silane functionalized graphene oxide nanosheets, Appl. Surf. Sci. 439 (2018) 45–59.
[43] Z. Yu, H. Di, Y. Ma, L. Lv, Y. Pan, C. Zhang, Y. He, Fabrication of graphene oxide–alumina hybrids to reinforce the anti-corrosion performance of composite epoxy coatings, Appl. Surf. Sci. 351 (2015) 986–996.
[44] Y.-H. Yu, Y.-Y. Lin, C.-H. Lin, C.-C. Chan, Y.-C. Huang, High-performance polystyrene/graphene-based nanocomposites with excellent anti-corrosion properties, Polym. Chem. 5 (2014) 535–550.
[45] K.-C. Chang, W.-F. Ji, M.-C. Lai, Y.-R. Hsiao, C.-H. Hsu, T.-L. Chuang, Y. Wei, J.-M. Yeh, W.-R. Liu, Synergistic effects of hydrophobicity and gas barrier properties on the anticorrosion property of PMMA nanocomposite coatings embedded with graphene nanosheets, Polym. Chem. 5 (2013) 1049–1056.
[46] Y. Zhan, J. Zhang, X. Wan, Z. Long, S. He, Y. He, Epoxy composites coating with Fe3O4 decorated graphene oxide: Modified bio-inspired surface chemistry, synergistic effect and improved anti-corrosion performance, Appl. Surf. Sci. 436 (2018) 756–767.
[47] M. Ramezanzadeh, B. Ramezanzadeh, M. Mahdavian, G. Bahlakeh, Development of metal-organic framework (MOF) decorated graphene oxide nanoplatforms for anti-corrosion epoxy coatings, Carbon 161 (2020) 231–251.
[48] S. Pourhashem, M.R. Vaezi, A. Rashidi, Investigating the effect of SiO2-graphene oxide hybrid as inorganic nanofiller on corrosion protection properties of epoxy coatings, Surf. Coat. Technol. 311 (2017) 282–294.
[49] H. Lu, S. Zhang, W. Li, Y. Cui, T. Yang, Synthesis of graphene oxide-based sulfonated oligoanilines coatings for synergistically enhanced corrosion protection in 3.5% NaCl solution, ACS Appl. Mater. Interf. 9 (2017) 4034–4043.

[50] S. Pourhashem, M.R. Vaezi, A. Rashidi, M.R. Bagherzadeh, Distinctive roles of silane coupling agents on the corrosion inhibition performance of graphene oxide in epoxy coatings, Prog. Org. Coat. 111 (2017) 47–56.

[51] V.K. Upadhyayula, D.E. Meyer, V. Gadhamshetty, N. Koratkar, Screening-level life cycle assessment of graphene-poly (ether imide) coatings protecting unalloyed steel from severe atmospheric corrosion, ACS Sustain. Chem. Eng. 5 (2017) 2656–2667.

[52] L. Yang, Y. Wan, Z. Qin, Q. Xu, Y. Min, Fabrication and corrosion resistance of a graphene-tin oxide composite film on aluminium alloy 6061, Corros. Sci. 130 (2018) 85–94.

[53] T. Monetta, A. Acquesta, F. Bellucci, Graphene/epoxy coating as multifunctional material for aircraft structures, Aerospace 2 (2015) 423–434.

[54] R.K. Gupta, M. Malviya, C. Verma, M. Quraishi, Aminoazobenzene and diaminoazobenzene functionalized graphene oxides as novel class of corrosion inhibitors for mild steel: experimental and DFT studies, Mater. Chem. Phys. 198 (2017) 360–373.

[55] R.K. Gupta, M. Malviya, C. Verma, N.K. Gupta, M. Quraishi, Pyridine-based functionalized graphene oxides as a new class of corrosion inhibitors for mild steel: an experimental and DFT approach, RSC Adv. 7 (2017) 39063–39074.

[56] R.K. Gupta, M. Malviya, K. Ansari, H. Lgaz, D. Chauhan, M. Quraishi, Functionalized graphene oxide as a new generation corrosion inhibitor for industrial pickling process: DFT and experimental approach, Mater. Chem. Phys. 236 (2019) 121727.

[57] K. Ansari, D.S. Chauhan, M. Quraishi, A. Adesina, T.A. Saleh, The synergistic influence of polyethyleneimine-grafted graphene oxide and iodide for the protection of steel in acidizing conditions, RSC Adv. 10 (2020) 17739–17751.

[58] A. Yabuki, A. Kawashima, I.W. Fathona, Self-healing polymer coatings with cellulose nanofibers served as pathways for the release of a corrosion inhibitor, Corros. Sci. 85 (2014) 141–146.

[59] A. Yabuki, T. Shiraiwa, I.W. Fathona, pH-controlled self-healing polymer coatings with cellulose nanofibers providing an effective release of corrosion inhibitor, Corros. Sci. 103 (2016) 117–123.

[60] P. Jain, B. Patidar, J. Bhawsar, Potential of nanoparticles as a corrosion inhibitor: a review, J. Bio- Tribo-Corros. 6 (2020) 1–12.

[61] M. Zheludkevich, S. Poznyak, L. Rodrigues, D. Raps, T. Hack, L. Dick, T. Nunes, M. Ferreira, Active protection coatings with layered double hydroxide nanocontainers of corrosion inhibitor, Corros. Sci. 52 (2010) 602–611.

[62] M.L. Zheludkevich, D.G. Shchukin, K.A. Yasakau, H. Möhwald, M.G. Ferreira, Anticorrosion coatings with self-healing effect based on nanocontainers impregnated with corrosion inhibitor, Chem. Mater. 19 (2007) 402–411.

[63] T.T.X. Hang, T.A. Truc, N.T. Duong, P.G. Vu, T. Hoang, Preparation and characterization of nanocontainers of corrosion inhibitor based on layered double hydroxides, Appl. Clay Sci. 67 (2012) 18–25.

Index

Page numbers followed by "*f*" and "*t*" indicate, figures and tables respectively.

A

Acidic electrolytes, 53
   anodic oxidation reaction, 49
   case studies, 54
      HCl-based electrolytes, 55
      H_2SO_4-based electrolytes, 56
   cathodic reduction reaction, 49
   corrosion protection in, 54
   hydrochloric acid, 50
   phosphoric acid, 52
   sulfuric/sulphuric acid, 51
Adsorption model, 16 *See also* Stress corrosion cracking (SCC)
Aliphatic corrosion inhibitors, 43–44
   acyclic compounds, 43–44
   alicyclic compounds, 43–44
Alkylamides, 236–237
Aluminum, 239
American Society of Metals (ASM), 11
Amino acids, 233, 241
   corrosion inhibitors, 234–236
   structure, 234*f*
2-aminobenzene-1,3-dicarbonitriles (ABDN), 61
Anodic reactions, 81, 117–118
Anodic-type corrosion inhibitors, 31
Anticorrosive coating formulations, 263–265
Aqueous phase corrosion inhibitors, 46
Atomic force microscope (AFM), 210

B

Back-donation, 54
Bioaccumulation, of a chemical, 32–33
Bis-phenol polymer, 154–155
Breakdown potential, 14

C

Carbohydrates, 225
   green corrosion inhibitors, 226
Carbon nanotubes (CNTs), 262–263
Carboxymethyl cellulose (CMC), 213
Cathodic corrosion inhibitors, 31, 41–42, 44–45
   cathodic poisons, 41–42
   hydrogen evolution reaction, 118
   poisons, 41–42
   precipitates, 41–42
   reactions, 175
   scavengers, 41–42
Cellulose, 213
Chemical corrosion, 22 *See also* Corrosion
   inhibitors, 46–47
   liquid metal corrosion, 24
   oxidative corrosion, 23
   oxidative dry-corrosion, 23
Chemical medicines, 193, 194
Chitosan, 208–209
   synthesis of, 209*f*
Coating phase corrosion inhibitors, 46
Copper-based alloys, 75–76
Corroion rate, 28
Corrosion, 1, 93
   adsorption mechanism, 177, 227
   adverse effects of, 6
   ASM classification of, 12*f*
   cost in major countries, 6*t*
   crevice, 15
   cycle, 22*f*
   defined, 1, 21
   direct costs of, 1, 2
   driving force for, 21
   economic costs of, 1
   erosion, 18
   factors affecting, 27
   galvanic, 14
   inhibition, 210
   inhibition effect, 156
   inhibitors, 31, 136–137, 194, 208
      application, 255
   intercrystalline, 12–13
   interdendritic, 12–13
   intergranular, 12
   monitoring, 11–12
   pitting, 13
   protection, 70
      in acidic electrolytes, 54
      in basic electrolytes, 60
   rate, 11, 27

Corrosion (*Continued*)
　related accidents in history, 7t
　stress corrosion, 16
　susceptibility, 11
　thermodynamics of, 21
　top-of-line, 18
　types of, 21
　　chemical corrosion or dry corrosion, 22
　　electrochemical or wet corrosion, 24
　uniform, 11
Corrosion inhibition/inhibitors
　in acidic electrolytes
　　hydrochloric acid, 50, 55
　　phosphoric acid, 52
　　sulfuric/sulphuric acid, 51, 56
　in alkaline electrolytes
　　basic electrolytes, 60
　　KOH-based electrolytes, 63
　　NaOH-based electrolytes, 61
　inorganic, 41
　organic, 43
　　based inhibition mechanism, 44
　　based molecular size, 45
　　based on mode of adsorption, 46
　　based on state of application, 46
　　based on the environment, 45
　　based on their nature, 43
　　based on the origin, 43
Crevice corrosion, 15
　factors influencing, 16
　mechanism of, 15–16
　minimize, 16

D

Dealloying. *See* Selective leaching
Decomposers, 32–33
Demetalification. *See* Selective leaching
Density functional theory (DFT), 103, 227
　based quantum chemical calculations, 120
　methods, 88–89
　parameters, 104
Derivatization, 200–201
Dipole moment, 106
Donor-acceptor mechanism (DFT), 244f
Dry corrosion, 22 *See also* Corrosion
　liquid metal corrosion, 24
　oxidative corrosion, 23
　oxidative dry-corrosion, 23

E

Electrochemical corrosion, 24
　anodic half-cell reaction, 25
　anodic sites, 24
　cathodic half-cell reaction, 25
　cathodic sites, 24
　electrical connection, 24
　electrolyte, 24
　exposed metal surface, 24
　inseparable anode/cathode type, 27
　interfacial anode/cathode type, 27
　potential difference, 24
　separable anode/cathode type, 27
Electrochemical impedance spectroscopy (EIS), 93, 177, 226–227
Electrochemical potential, 93
Electrochemical techniques, 97, 101
Electrochemical testing practice, 93
Electron-donating substituents, 104–105
"Electron-donation ability,", 108–109
Embrittlement model, 16 *See also* Stress corrosion cracking (SCC)
Erosion corrosion, 18

F

Faraday, 21
Fatty acids, 246f
Ferric ions, 69
Film-forming corrosion inhibitors, 4–5
Film rapture model, 16 *See also* Stress corrosion cracking (SCC)
Fourier-Transformed Infra-Red (FT-IR) spectroscopic analyses, 120
Fraction of electron, 108
Frontier molecular orbitals, 104

G

Galvanic corrosion, 14
　of common metals and alloys, 15f
　prevention of, 14–15
Galvanic couple, 14–15
General corrosion. *See* Uniform corrosion
Gibbs free energy, 21, 71
Green corrosion inhibitors, 31, 135, 165
　assessment of, 32
　green chemistry principles and, 33
　　derived through natural resources, 37
　　inhibitors through proper designing, 35
　　practices, 38
　　synthetic green corrosion inhibitors, 34
　microwave and ultrasound irradiations, 147
　microwave irradiation, 151
　plant extracts, 173
　polyethylene glycol, 169
　ultrasound irradiation, 156

H

Hammett constant, 136–137
Hardness, 108
　absolute, 108

Heterocyclic compounds, 148
Highest occupied molecular orbital (HOMO), 38
High-frequency response, 94
High temperature corrosion, 255, 258, 258t
Hot corrosion, 255
Hydrochloric acid, 49, 50, 52
    solutions, 118–119, 176–177
Hydroxymethyl cellulose, 213

I
Imidazoline-based fatty acid derivatives, 247
Inorganic corrosion inhibitors, 41
    anodic, 41
    cathodic, 41–42
    classification of, 42f
Intergranular attack. See Intergranular corrosion (IGC)
Intergranular corrosion (IGC), 12, 13
Ionic liquids (ILs), 35–36, 103, 115
    anions in, 36t
    application, 115
    applications of, 36–37
    biological applications, 116f
    classification, 116f
    corrosion inhibitors, 118–119
    imidazolium-based, 104
    industrial applications, 115, 116f
    mechanism, 117
    property, 115

J
Janak's theorem, 108

K
Koopman's theorem, 104

L
Langmuir adsorption isotherm, 120, 125, 153–154, 227–228, 239
    model, 72, 152–153, 156, 198, 209, 248
Liquid metal corrosion, 24 See also Dry corrosion
Liquid metal cracking. See Liquid metal corrosion
Liquid metal embrittlement. See Liquid metal corrosion
Lowest unoccupied molecular orbital (LUMO), 38

M
Macromolecular corrosion inhibitors, 45
Mechanochemical mixing (MCM), 31–32
Metal-inhibitor interaction, 110
Metal oxides, 262
Microbubbles formation, 151f
Mixed-type adsorption mechanism, 136–137
Monte Carlo simulations, 109

Multicomponent reactions (MCR), 31–32, 135
    advantages, 135–136
    one-step, 143–145
MW-mediated reaction, 148–149

N
Nanofibers, 266
Nanomaterials (NMs), 261
    corrosion inhibitors, 262
Naphthenic acid inhibitors, 257–258
National Association of Corrosion Engineers, 1
Natural polymeric corrosion inhibitors, 45
Natural polymers, 208, 217
Neutral electrolytes, 69, 70
Non-polymeric carbohydrates, 231
Nyquist curves, 97, 101, 247
Nyquist plot, 95f

O
Oleic hydrazide salicylate (OHS), 247
Oleochemicals, 243, 252
Organic compounds, 71, 74, 135, 137, 194, 209, 239
Organic corrosion inhibitors, 43, 43f, 71–72, 147
    aluminum alloys, 74
    based inhibition mechanism, 44, 45f
    based molecular size, 45, 46f
    based on mode of adsorption, 46
    based on state of application, 46
    based on the environment, 45
    based on their nature, 43, 44f
    based on the origin, 43
    copper alloys, 75–76
    iron alloys, 72
    nanosized, 45
Oslo Paris commission (OSPAR), 32
Oxidative dry-corrosion, 23

P
Palmitic acid imidazole, 244f
Partition coefficient, 32–33
Passivators, defined, 31, 41
Phenyl sulfonylacetophenoneazo derivatives (PSAAD), 55–56
Phosphoric acid (H_3PO_4), 52, 52t, 53t
Phytochemicals adsorption, 176f
Pilling-Bedworth ratio, 22–23, 256
Pitting corrosion, 13
    breakdown potential, 14
    electrochemical cyclic polarization, 14
    electrochemical noise testing in, 14
    measurement of, 13–14
    repassivation potential, 14
    testing, 14

Plant extracts, 175, 177, 181, 183–184
  diagrammatic illustration of, 174f
  mechanism, 175
Polymers, 207
Potentiodynamic polarization, 95, 156, 248
Potentiodynamic polarization (PDP), 194–195, 226–227
  studies, 142, 167
Pyrimidine derivatives, 137–142
Pyrimidine-fused heterocycles, 142

Q
Quantum chemical calculations (QCCs), 227
Quercetin-3-glucuronide, 178f

R
Randles, equivalent circuit, 94f
Reactant molecules, 149f
Registration, evaluation, authorization and restriction (REACH), 32
Repassivation potential, 14
Retro-donation, 54

S
Scanning electron microscopy (SEM), 136–137, 210
  method, 55
Selective attack. *See* Selective leaching
Selective corrosion. *See* Selective leaching
Selective leaching, 17
  mechanisms of, 17
  prevention of, 17
Sensitization, defined, 12–13
Sodium chloride based solutions, 69, 76–77
Solid state reactions (SSR), 31–32
Solvents, 168
Sour corrosion, 81
Stress corrosion cracking (SCC), 16, 198
  adsorption model, 16
  embrittlement model, 16
  film rapture model, 16
  impact of, 16
  mechanisms for, 16
  occurrence of, 16
  prevention of, 17
Sulphuric acid based electrolytic media, 125
Supercritical carbon dioxide, 167
Sweet corrosion, 79
  mechanism of, 79, 80f, 82f
  protection for, 82–83
Synthetic corrosion inhibitors, 43
Synthetic green corrosion inhibitors, 34
Synthetic polymers, 207

T
Tafel constant, 226–227
Tafel curves, 226–227
Tafel polarization, 96–97
Tafel reactions, 99
Top-of-line corrosion (TLC), 18
  prevention of, 18
Triazolyl bis-amino acid derivatives, 152f
Triglyceride, 244f
Two-metal corrosion. *See* Galvanic corrosion

U
Ultrasound assisted chemical transformations, 150f
Ultrasound heating, 150
Uniform corrosion, 11
  corrosion rate, 11
  methods, 12
  susceptibility, 11
United States
  corrosion effects loss in, 1
  cost of corrosion, 2f
    annual, 1
    infrastructure (upper) and utilities (lower) sectors, 5f
    production and manufacturing sectors, 4f
    shipping industry, 3
    transportation sector, 3f
  economy, 2
"Universal solvent,", 165

V
Valence state parabola model, 108
Volatile compounds (VCIs), 246

W
Weight loss technique, 86–87
  advantages of, 86
  disadvantages of, 87
Weight loss (WL) technique, 85–86
  experiment, 88
Well acidization, 256–257
Wet corrosion, 24
  anodic half-cell reaction, 25
  anodic sites, 24
  cathodic half-cell reaction, 25
  cathodic sites, 24
  electrical connection, 24
  electrolyte, 24
  exposed metal surface, 24
  potential difference, 24

Printed in the United States
by Baker & Taylor Publisher Services